Analytical Chemistry - Advancement, Perspectives and Applications

Edited by Abhay Nanda Srivastva

Published in London, United Kingdom

IntechOpen

Supporting open minds since 2005

Analytical Chemistry - Advancement, Perspectives and Applications
http://dx.doi.org/10.5772/intechopen.87743
Edited by Abhay Nanda Srivastva

Contributors
Serge-Bertrabd Adiko, Rifat Radisovich Mingazov, Muhammad Sagir, Myung-Hoon Kim, Xiaoying Wang, Mingqiang Ma, Xueliang Wang, Shoujuan Wang, Noboru Sasagawa, Sergiy Andreevich Kurta, Khatsevych Olha, Syed Amir Ashraf, Zulfiqarur Rashid Azaz Ahmad Azad, Sadaf Nazir, Mohd Adnan, Mahesh Deshpande, Francisco J.A. Loureiro, Laura I.V. Holz, Vanessa C.D. Graça, Duncan P. Fagg, Susan Marie Marie Viet, Maire S.A. Heikkinen, Michael Dellarco, Theopoline Omagano Itenge, Abhay Nanda Nanda Srivastva, Sunil Kumar, Muhammad Bilal Tahir, Muhammad Pervaiz, Muhammad Hassan Qasim, Sami Ullah, Reema Ansar

First published in London, United Kingdom, 2021 by IntechOpen
IntechOpen is the global imprint of INTECHOPEN LIMITED, registered in England and Wales, registration number: 11086078, 5 Princes Gate Court, London, SW7 2QJ, United Kingdom
Printed in Croatia

British Library Cataloguing-in-Publication Data
A catalogue record for this book is available from the British Library

Additional hard and PDF copies can be obtained from orders@intechopen.com

Analytical Chemistry - Advancement, Perspectives and Applications
Edited by Abhay Nanda Srivastva
p. cm.
Print ISBN 978-1-83968-020-5
Online ISBN 978-1-83968-021-2
eBook (PDF) ISBN 978-1-83968-022-9

For EU product safety concerns:
IN TECH d.o.o., Prolaz Marije Krucifikse Kozulić 3, 51000 Rijeka, Croatia, info@intechopen.com or visit our website at intechopen.com.

Meet the editor

Dr. Abhay Nanda Srivastava has been appointed as Assistant Professor in the chemistry department of Babasaheb Bhim Rao Ambedkar Bihar University, Muzaffarpur-India, and then posted as Head of Chemistry Department at Nitishwar College (A constituent unit of B.R.A. Bihar University), Muzaffarpur-India in 2017. Along with this, Dr. Srivastva has also been entrusted with additional academic and research activities in the University Department of Chemistry, B.R.A. Bihar University, Muzaffarpur-India. Before joining BU, Dr. Srivastava was an assistant professor of chemistry at the Engineering College, Delhi, NCR, affiliated to Dr. A. P. J. Abdul Kalam Technical University, Lucknow, India. He obtained his master's and doctorate degrees in Chemistry from Chaudhary Charan Singh University, Meerut. His areas of interest include coordination chemistry, medicinal chemistry, green chemistry, and bioinorganic/organic chemistry. Dr. Srivastva supervises Ph.D. scholars in these areas. He has authored many research papers, edited books, and contributed chapters to reputed international and national journals and book publishers. He is the associate editor of the *Journal of Transition Metal Complexes* and an editorial board member for several international journals. Dr. Srivastva reviews articles, delivers invited talks and lectures, and presents research papers at national and international conferences, seminars, webinars, and workshops.

Contents

Preface

Analytical chemistry deals with the identification and structural verification of materials developed through any discipline of chemistry or material sciences. The versatile application of this branch of chemistry provides researchers with a vast field of analytical instruments and methodologies for both qualitative and quantitative analysis. The suitability of chemicals and materials in different areas of technological developments can also be determined using analytical chemistry. The field of analytical chemistry is improving daily due to the refinement of instruments and related principles and theories with a practical approach. This advancement makes it easier for chemists to analyze their studies deeply for optimum utilization of materials in various areas.

This edited book is for researchers, postgraduate students, and all those associated with analytical chemistry. It presents the latest advancements and applications of analytical chemistry in a systematic manner. It is an anthology of scientific findings and views of researchers from various research centers across the globe on emerging topics of instrumentation, energy, environment, biotechnology, and synthetic enhancement analysis techniques related to analytical chemistry. The volume includes twelve chapters containing discussion, analogies, and graphics for a better understanding of the presented concepts.

I am grateful to the authorities of the institution affiliated with me for their encouragement throughout the preparation of this book. I am thankful to the faculty members of the University Department of Chemistry, B.R.A. Bihar University, Muzaffarpur, India, for productive discussions about the content of the book. I am most grateful for the assistance of Dr. Netra Pal Singh, Associate Professor, Department of Chemistry, D.D.U. Gorakhpur University, Gorakhpur, India. I am also appreciative of my family members for their support and care, especially my wife Shalini Srivastava. I extend my sincere thanks to all the authors who contributed chapters and the team at IntechOpen for publishing this book with all their best efforts.

This book is dedicated to my respected grandfather Shri Vishwanath Prasad Srivastava (in loving memory) and father Shri Sunil Kumar Srivastava.

Abhay Nanda Srivastva
Department of Chemistry,
Nitishwar Mahavidyalaya (B.R.A. Bihar University),
Muzaffarpur, India

Chapter 1

Advances in Derivative Voltammetry - A Search for Diagnostic Criteria of Several Electrochemical Reaction Mechanisms

Myung-Hoon Kim

Abstract

New methods for analysis of current-potential curves in terms of their derivatives are presented for studying various types of electrode processes – such as simple electron transfer reactions (reversible, quasi-reversible, and irreversible electron transfer) as well as chemically coupled electron transfer reactions along with a diagnostic scheme for differentiating these various types of electrochemical reaction mechanisms. Expressions for first- and higher order derivatives are derived from theoretical analytical solutions for currents for the different types of electrode mechanisms. The derivative curves are analyzed in terms of various parameters which characterize *peak shape or peak symmetry* with an emphasis on *the second derivatives* with well-defined anodic and cathodic peaks. Second derivatives can yield, in a simpler manner, the symmetry ratios; i.e., a ratio of anodic to cathodic peak-currents ($\mathbf{i_p{}^a/i_p{}^c}$), and a ratio of anodic to cathodic peak-widths ($\mathbf{W_p{}^a/w_p{}^c}$) and a ratio of anodic to cathodic peak potential differences ($\mathbf{\Delta E_p{}^a/\Delta E_p{}^c}$) or a peak separation ($\mathbf{E_p{}^a\text{-}E_p{}^c}$) are evaluated, and these ratio can be related to kinetic parameters associated with a particular types of electrode mechanisms. Peaks are found to be symmetrical for a simple reversible electron transfer process (E_r). However, peaks *become asymmetrical* when the electron transfer become slower (namely, irreversible, E_{irr}) or e^- transfer reaction is coupled with homogeneous chemical reactions such as a prior reaction (CE_r) or a follower-up reaction (EC_r). From measured values of such symmetry ratios above, one can gain insight to the nature of the electrochemical systems enabling us to determine various kinetic parameters associated with a system. A diagnostic criteria for assigning an electrode mechanism is devised based on the values of asymmetry parameters measured, which are unity for a simple reversible electron transfer process.

Keywords: Derivative Voltammetry, Electrochemical Reaction Mechanism, Electrode Kinetics, Reversible, Quasi-reversible, and Irreversible Electron Transfer, Chemically Couple Electron Transfer, Diagnostic Criteria for Electrode Processes

1. Introduction

The aim of this Chapter is several folds (1) to provide a brief background on the derivative approach in measurements and its various applications in science and engineering fields, (2) to review some of our research work on advances on derivatives voltammetry (DV) which have been scattered among dozens of different journls and conference presentations/proceedings in the past, together with additional new findings and development which have not been published yet, (3) to analyze and compare those characteristic results from each of different electrochemical reaction mechanisms, and to construct a master table of diagnostic criteria for differentiating various types of the electrode mechanisms. Namely, as the subtitle indicates, the primary goal of the project is, utilizing derivative voltammetry, to invent a general criteria for differentiating various types of electrode processes, in order to complement other methods such as cyclic voltammetry(CV), which is the most popular one, or Cyclic Square-Wave Voltammetry (CSWV).

1.1 Backgrounds

Derivatives approaches in analyzing data signals, have attracted much interests producing numerous research articles in the past [1–32]. Although earlier pioneering reports on the derivative approach in electrochemistry are found in the reference sections in our previous reports in electronalyticalchemistry [11–15], the approach with derivatives is not only limited to electrochemical methods but also have extended to other various fields of science and engineering [23–27, 33]. These reviews and earlier articles introduces basics of the method and techniques of signal processing [2–4, 28, 34], as applied to spectroscopic and spectrophotometric data [2–9, 34], electrochemical data [10–21], and other signals from seismic data [23] and data from biomechanics measurements [24–27, 33]. Several group [3, 6] extended the derivatives to as high as fourth order for enhanced peak resolutions. For example, a single broad peak of UV absorption spectrum from a ternary mixture of amino acids (phenylalanine, tyrosine and tryptophan), were separated and resolved in its second and fourth derivatives [3]; the physiological states of certain algae have been assessed from an analysis of 4th derivative of absorption spectra of chlorophyll a and b [6]; a second order derivatives of an IR spectra of a DNA in malignant tissues exhibits some differences from those of a normal one [7]. The ever increasing interests in the derivative approaches in analysis of signals stem from (a) the information hidden behind in the raw data can be easily exposed from analysis of finer features emerged in the shapes of derivatives, the higher order the more the revelation, and (b) the technical advances in digital signal enhancing methods [2, 14, 15, 23–29, 34] that improve the S/N ratio which was not possible from analog instruments. A review of signal processing in electrochemistry with an introduction of software packages and extensive references therein by Jakubowska [29] may provide further assistance for interested readers.

In the field of electrochemical analysis, the derivative approach has been applied to potentiometric [10, 16] as well as to voltammetric measurements [10–22] with the enhanced resolution and better signal-to-noise (S/N) ratios.

More recent applications for derivative techniques have also been reported for analysis of oxidation of methanol [21], for improved quantification of ill-defined peaks in determining Pt in water and sediments [35], for analysis of alloxan [36], and for quantitation of naringin, an antioxidant [37].

1.2 Application to study of electrochemical reaction mechanisms

Most reports of derivative approach on electrochemical system has been for its application in chemical analysis [10, 19, 20] while only few groups have attempted to apply the approach to the study of kinetics of electrode reactions [11–18, 21, 22]. Elucidation of electrochemical reaction mechanisms on electrode surface is not only important in developing electroanalytical methods for determining an analyte in a system but also devising effective electrochemical energy storage cells and other applied fields of engineering.

1.2.1 Drawbacks of cyclic voltammetry

As far as method of studying electrode reaction mechanisms concerned [1, 38], the cyclic voltammetry (CV) has been dominant in the field [39–41]. However, we have developed this new derivative approach to investigate electrode kinetics [11–15] in order to help in elucidating electrochemical reaction mechanisms: this is because CV has drawbacks of complications arising from the interference of the capacitive current to charge the electrical double layer at the interface; as results those peaks from CV are not purely Faradaic, and giving problems associated with defining/evaluating peak currents (i_p), and peak potentials (E_p) and peak widths (W_p). In present work, currents are taken in normal pulse voltammetric (NPP) mode by sampling the current near the end of a pulse in order to minimize the non-Faradaic currents: then, first-, second-, and third- derivatives are taken from the NPP data.

1.2.2 Several common types of electrochemical systems

The electrochemical systems under present investigation includes the simple electron transfer process; the reversible electron transfer reaction (**E_r** type of mechanism), and quasi-reversible (**$E_{quasirev}$** type) and irreversible (**E_{irr}** type) electron transfer [11, 15], and chemically coupled e^- transfer processes with a prior chemical equilibrium (**CE_r** type) [12, 13, 42] as well as those with a post-kinetic process (**E_rC** type) [14] all on a planar electrode. The observed currents may be influenced by the heterogeneous kinetic constants for an electron transfer reactions (*i.e.*, the transfer coefficient, α, and heterogeneous rate constant, k^o) for E_{irr} and $E_{quasirev}$ type of processes [15], as well as the homogeneous rate constant (k) and equilibrium constant (K_{eq}) for E_rC or CE_r type of processes [13, 14]. In this study, the effects of these various kinetic variables on the current and it's derivatives are also investigated.

1.2.3 Experimental measurements of derivatives

The theoretical voltammogram (zeroth-order derivative) is basically to emulate a normal pulse voltammogram (NPV) obtained with a normal pulse mode from an electrochemical analyzer with a DME or HMDE or SMDE (static mercury drop electrode) [11]; typically, with a pulse period of 1.000 s, pulse width of 50 ms, current sampling time of 33 ms, and scan rate of 1 mV/sec or 2 mV/s with a commercial electrochemical analyzer such an instrument as BAS-100 and Tacussel PRG-5 Pulse Polarograph. In calculation of theoretical currents, the original currents are normalized by dividing them with the diffusion controlled current (i_d) to obtain a dimensionless currents.

1.3 Basic scheme with symmetry in the peaks associated with derivative voltammetry and definitions of various parameters from the plots of currents and it's derivatives

Please refer to the **Figure 1** for the definitions of all the dependent variables associated with the current-potential curve (zeroth order derivative) and it's first, second and third derivatives. All measured variable and derived variables associated

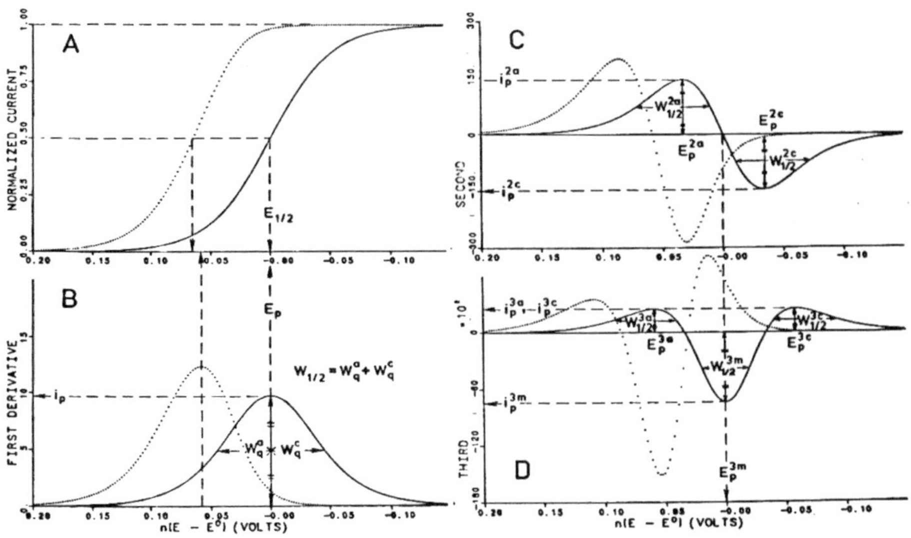

Figure 1.
Normal pulse voltammograms (A) and their higher order derivatives: 1st order (B), 2nd order (C), and 3rd derivatives (D). Several important experimental variables (parameters) such as peak-potentials, peak-currents and peak-widths are defined on the graph,, for two different electrode mechanisms; (1) for a simple E_{rev} type mechanism (solid lines), and (2) for a reversible electron transfer coupled with a follow-up chemical reaction ErC type mechanism (dotted line). Potentials (x-axis) are with respect to a formal potential, E^o, or $E_{1/2}$ for the E_{rev} reaction. Calculated for T = 298 K and t = 0.952 s. the currents are normalized with respect to the diffusion currents and dimensionless; thus, the unit of first derivatives is V^{-1}, second derivatives is V^{-2}, third derivatives is V^{-3}.Note: For the original currents (A), only the half-wave potential ($E_{1/2}$) is graphically defined. Graphic definitions of the quarter-wave potential ($E_{1/4}$) and three-quarter-wave potential ($E_{3/4}$), are omitted in order to make the figure simple.

with current and the derivatives are listed in **Table 1** along with those values for a reversible e- transfer process, which is the most simple type [15].

1.3.1 Symmetry parameters

A quantitative measure of symmetry *in the original current are not readily available*, but can be found indirectly (and inconveniently) from a ratio of difference in several potentials defined from an original voltammogram (zeroth order deivatives): namely, a quarter-wave potentials ($E_{1/4}$, at which a current reaches a quarter of the diffusion current, i_d), half-wave potentials ($E_{1/2}$ or E_h) and a three-quarter-wave potentials, $E_{3/4}$, at which a current becomes $(3/4)i_d$); then a ratio of an anodic to cathodic quarter-wave potential differences from a half-wave potentials (i.e., $\mathbf{\Delta E_q{}^a/\Delta E_q{}^c}$ = $(E_{1/2}-E_{1/4})/(E_{1/2}-E_{3/4})$ = $\mathbf{R_q{}^o}$) is calculated. It is very limited.

However, measurements of symmetry parameters from *the first derivatives* are much easier than those from the oignal current, yielding a variety of (half a dozen) of variables. Namely, in addition to that based on peak potential differences (with respect to E^o), a peak current, half-peak width and anodic and cathodic and quarter-peak widths, and a ratio of anodic to cathodic quarter-peak width ($\mathbf{R_w{}^1 = W_q{}^a/W_q{}^c}$) are also available. *The second derivative* can be characterized by more (about ten) variables, such as peak potentials and their ratio ($\mathbf{R_{\Delta E}{}^2 = \Delta E_p{}^{2a}/\Delta E_p{}^{2c}}$), peak currents and their ratio ($\mathbf{R_i{}^2 = |i_p{}^{2a}/i_p{}^{2c}|}$), and half-peak widths and a ratio of anodic to cathodic half-peak widths ($\mathbf{R_w{}^2 = W_{1/2}{}^{2a}/W_{1/2}{}^{2c}}$). For *the third derivative*, with appearance of an additional third peak, the number of variables doubles to about twenty: three peak potentials, the differences among them, and

	Parameters	Definitions	E_r		Parameters	Definitions	E_r
Current				*3rd Der.*			
	$E_{1/4}$ (mV)		+28		E_p^{3a} (mV)		59
	$E_{1/2}$ (mV)		0		E_p^{3m} (mV)		0.00
	$E_{3/4}$ (mV)		-28		E_p^{3c} (mV)		-59
	ΔE_q^a (mV)	$E_{1/4}$-$E_{1/2}$	28		ΔE^3_p (mV)	E_p^{3a}-E_p^{3c}	118
	ΔE_q^c (mV)	$E_{1/2}$-$E_{3/4}$	28		ΔE^3_a (mV)	E_p^{3a}-E_p^{3m}	59
	ΔE_h^0 (mV)	$E_{1/4}$-$E_{3/4}$	56.4		ΔE^3_c (mV)	E_p^{3m}-E_p^{3c}	59
	$\mathbf{R_q^0}$	$\mathbf{\Delta E_q^a/\Delta E_q^c}$	**1.0**		$\mathbf{R_{\Delta E}^3}$	$\mathbf{\Delta E^{3a}/\ \Delta E^{3c}}$	**1.00**
1st Der.					i_p^{3a}		2461
	ΔE^1_p (mV)	E_p-E^o	0.0		i_p^{3m}		-7382
	R_i^1	i_p/i_d	9,7		i_p^{3c}		2461
	$W_{1/2}$ (mV)		90.5		$\mathbf{R_i^3}$	$\mathbf{i_p^{3a}/i_p^{3c}}$	**1.00**
	W_q^a (mV)		45.3		$\mathbf{R_i^{3a}}$	$\mathbf{\lvert i_p^{3a}/i_p^{3m}\rvert}$	**0.32**
	W_q^c (mV)		45.3		$\mathbf{R_i^{3c}}$	$\mathbf{\lvert i_p^{3c}/i_p^{3m}\rvert}$	**0.32**
	$\mathbf{R_w^1}$	$\mathbf{W_q^a/W_q^c}$	**1.00**		$W_{1/2}^{3a}$ (mV)		54
2nd Der.					$W_{1/2}^{3m}$ (mV)		41
	E_p^{2a} (mV)		34		$W_{1/2}^{3c}$ (mV)		54
	E_p^{2c} (mV)		-34		$\mathbf{R_w^3}$	$\mathbf{W_{1/2}^{3a}/W_{1/2}^{3c}}$	**1.00**
	E_p^2 (mV)	E_p^{2a}-E_p^{2c}	68		$\mathbf{R_w^{3a}}$	$\mathbf{W_{1/2}^{3a}/W_{1/2}^{3m}}$	**1.32**
	$\mathbf{R_{\Delta E}^2}$	$\mathbf{\Delta E_p^{2a}/\Delta E_p^{2c}}$	**1.00**		$\mathbf{R_w^{3c}}$	$\mathbf{W_{1/2}^{3c}/W_{1/2}^{3m}}$	**1.32**
	i_p^{2a}		14				
	i_p^{2c}		-14				
	$\mathbf{R_i^2}$	$\mathbf{\lvert i_p^{2a}/i_p^{2c}\rvert}$	**1.00**				
	$W_{1/2}^{2a}$ (mV)		64				
	$W_{1/2}^{2c}$ (mV)		64				
	$\mathbf{R_w^2}$	$\mathbf{W_{1/2}^{2a}/W_{1/2}^{2c}}$	**1.00**				

Table 1.
Definitions of original and derived parameters and the values for the simple reversible e^- transfer reaction. Calculated for $n = 1$ and at 298 K.

three peak currents an their ratios, and three half-peaks widths and their ratios. Among the ratios from third derivative, the ratio of anodic to cathodic peak potential difference ($\mathbf{R_{\Delta E}^3 = \Delta E^{3a}/\Delta E^{3c}}$), the anodic to cathodic peak currents ratio ($\mathbf{R_i^3 = i_p^{3a}/i_p^{3c}}$), and a ratio of anodic to cathodic half-peak widths ($\mathbf{R_w^3 = W_{1/2}^{3a}/W_{1/2}^{3c}}$) are most notable. All of these ratios are particularly noticeable being utilized as symmetry parameters: and *all of these values of symmetry parameters are unity (1.00) for an E_r type*. Unlike E_r that exhibits a well-defined perfect symmetry in the derivative curves, these parameter values become no longer unity for other type of slower electron transfer mechanisms (E_{irr}, and $E_{quasirev}$) or a fast e^- transfer reaction (E_{rev}) is chemically coupled homogeneously (namely, CE_r and E_rC type), introducing *asymmetry* in those curves.

As far as analysis of *the derivatives* concerned, most of the work have dealt with only the linear diffusion (on a planar electrode) or semi-infinite liner diffusion (on DME, or SMDE), not with spherical diffusion (on a spherical electrode), even though there are many studies on the original currents (not derivatives) with spherical electrodes from earlier days of polarographic/voltammetric works as given in general monographs [30, 38, 43, 44] as well as in articles on particular systems and methods [45–55]. These include studies on EC mechanisms [45, 46], CE mechanisms [46, 47, 54], on various pulse polarographic methods [48, 55], on DC polarography [49], on AC polarography [72], on cyclic voltammetry [51], on catalytic EC mechanism with the square-wave voltammetry [52], on a double potential method [53] and on DPV at spherical and microelectrodes [56].

1.3.2 Effects of electrode sphericity

Current-potentials curves depends not only on types of the electrode mechanisms but also shape (geometry) of electrodes. However, present study focuses on a planar electrode with a linear diffusion. The effect of electrode sphericity can be minimized on planar electrode in pulse or ultrafast voltammetry because the diffusion layer thickness near the electrode is also depend on the duration of the applied potentials [43]: with application of short pulses, diffusion onto a spherical electrode (for cases with DME, HMDE, and SMDE) is approximated to a linear diffusion by minimizing the diffusion layer thickness, Nevertheless, effects of electrode sphericity *on the derivatives* of currents have been addressed by other groups [18, 54, 56] and author's group [31, 32]. Recently, Molina and coworkers [18] took a different approach that examine the dependence of peak-potentials (E_p) and peak-currents (i_p) in first derivatives on the dimensionless rate constant ($\chi = k\tau$) and on the dimensionless electrode sphericity ($\xi = (Dt)^{1/2}/r$); and found that, as the sphericity increases, the peak-potentials (E_p) moves towards more negative potentials for EC and moves oppositely towards more positive potentials for CE mechanism, while peak currents (i_p) for both types of mechanisms decreases. As far as DV concerned, our recent work [31, 32] show that even slower electron transfer processes exhibits symmetry when electrode sphericity increases: this suggess that planar electrode is much more effective than the spherical electrode with the present DP method.

1.3.3 Similar analysis from differential pulse polarography (DPP)

Our group has adopted a different approach focusing on the analysis of *peak shapes in terms of symmetry parameters* as reported in our earlier studies on the first- and higher- order derivatives [12–15, 42], instead of examining shifts in peak potentials and changes in the magnitude of peak currents which is adopted by the other group [18, 54, 56]. Namely, in our previous studies [57–61], we have introduced the approach of analyzing peak asymmetry *found in the differential pulse polarography/voltammetry (DPP/DPV)*: namely, the differential pulse voltammograms can be viewed as a *pseudo-derivative of the i-E curve* which is emulated mechanically and/or electronically with the voltammetric/polarographic analyzer. In the earlier works, two peaks are generated from DPP; namely, a cathodic peak generated from a reduction process with a pulse going with the same direction as the scan direction (with a pulse amplitude of ΔE = -50 mV), and an anodic peak generated with a pulse going against the scan direction (with ΔE = +50 mV) to drive oxidation of reduced species back to the oxidized form, then the two peaks are compared and analyzed. In the original normal pulse program, the currents are assigned as positive for the reduction process. In the derivative voltammetry, however, if values of the derivative are positive or the peak appears in more positive potentials, it is labeled as "anodic", if the values are negative, or the peak appears in more negative potentials, it is labeled as "cathodic" regardless of actual redox reaction occurring on the electrode surface, following the previous convention [12–15, 42]. In addition to the analysis of peak in term of peak-currents and peak-potentials [57–60], an analysis in terms of half-peak-widths was introduced later [61]. In present study a focus is given more to the *second derivative* because (a) it generates two well-defined peaks of anodic and cathodic that parallel to two peaks of DPP, (b) it is more sensitive and practical than first derivative which has only one peak and fewer visible parameters, and (c) it is less complicated and more practical than third derivative even though it is less sensitive than the third derivative.

1.3.4 Other approaches for diagnostic criteria with cyclic square-wave voltammetry (CSWV)

Cyclic square-wave voltammetry (CSWV), which is an extension of SWV [62–65], has been adopted by Bottomley group [66–70]. Like pulse voltammetric techniques, SWV is effective in removing the interference from the capacitive current that is inherent in cyclic voltammetry(CV). SWV, like DPV, is a pseudo-derivative technique in which a derivative-shaped peak is obtained from the current with electronic control of input of square-wave potentials and current sampling times [62]. Diagnostic criteria as a tool for studying electrode mechanisms based on CSWV scheme have been investigated by the group for reversible [66] and quasi-reversible [67] processes, EC type [68], ECE type [69], CE type [70] mechanisms, and other non-unity stoichiometric cases [71]. Basically, the impact of experimental SW parameters - such as pulse heights (or amplitude), step heights (or potential increments), switching potentials, and period – on the peak-currents, peak-potentials and separations, peak-widths were examined. It was found that those observed peak-related quantities changes characteristically according to each particular types of the electrochemical mechanisms, which make it feasible to be utilized as diagnostic parameters.

2. Theory

2.1 Advantages of derivative approach

In general, original signals (y) from a measurement can be given in a polynomials of a dependent (x) variable:

$$y = \mathrm{a}x^{\mathrm{n}} + \mathrm{b}x^{(\mathrm{n}-1)} + \mathrm{c}x^{(\mathrm{n}-2)} + - - - + \mathrm{f}x^2 + \mathrm{g}x + \mathrm{h}$$

Taking its firstst derivative of a signal above removes a constant component (h) which represent a background process yielding the signals, and taking 2nd derivative removes a linear component (gx term) which represent a secondary process for a signal component, thus revealing the hidden higher order processes. In other words, from these operations, (a) undesired components in the signal (a constant term, a linier component, and quadratic component) can be removed successively upon repeated differentiations) leaving higher order terms only. Thus, upon differentiations, changes (in- or de-creases) in the signal can be developed to yields peaks which were hidden behind the original signal; hence the better-defined peaks, makes the analysis of curve much easier than the original curves; thus revealing concealed features in the original function. Basically, higher order non-linear components in the function (original signals) transforms into peaks in the process of differentiations.

2.2 Peak currents, peak potentials, and peak widths and the ratios

Theoretical expressions for the first- second- and third- derivatives of current-potential curves were analytically derived from successive differentiations of the original current expression (i.e., zeroth order derivative) obtained; then the derivatives as function of potential were computed and plotted. The raw and derived parameters (**Figure 1**) for the simple reversible process (E_{rev}), calculated from the theoretical equations (Eq. 1–21; refer to Section 3.1.1) for lower order derivatives and those values given in the last column of **Table 1**; this analytical approach requires only a pencil and papers. However, for third derivatives for E_{rev} and all derivatives for other types (E_{iir}, CE_r, and E_rC), it is difficult to derive analytical solutions (equations) to calculate values of the parameter. Thus, numerical approaches had to be employed; namely, the values of a parameter as a function of an independent variable (such as a rate constant, k) had to be plotted on a graphs first, then a possible value or ranges of values were read from the graphs, and an equation for the relationships to calculate the parameters had to be found from such graphs, and this numerical approach requires much of computations, graph papers and a ruler. Those values of parameters, found analytically or numerically/graphically are summarized (refer to the results sections for details) in **Table 2** for four types of the mechanisms for comparisons.

2.3 Obtaining theoretical derivatives for various types of electrode mechanisms - analytical method vs. numerical (digital) method

Theoretical equations for first, second and third derivatives for irreversible (and quasi reversible) processes are obtained with successive differentiations of analytical expressions of the current [15]. The derivatives of the CE type of mechanism are obtained with successive differentiations in a similar fashion [12, 13, 42]. Whenever possible, analytical solutions are sought after first because it is computationally less expensive. Numeral differentiations had to be adopted when analytical solution cannot be found. Thus, the analytical solutions of current expression and it's first derivative for the EC type of mechanisms are so complicated that analytical differentiations for the second and third derivatives were practically impossible [11, 14], numerical differentiations had to be adopted [14]. A value of $\Delta E = 1.0$ mV was used for a finer resolution for all curves. The differentiation methods are summarized in **Table 3**.

3. Results and discussions

3.1 Derivation of the original current expression for various types of electrode mechanisms

3.1.1 Simple reversible electron transfer reaction (E_{rev} or E_r)

A theoretical current (i) as a function of the applied potential (E), under semi-infinite linear diffusion, is expressed by the following relationship for a simple reversible electron-transfer process [1, 38]

$$O + ne^- \rightarrow R \text{ (Nernstian)} \tag{1}$$

$$i(E) = id\,[1/(1 + e)] \tag{2}$$

where

	Parameters	Definitions	E_r	E_{irr}	E_rC	CE_r
Current	$E_{1/4}$ (mV)		+28	<28 (-205)		
	$E_{1/2}$ (mV)		0	<0 (-253)	>0	<0
	$E_{3/4}$ (mV)		-28	<-28 (-295)		
	ΔE_q^a (mV)	$E_{1/4}$-$E_{1/2}$	28	>28 (48)	<28	>28
	ΔE_q^c (mV)	$E_{1/2}$-$E_{3/4}$	28	>28 (42)	<28	>28
	ΔE_h^0 (mV)	$E_{1/4}$-$E_{3/4}$	56.4	>56.4 (90)	<56.4	>51.5, <56.4
	$\mathbf{R^0_{qw}}$	$\mathbf{\Delta E_q^a/\Delta E_q^c}$	**1.0**	**>1.00 (1.14)**	**>1.00**	**<1.00**
	R_i^0 (normalized o Er)		1.0	<1.0 (0.64)	1.0	<1.0
1st Der.	ΔE^1_p (mV)	E_p-E^o	0.0	<0.0 (-265)	>0	<0.0
	i_p		9.6	<9.6 (6.4)	>9.6	<9.6
	R_i^1 (normalized o Er)		1.00	<1.00 (0.67)	>1.00	<1.00
	$W_{½}$ (mV)		90.5	>90.5 (143)	<90.5	<90.5
	W_q^a (mV)		45.3	>45.3 (80)	<45.3	>42, <45.3
	W_q^c (mV)		45.3	>45.3 (63)	<45.3	>40, <45.3
	$\mathbf{R_w^1}$	$\mathbf{W_q^a/W_q^c}$	**1.00**	**>1.00 (1.27)**	**> 1.00**	**<1.00**
2nd Der.	E_p^{2a} (mV)		34	<34 (-210)	< 34	> 34
	E_p^{2c} (mV)		-34	<-34 (-319)	<-34	>-34
	ΔE_p^2 (mV)	E_p^{2a}-E_p^{2c}	68	>68 (108)	< 68	>59, <68
	$\mathbf{R_E^2}$	$\mathbf{\Delta E_p^{2a}/\Delta E_p^{2c}}$	**1.00**	**<1.0 (0.66)**	**>1.0**	**>1.0**
	i_p^{2a}		14	>14 (51)	> 14	>10
	i_p^{2c}		-14	<14 (-72)	>-14	>10
	$\mathbf{R_i^2}$	$\mathbf{\lvert i_p^{2a}/i_p^{2c}\rvert}$	**1.00**	**<1.00 (0.71)**	**>0.71, <1.00**	**>1.0, <1.25**
	$W_{½}^{2a}$ (mV)		63	>63 (111)	>50, <63,	<63
	$W_{½}^{2c}$ (mV)		63	>63 (81)	> 35, <63	<63
	$\mathbf{R_w^2}$	$\mathbf{W_{½}^{2a}/W_{½}^{2c}}$	**1.00**	**>1.00 (1.37)**	**>1.00, <1.45**	**<1.0**
3rd Der.	$\mathbf{E_p^{3a}}$ **(mV)**		**59**	**<59 (-165)**	**>59**	**>-180**
	$\mathbf{E_p^{3m}}$ **(mV)**		**0.00**	**<0.00 (-275)**	**>0.00**	**>-230**
	E_p^{3c} (mV)		-59	<-59 (-353)	>-59	>-300
	ΔE^3_p (mV)	E_p^{3a}-E_p^{3c}	118	>118 (188)	<118	<118
	ΔE^3_a (mV)	E_p^{3a}-E_p^{3m}	59	>59 (110)	<59	<59
	ΔE^3_c (mV)	E_p^{3m}-E_p^{3c}	59	>59 (78)	<59	<59
	$\mathbf{R_{\Delta E}^{\ 3}}$	$\mathbf{\Delta E^{3a}/\ \Delta E^{3c}}$	**1.00**	**>1.00 (1.41)**	**>1.0, <~1.6**	**<1.00**
	i_p^{3a}		2461	<2641 (449)	<3692	<2461
	i_p^{3m}		-7382	>-7382 (-1961)	>-15500	>-7382
	i_p^{3c}		2461	<2461 (1061)	< 8367	<2461
	$\mathbf{R_i^3}$	$\mathbf{i_p^{3a}/i_p^{3c}}$	**1.00**	**<1.00 (0.42)**	**>0.4, <1.0**	>1.0, <1.27
	$\mathbf{R_i^{3a}}$	$\mathbf{\lvert i_p^{3a}/i_p^{3m}\rvert}$	**0.32**	**<0.32 (0.22)**	**>0.2, <0.32**	>0.32, <0.36
	$\mathbf{R_i^{3c}}$	$\mathbf{\lvert i_p^{3c}/i_p^{3m}\rvert}$	**0.32**	**>0.52 (0.52)**	**>0.32,<0.60**	>0.28,<0.32
	$W_{½}^{3a}$ (mV)		54	>54 (102)	>50, <54	>42.5, <54
	$W_{½}^{3m}$ (mV)		41	>41 (64)	>32, <54	>35, <41

Parameters	Definitions	E_r	E_{irr}	E_rC	CE_r
$W_{½}^{3c}$ (mV)		54	>54 (63)	>32, <42	>50, <54
R_w^{3}	$W_{½}^{3a}/W_{½}^{3c}$	1.00	>1.00 (1.62)	>1.0, <1.7	>1.00,<1.19
R_w^{3a}	$W_{½}^{3a}/W_{½}^{3m}$	1.32	>1.32 (1.60)	>1.3, <1.6	>1.19,<1.32
R_w^{3c}	$W_{½}^{3c}/W_{½}^{3m}$	1.32	<1.32 (0.98)	>1.0, <1.3	>1.32, <1.43

Notes: (a) Calculated for n=1and T=298°K for the reversible e^- transfer rection (Third Column).
(b) All values for the irreversible case were calculated for k^o=1.0x10^{-5}cm/s and αn=0.50;
For the case with αn ≠ 0.50, these values are not restricted to the ranges given above. those values the parentheses are calculated with k_s=1.0x10^{-5}cm/s
(c) Values for quasi-reversible processes, calculated for αn=0.5, lie in between those for reversible processes and those for irreversible processes, hence the column for quasi-reversible cases is omitted.

Table 2.
Comparisons of values independent variable and derived parameters for various mechanisms.

$$i_d = nFAD_o^{1/2}C_o/(\pi t)^{1/2} \text{ (the diffusion-controlled current)} \quad (3)$$

$$e = \exp\ (nFE/RT) \quad (4)$$

and E is an applied potential with respect to the reversible polarographic half-wave-potential, $E_{1/2}$ (rev) or E_h. Other parameters and variables have their usual meanings. Eq. 2 is derived assuming $D_O = D_R$. Successive differentiations of Eq. 2 with respect to E yields [15].

$$i'(E) = -i_d(nF/RT)\left[e/(1+e)^2\right] \quad (5)$$

$$i''(E) = -id(nF/RT)^2\left[e(e-1)/(1+e)^3\right] \quad (6)$$

$$i'''(E) = -id(nF/RT)^3\left[e(e^2-4e+1)/(1+e)^4\right] \quad (7)$$

Examination of Eq. 2 reveals that

$$i(-E) = i_d - i(E) \quad (8)$$

$$i(0\ ) = i_d/2 \quad (9)$$

This implies that *the i-E curve (solid line, **Figure 1A**) is symmetrical with respect to its inflection point (0, i_d/2).*

Types of Mechanisms	Current i	1^{sit} Der. i'	2nd Der. i"	3rd Der. i"'
		di'/dE	di"/dE	di"'/dE
E_r(rev)	i	di'/dE	di"/dE	di"'/dE
E_{irr}(irreversible)	i	di'/dE	di"/dE	di"'/dE
CE_r (Prior Chemical)	i	di'/dE	di"/dE	di"'/dE
E_rC(Post Chemical)	i	di'/dE	***Δi"/ΔE*** *(numerical)*	***Δi"'/ΔE*** *(numerical)*

Table 3.
Methods of differentiation for various electrode mechanisms. Derivatives are obtained from the analytical solutions of currents for each cases, except for i", i"' with the post kinetics system (E_rC).

3.1.1.1 For the first derivatives

Eq. 5 shows

$$i'(-E) = i'(E) \tag{10}$$

This means that the *first derivative is symmetrical with respect to an axis E = 0* (**Figure 1B**, solid line). The value of the peak-width at a half-height for the reversible process proved to be 90.5/n mV at 298 K [18]. A shape parameter from the first derivative can be defined; the peak widths are divided into positive (anodic) and negative (cathodic) parts to define two semi-peak widths ($W_q{}^a$ and $W_q{}^c$, **Figure 1B**) and their ratios. For a reversible process it was found [14].

$$W_{1/2} = 90.5\text{mV} \tag{11}$$

$$W_q{}^a = W_q{}^c = (1/2)W_{1/2} = 45.3/\text{n mV at 298 K} \tag{12}$$

$$W_q{}^a/W_q{}^c = 1 \text{ at any temperature} \tag{13}$$

3.1.1.2 For the second derivatives

One can examine the second derivative (i”), Eq. (6), to find

$$i''(-E) = -i''(E) \text{ and } i''(0) = 0 \tag{14}$$

Thus, the second derivative (**Figure 1C**, solid line) *has the same symmetry as the current with respect to its inflection point*. One can define shape parameters $W_{1/2}{}^{2a}$ and $W_{1/2}{}^{2c}$ which are the peak-widths for the anodic (positive) and cathodic (negative) peaks in the second derivative. It should be pointed out that the adjectives “anodic” and “cathodic” in this context are *not* related to the actual redox processes associated with the peaks. The two peak heights for the anodic, $i_p{}^{2a}$, and the cathodic, $i_p{}^{2c}$, peaks are defined at two peak potentials, $E_p{}^{2a}$ and $E_p{}^{2c}$, respectively (**Figure 1C**).

Following relationships are found for E_r type.

1. Peak Potentials

$$E_p{}^{2a} - E_p{}^{2c} = 68/\text{n mV (at 298 K)} \tag{15}$$

$$|E_p{}^{2a}/E_p{}^{2c}| = 1.00 \tag{16}$$

2. Peak Currents

$$i_p{}^{2a} = -i_p{}^{2c} \tag{17}$$

$$|i_p{}^{2a}/i_p{}^{2c}| = 1.00 \tag{18}$$

3. Peak Widths

$$W_{1/2}{}^{2a} = W_{1/2}{}^{2c} = 62.5/\text{n mV (at 298 K) or} \tag{19}$$

$$W_{1/2}{}^{2a}/W_{1/2}{}^{2c} = 1.00 \tag{20}$$

The peak-potentials in Eq. 15 can be found by solving i’“(E) = 0 (Eq. 7) for E.

$$-\mathrm{id}(\mathrm{nF/RT})^3\left[\mathrm{e}\left(\mathrm{e}^2-4\mathrm{e}+1\right)/(1+\mathrm{e})^4\right]=0 \tag{21}$$

As given in the value of the unity (1.0) for those symmetry parameters (i.e. anodic to cathodic peak-current ratios, anodic to cathodic half-peak-width ratios, and anodic to cathodic peak potential ratios) for first or second derivatives, the derivative curves are symmetrical.

3.1.1.3 For the third derivatives

It's symmetry is the same as that of the first derivative, *being symmetrical with respect to E = 0*. One can find the ranges of parameter values in similar fashion for the third derivatives. Details of the work can be found elsewhere [15]: the key findings on the peak potentials, peak current and half peak widths and the various ratio for symmetry are given in **Tables 1** and **2**. It should be noted that all those values for the anodic to cathodic symmetry parameters (those ratios Rs) are unity, the same as in lower order derivatives.

As shown above, the values of parameters above for E_r process are derived analytically from the equations which are much simpler than others. However, for an irreversible electron process (E_{irr}, or $E_{quasirev}$) and chemically coupled electron transfer reactions such as E_rC and CE_r, the expressions for the current and their derivatives are so complicated that it is practically impossible to derive the values of those parameters analytically by solving algebraic equation (Refer to later section). Therefore, a numerical approach had to be adopted. Namely, (a) about 10 curves for derivatives vs. potentials at various values of independent variables (the heterogeneous rate constant k^o and αn for simple electron transfer; the homogeneous constant K_{eq}, k_f, k_b and k for chemically coupled electron transfer) had to be calculated, (b) then, all those curves are plotted out, and (c) those plots are analyzed graphically by examining the curves carefully in order to obtain values or equations (relationships) for those parameters.

3.1.1.4 Simple values of symmetry parameters of reversible electrochemical process (E_r)

The symmetric relationship among the symmetry parameters are found from the simple fast electron transfer process; namely, a reversible electron transfer (E_{rev}, or E_r) type in which the electrode process is basically Nernstian. What will happen to the shapes of voltammograms and its derivatives if the system involve non-Nernstian behavior with a slower irreversible electron transfer rate (E_{irr} type), or the electron transfer is coupled with a prior chemical equilibrium (CE_r Type) or an E_r coupled with a follow-up chemical reactions (E_rC). These studies have been already done mostly and discussed in detail by our group [14]; to summarize, the symmetry exhibited in the derivatives in E_r disappears for the non-Nernstian electron transfer (i.e., irreversible) system. Namely, the asymmetry strongly depends on those the electron transfer parameters (α and k^o) for E_{irr} type, and the kinetic parameters (i.e., k and K_{eq}) for chemically coupled processes (C_rE_r and E_rC_{irr} Type). In this work, comparisons are made for the symmetry parameters for the different electrode mechanisms. Details are given in later sections for each of corresponding mechanisms.

3.1.1.5 Reversibility in electrochemical reactions

In general, reversibility in electrochemical reactions is divided into three kinds depending on the magnitude of the standard heterogeneous rate constant k^o (or k_s)

with respect to the diffusion coefficient (D): an electron transfer reaction is considered reversible if e^- transfer is much faster the diffusion (i.e., $k^o >> (D/t)^{1/2}$), and irreversible if e^- transfer is much slower the diffusion ($k^o << (D/t)^{1/2}$) and quasi-reversible if those competing rates are comparable ($k^o \sim (D/t)^{1/2}$).

In terms of values of k^o, by adopting typical values of D (=5x10^{-5} cm^2/s) and t (=1.000 s).

reversible for $k^o > 0.020$ cm/s,
quasi-reversible for 0.020 cm/s > k^o > 5x10^{-5}cm/s,
irreversible for k^o < 5x10^{-5}cm/s.

In the following sections, other types of processes than the simple reversible process will be treated.

3.1.2 Theoretical expression for and the derivatives for $E_{quasirev}$ and E_{irr} types of electron transfer

Currents (i) as a function of the applied potential (E) for non-Nernstian (i.e., irreversible and quasi-reversible) system with a slower electron transfer, have been previously derived and given in several references [44, 54–56, 72, 73], and has final forms of the following equations for a planar diffusion:

$$I = nFAC_o k_f \left[\exp\left(Q^2 t\right)\text{erfc}\left(Qt^{½}\right)\right] \tag{22}$$

assumed $D_R = D_O = D$.

All symbols have their usual meanings and may be refer to the reference for details,

where,

$$Q = (k_f + k_b)/D^{½} \tag{23}$$

$$k_f = k^o \exp\{-\alpha nF(E - E^o)/RT\} \tag{24}$$

$$k_b = k^o \exp\{(1 - \alpha)nF(E - E^o)/RT\} \tag{25}$$

$$E_{1/2} = E^o + (RT/nF)\ln(D_R/D_O)^{½} = E^o \tag{26}$$

normalizing the current, Eq. (22), to the diffusion current, Eq. (3), yields

$$i_n = (\pi t/D)^{1/2} k_f \left[\exp\left(Q^2 t\right)\text{erfc}\left(Qt^{½}\right)\right] \tag{27}$$

where,

$$F\left(Qt^{1/2}\right) = \exp\left(Q^2 t\right)\text{erfc}\left(Qt^{½}\right) \tag{28}$$

This normalized current, which is independent of the concentration and electrode surface area, mostly is dependent on the heterogeneous rate constants (k^o), the transfer coefficient (α) and n, and the curve found to exhibit an asymmetry [15].

Successive differentiations of the current yield the following expressions for the first-, second- and third- derivatives respectively [15]. Taking derivative of the current with respect to potential (E) yields the first derivative,

$$i' = (\pi t/D)^{1/2} k_f \left[(-nF/RT)F\left(Qt^{1/2}\right) + 2Q'\left[QtF\left(Qt^{1/2}\right) - (t/\pi)^{1/2}\right]\right] \tag{29}$$

where

$$Q' = (-\alpha nF/RT)k_f/D_o^{1/2} - ((1-\alpha)nF/RT)k_b/D_r^{1/2} \quad (30)$$

The analytical differentiations of i' and expressions for the second- and third-derivatives are more involved and given elsewhere [15]. Alternatively, it is also possible to obtain second and third derivative curves from successive *numerical* differentiations starting from the values of first derivative.

As shown above, the current expression for or an irreversible e^-transfer and quasi-reversible much complicated than that of reversible case, because the current depends not only on the applied potential but also k^o and αn which introduces much differences in the peak shapes and peak symmetry. Comparisons of the cases are made in **Figure 2** for a contrast for E_r and E_{irr} processes.

It can be readily seen from the curves that as the reduction becomes more difficult (i.e., with smaller k^o) (a) all curves shifts to negative (cathodic) direction for the quasi-reversible and irreversible process, (b) the curves and peaks become widened, and (c) the curve becomes asymmetric which is much pronounced in the second and third derivatives although it is not readily seen from the current and its first derivative. By analyzing the formula and the curves graphically, one can find possible values and their ranges for the following variables or asymmetric parameters listed on **Table 1**: such values of the irreversible process are summarized in **Table 2** along with those of the reversible process for comparisons.

3.1.2.1 For the first derivative

First derivative is obtained by differentiating Eq. (27) and (28), the details of differentiation and the results were given elsewhere in original work [15] of the author's group: for the irreversible or quasi-reversible process, following relationships are found from plots

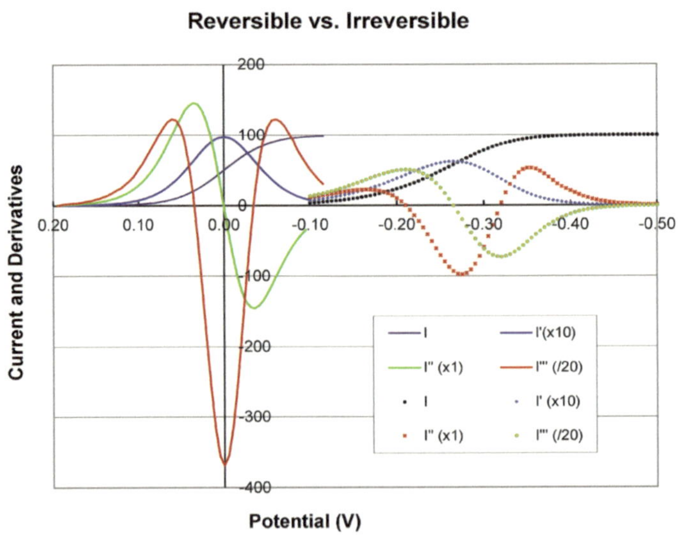

Figure 2.
Comparison of currents (normalized at 100, black), and their first (blue), second (green), and third derivatives (red). Those for a simple reversible electron transfer are given in solid lines on the left, and those for an irreversible electron transfer reaction (with k^o = 10^{-5} cm/s, n = 1, and α = 0.5 at T = 298 K) are in dotted lines on the right. The derivatives were scaled appropriately in order to bring all of them in the same plotting area: Scaling factors are given in the parentheses.

1. Peak Potentials
   It is found that the peak potential for totally irreversible process ($k^o < 10^{-5}$ cm/s) is directly proportional to log (k^o), shifting towards cathodic direction as k^o decreases for a given αn, and is inversely proportional to αn at a constant k^o [15].

$$E_p = E^o + [(60.3)\log(k^o) + 165)](1/\alpha n)\ (mV) \quad (31)$$

<0 (because log(k^o) is always negative for E_{irr}).

2. Peak Currents
   The normalized peak heights is directly proportional to αn, decreasing from 9.6 (for E_r) to 6.2 as k^o decreases [15]

$$i_p/i_d = 12.3(\alpha n) \sim 6.2 < 9.6 \text{ for } \alpha n = 0.5 \quad (32)$$

3. Peak-Widths.
   The half-peak width ($W_{½}$), the anodic half-peak width ($W_q{}^a$), and the cathodic half-peak width ($W_q{}^c$) at 298 K for a totally irreversible reaction are independent of k^o, but inversely proportional to αn [15]

$$W_{½} = 2.80(RT/F)(1/\alpha n)(V) = 72(1/\alpha n)(mV) = 144\ (mV) > 90.4\ (mV) \quad \text{for } \alpha n = 0.5 \quad (33)$$

$$W_q{}^a = 1.62(RT/F)(1/\alpha n)(V) = 41.7(1/\alpha n)(mV) = 83.4\ (mV) > 45.3(mV) \text{ for } \alpha n = 0.5 \quad (34)$$

$$W_q{}^c = 1.26(RT/F)(1/\alpha n)(V) = 32,4(1/\alpha n)(mV) = 64.8\ (mV) > 45.3(mV) \text{ for } \alpha n = 0.5 \quad (35)$$

$$W_q{}^a/W_q{}^c = 1.28 > 1.00 \text{ for all } \alpha n \quad (36)$$

3.1.2.2 For the second derivative

Second derivative was obtained by differentiating the first derivative, Eq. (29). The details of the derivation and the results are given elsewhere and following relationships were found [15]:

1. Peak-Potentials and Peak Separation
   Both peak potentials (anodic and cathodic) of the second derivatives for totally irreversible processes ($k^o < 10^{-5}$ cm/s) are directly proportional to log (k^o) for a given αn, moving to cathodic direction as k^o decreases, which is the same trend as for E_p in the first derivative.

$$E_p{}^{2a} = E^o + \{(60.3)\log(k^o) + 201\}(1/\alpha n)\ (mV) \quad (37)$$

$$E_p{}^{2c} = E^o + \{(60.3)\log(k^o) + 67\}(1/\alpha n)\ (mV) \quad (38)$$

and $E_p{}^{2a}$-$E_p{}^{2c}$ = 2.3(RT/F)(1/αn) = 0.0591(1/αn) (mV)

$$= 118\ mV > 68 mV \text{ for } \alpha n = 0.5 \quad (39)$$

Both peak potentials (E_p^{2a} and E_p^{2a}) are dependent on k° as well as αn values yielding −210 mV and -319 mV respectively, with $k^o = 10^{-5}$ cm/s and αn =0.5. However, the peak potential difference depends only on αn values, not on k°.

$$\Delta E_p^{2a}/\Delta E_p^{2c} = 210\text{mV}/319\text{mV} = 0.66 < 1.00 \quad (40)$$

2. Peak-Heights and Their Ratios.
   A plot of the two peak-heights (i_p^a for anodic and i_p^c cathodic side) of the second derivative depend heavily on αn, a magnitude of both heights increases with increasing αn. However, i_p^c is greater than i_p^a for most of the case unless αn = 0.3 which is strange.

$$|i_p^a/i_p^c| = 0.78 < 1.00 \text{ for } \alpha n = 0.5 \quad (41)$$

3. Peak-Widths and Their Ratio.
   Two half-peak-width (anodic and cathodic) of the second derivatives, and the ratio for E_{irr} processes are given below

$$W_{1/2}^{2a} = 2.10(RT/F)(1/\alpha n)\ (V) = 54.0(1/\alpha n)\ (mV) = 108\ mV > 64\ mV \text{ for } \alpha n = 0.5 \quad (42)$$

$$W_{1/2}^{2c} = 1.53(RT/F)(1/\alpha n)\ (V) = 39.4(1/\alpha n)\ (mV) = 78.8\ mV > 64\ mV \text{ for } \alpha n = 0.5 \quad (43)$$

$$W_{1/2}^{2a}/W_{1/2}^{2c} = 1.37 > 1.00 \text{ for all } \alpha n \quad (44)$$

Here again, it should be noted that although the two half-peak-widths depends on αn, their ratio is invariant.

3.1.2.3 For the third derivatives

Third derivative are obtained by differentiating the second derivative. The details of the derivation and results were given elsewhere [15]. The number of parameters increases, and results on the parameters increase as three peaks are available, and those parameters (peak separations, peak-current ratios, and half-peak ratios were analyzed from various plots. The results with possible ranges of the values are summarized in **Table 2**.

3.1.2.4 Sensiivity increases with increasing order of derivatives

In **Table 2**, the key results from irreversible process (as well as two chemically coupled processes) are juxtaposed here for a ready comparison of the differences observed among the types of mechanisms. Basically, the symmetry observed in the curves for the reversible process, disappears as the electron transfer rates become slower (i.e., irreversible type): the unity (1.00) values of the ratio parameters becomes no longer 1.00 for irreversible (E_{irr}) process and for other types of electrode mechanisms. Namely, the ratio of anodic to cathodic peak-widths ($W_{1/2}^a/W_{1/2}^c$) increases to 1.27 (an increase of 27%) for the first derivative, and it increases to 1.37 (an increase of 37%) for the second derivative, while it increases to1.62 (an increase of 62%) for the third derivative. In general these ratios increases systematically, as an order of a derivative increases; the higher the order is, the larger the changes are.

These changes (or sensitivity) of symmetry parameter values with increasing order of a derivative are also observed in CE_r as well as E_rC. In short, the values of the symmetry parameters with a ratio of anodic values to cathodic values (all **R's**) are 1.00 for reversible electron transfer reactions; however it will deviates from the unity (<1.00 or > 1.00) for other types of processes as shown in later sections.

3.1.2.5 Notes on computations for quasi-reversible and irreversible processes

All computations are carried out with $k^o = 1.0x10^1$cm/s for a reversible case, $k^o = 1.0x10^{-3}$cm/s for a quasi-reversible case, $k^o = 1.0x10^{-5}$ cm/s for an irreversible case for simple e^- transfer reactions. A value of *αnof* 0.50 is used for all quasi-reversible and irreversible cases. For quasi-reversible processes, only the case with $\alpha n = 0.5$ is reported. The values for all parameters for quasi-reversible process (with $\alpha n = 0.5$) lies in between those for reversible and those for irreversible case. Those values for quasi-reversible case with $\alpha n \neq 0.5$ exhibits much complicated behaviors depending on values of αn [15]. This is because depending on values of the kinetic parameters (k^o and αn) the changes in some dependent variables become not monotonic,, often exhibiting discontinuity at around $\alpha n = 0.3$.

Propagations of round-off errors, associated with differentiations (or subtractions) which result in a loss of significant figures, encountered in evaluating the derivatives (Eq. 27–30): the error were so severe that an extended quadruple-precision mode (32 significant digits) had to be employed for all calculation: namely, the function $\exp.(x^2)\text{erfc}(x)$: namely $F(Q_s t^{1/2}) = \exp.(Q_s^2 t)\text{erfc}(Q_s t^{1/2}))$ (Eq. 28) had to be evaluated to the 32 digit-precision [13].

3.1.3 Reversible electron transfer coupled with a prior chemical equilibrium (CE_r)

3.1.3.1 Theoretical expressions for and the derivatives

Full derivations of the derivatives from a current expression are given elsewhere [1, 42]: the basic rection scheme and final forms of formula are presented here.

This mechanism that involves pre-chemical step can be given as follows.

$$Y \underset{k_b}{\overset{k_f}{\rightleftharpoons}} O \quad K_{eq} = k_f/k_b \quad \text{(in solution)} \tag{45}$$

$$O + ne^- \rightarrow R \quad \text{Nernstian} \quad \text{(at the electrode)} \tag{46}$$

At a more interesting case of $K_{eq} << 1$, the currents [1, 13] is given by

$$i = i_d\left[(\pi t)^{1/2}\left(K_{eq}k_f\right)^{1/2}\exp\left(K_{eq}k_f(1+e)/K_{eq}\right)^2 t\right)\text{erfc}\left(\left(K_{eq}k_f\right)^{1/2}(1+e)/K_{eq}\right)t^{1/2})] \tag{47}$$

Successive differentiations of the current yield the first-, second- and third-derivatives, respectively.

$$i' = i_d(nF/RT)(2k_f\,t)e\left[(1+e/K_{eq})i - 1\right] \tag{48}$$

Second derivative becomes

$$i'' = i_d(nF/RT)^2(2k_f\,t)e\left[(1+e/K_{eq})i + (RT/nF)(1+e/K_{eq})i' - 1\right] \tag{49}$$

Third derivative becomes

$$i''' = i_d(nF/RT)^3(2k_f\ t)e\Big[(1+4e/K_{eq})i + (2RT/nF)(1+2e/K_{eq})i' + (RT/nF)^2(1+e/K_{eq})i'' - 1\Big] \tag{50}$$

Such curves according to the derivative formula are given in **Figure 3**. Full details of derivations of the three derivatives above and the analysis of the all parameters from those graphs were presented elsewhere [12, 15, 42], and only important key results are presented here, and summarizedin the **Tables 2** and **4**.

Typical i-E curves at various k_f value for CE process are given **Figure 3A** in two dimensions, and **Figure 4A** in three dimensions with an additional axis for $\log(k_f)$. The plots of the original currents and three derivatives were calculated according to the Eq. (47), (48), (49), and (50). As it can be expected the entire i-E curve moves upward with the diffusion-controlled current as k_f increases. Asymmetry is not readily noticeable in the current and first derivatives, but it becomes more visible in the second and third derivatives: namely, anodic side of the peak height is larger than the cathodic side in both second and the third derivatives, This is more evident for larger k_f where currents are larger in all potentials. All the relationships are obtained graphically [42], because finding the analytical solution of the dependent variables and parameters from the theoretical equations are too involved to be calculated. Graphic analysis of the curves yields the following results: the full account for the analysis of the derivative curves can be found in Ref. [12, 42].

3.1.3.2 For the first derivative

For a CE_r process, following relationships can be obtained from analysis of the graphs for the first derivative.

1. Peak-Potentials

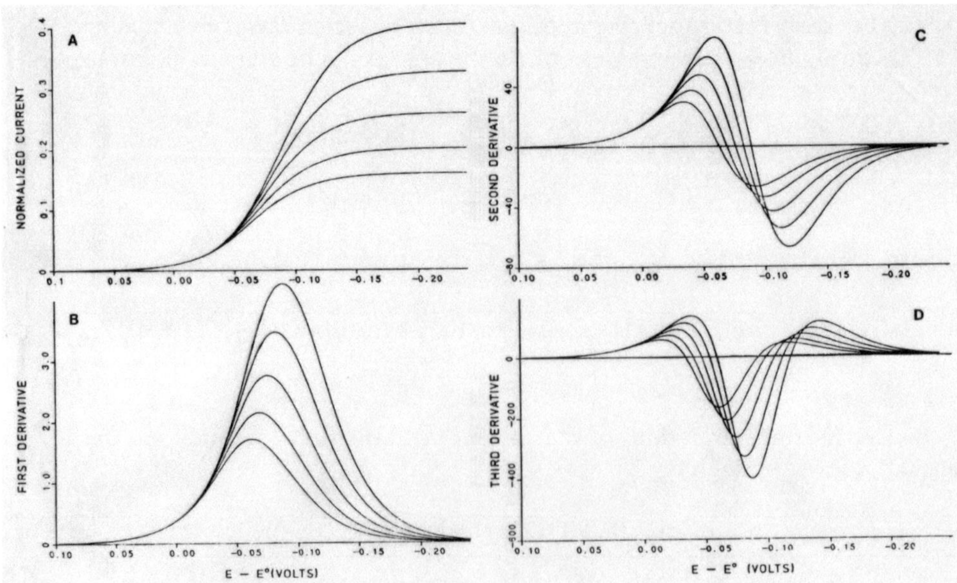

Figure 3.
Normalized currents (a), their first (B), second (C) and third (D) derivatives at various values of forward rate constant at Keq = 10^{-2} *for* CE_r *type mechanism: (a)* $k_f = 10\ s^{-1}$ *(for the lowest current and derivatives), (b) 5.62, (c) 3.16, (d) 1.78, (e) 1.00 (for the highest current and derivatives), calculated for* $T = 298$, $t = 1{,}0$, *and* $n = 1$.

	Parameters	Definitions	E_r	E_{irr}	E_rC	CE_r
Current						
	ΔE_h^0 (mV)	$E_{1/4}$-$E_{3/4}$	56.4	>56.4 (90)	<56.4	>51.5, <56.4
	R_q^0	**$\Delta E_q^a/\Delta E_q^c$**	**1.00**	**>1.00 (1.14)**	**>1.00**	**<1.00**
1st Der.						
	$W_{½}$ (mV)	90.5	>90.5 (143)	<90.5	>90.5	
	R_w^1	**W_q^a/W_q^c**	**1.00**	**>1.00 (1.27)**	**>1.00**	**<1.00**
2nd Der.						
	ΔE_p^2 (mV)	E_p^{2a}-E_p^{2c}	68	>68 (108)	< 68	>59, <68
	$R_{\Delta E}^2$	**$\Delta E_p^{2a}/\Delta E_p^{2c}$**	**1.00**	**<1.0 (0.66)**	**>1.00**	**>1.00**
	R_i^2	**$\lvert i_p^{2a}/i_p^{2c}\rvert$**	**1.00**	**<1.00 (0.71)**	**>0.71, <1.00**	>1.0, <1.25
	R_w^2	**$W_{½}^{2a}/W_{½}^{2c}$**	**1.00**	**>1.00 (1.37)**	**>1.00, <1.45**	**<1.00**
3rd Der.						
	$R_{\Delta E}^3$	**$\Delta E^{3a}/\Delta E^{3c}$**	**1.00**	**>1.00 (1.41)**	**>1.0, <~1.6**	**<1.00**
	R_i^3	**i_p^{3a}/i_p^{3c}**	**1.00**	**<1.00 (0.42)**	**>0.4, <1.0**	>1.0, <1.27
	R_i^{3a}	**$\lvert i_p^{3a}/i_p^{3m}\rvert$**	**0.32**	**<0.32 (0.22)**	**>0.2, <0.32**	>0.32, <0.36
	R_i^{3c}	**$\lvert i_p^{3c}/i_p^{3m}\rvert$**	**0.32**	**>0.52 (0.52)**	**>0.32,<0.60**	>0.28, <0.32
	R_w^3	**$W_{½}^{3a}/W_{½}^{3c}$**	**1.00**	**>1.00 (1.62)**	**>1.0, <1.7**	**>1.00, <1.19**
	R_w^{3a}	**$W_{½}^{3a}/W_{½}^{3m}$**	**1.31**	**>1.31 (1.60)**	**>1.31, <1.6**	**>1.19, <1.31**
	R_w^{3c}	**$W_{½}^{3c}/W_{½}^{3m}$**	**1.31**	**<1.31 (0.98)**	**>1.0, <1.31**	**>1.31,<1.43**

*Notes: All values calculated with the same conditions as in **Table 2**.*

Table 4.
Summary of asymmetry parameters which are more sensitive to the kinetics.

Peak potentials always shifts to more cathodic direction as the equilibrium constant for the prior chemical step, K_{eq} (or k_f), becomes smaller. Namely, E_p is more negative than E°,

$$E_o - E_p > 0.0 \tag{51}$$

2. Peak Current
   Normalized peak currents found to be smaller than 9.6.
   As expected, i_p deceases as K (hence k_f) decreases from the reversible value of 9.6

$$i_p < 9.6 \tag{52}$$

3. Peak Width
   Half-peak-withs become less than 90.4/n (mV), approaching to 80/n (mV) as K (hence k_f) decreases

$$W_{1/2} < 90.5/n\ (\text{mV}) \tag{53}$$

The anodic part of a half-peak is always smaller than the cathodic part.

$$W_q^{\ a} < W_q^{\ c} < W_q^{\ a}\ (\text{rev}) < W_q^{\ c}\ (\text{rev}) = 90.5/n\ (\text{mV}) \tag{54}$$

$W_q^{\ a}/W_q^{\ c} < 1.00$.

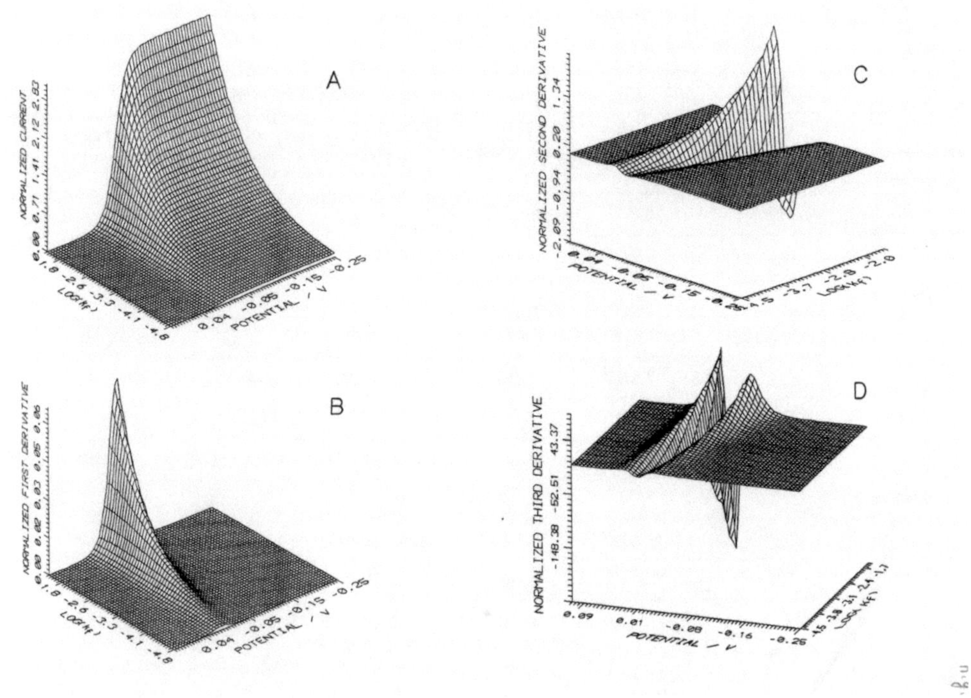

Figure 4.
Three-dimensional plots of normalized current-potential (i-E) curves at various forward rate constants (k_f) with $K_{eq} = 10^{-4}$. The perspectives of the current (A) and first derivative(B) is different from those of the second (C) and third derivatives (D) for a better view of valleys present in second and third derivatives.

3.1.3.3 For the second derivative

Following relationships can be obtained from analysis of the graphs for the second derivative.

1. Peak-Potentials and Peak Separation and the Ratio
   The separation of the two peaks decreases as K (hence k_f) from 68.0/n (mV), the value for E_r.
   E_p^{2a}-E_p^{2c} < 68.0/n (mV)

$$\Delta E_p^{2a}/\Delta E_p^{2c} > 1 \qquad (55)$$

2. Peak Currents and the Ratio
   It is observed that the anodic peak currents are always larger than the cathodic ones, but smaller than 1.25.

$$1.00 < |i_p^{2a}/i_p^{2c}| < 1.25 \qquad (56)$$

3. Peak Widths and the Ratio
   The anodic half-peak is always smaller than the cathodic one

$$W_{1/2}^{a} < W_{1/2}^{c} < W_{1/2}^{a}\ (\text{rev}) < W_{1/2}^{c}\ (\text{rev}) = 64/n\ (\text{mV}) \qquad (57)$$

$W_q^a/W_q^c < 1.00$

3.1.3.4 For the third derivative

Parameters associated with the third derivatives can be graphically analyzed in a similar fashion, and the results are given in the **Table 2** for the various values for three peaks, peak separations, peak-current ratios, and half-peak ratios. Refer to Ref. [42] for full details.

3.1.4 Reversible Electron transfer coupled with a follow-up chemical reaction ($E_r C$)

3.1.4.1 The mechanism and derivation of the current and it's derivatives

This e^- transfer reaction with a post kinetic process mechanism can be given as follows [11].

$$O + ne^- \rightarrow R \quad \text{Nernstian} \quad (\text{at the electrode}) \tag{58}$$

$$R \xrightarrow{k} P \quad (\text{in solution}) \tag{59}$$

The closed form of the analytical solution of the concentration gradient and the current at the specific boundary conditions for this $E_r C$ mechanism can be found elsewhere, but the solutions for the current include Dawson Integrals are too lengthy and complicated be reproduced here and can be found elsewhere [11]. Nonetheless, we managed to obtain the first derivatives by analytically differentiating the current equation. However, further analytical differentiation of the first derivative in order to obtain second and third derivatives were nearly impossible; thus, a numerical approach had to be adopted; namely, the two higher order derivatives i.e., (second and third derivatives) were obtained with ΔE being as small as 1 mV in order to increase a resolution of peaks [14]. Such derivatives thus obtained are given in **Figure 5**. In general, all curves shift to the left (*i.e.*, anodic direction) as the rate constant (k) increases.

3.1.4.2 For the first derivative

For an $E_r C$ type mechanism, following relationships have been observed from the analysis for the first derivative.

1. Peak Potentials
   Peak potentials always shift to anodic direction as the rate constant (k) for the follow up chemical reaction increases. Namely, E_p is more positive than E^o,

$$E_o - E_p < 0.0 \tag{60}$$

2. Peak Currents
   Peak currents increase from 9.6 (the value for E_r) as the rate constant (k) increases to a value of 12.3, thus

$$i_p > 9.6 \tag{61}$$

   i_p (normalized) > 1.00.
   which is opposite of CE_r.

3. Peak Width
   Half peak-with decreases from 90.5/n (mV) as the rate constant (k) increases,

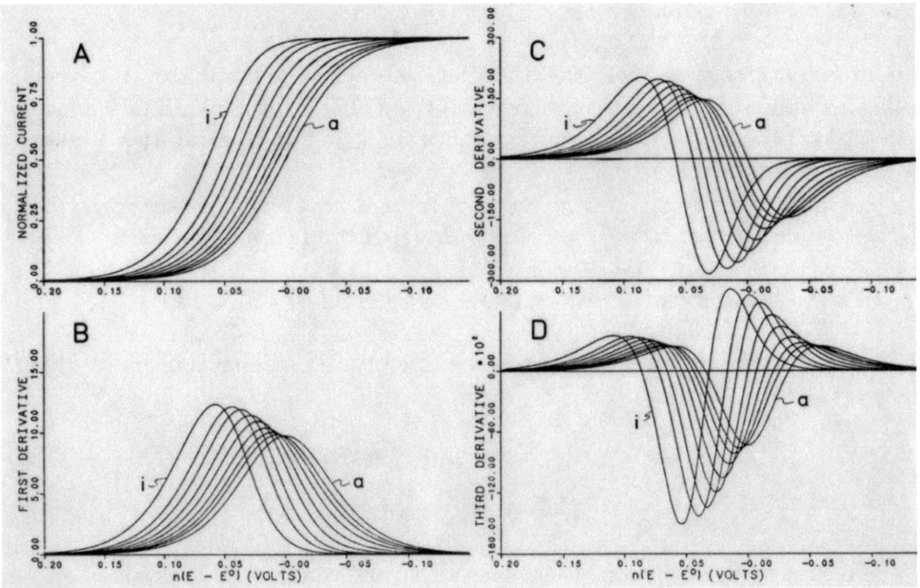

Figure 5.
Theoretical normalized current (A), and its first (B), second (C), and third derivatives (D) at various values of the homogeneous rate constant(k) for the E_rC type of electrode reaction. Calculated for k values of (a) 0 (b) 0.3, (c) 0.562, (d) 1.0, (e) 1.78, (f) 3.0, (g) 5.62, (h) 10.0 (i) 30.0 s^{-1}, and T = 298 K and t = 0.952 s. the normalized currents are dimensionless; thus, the first derivatives are in a unit of V^{-1}, second derivatives in V^{-2}, third derivatives in V^{-3}.

$$W_{1/2} < 90.5\ (\text{mV}) \tag{62}$$

The anodic part of a half-peak is always larger than the cathodic part.

$$W_q{}^c < W_q{}^a < W_q{}^a\ (\text{rev}) = W_q{}^c\ (\text{rev}) = 45.3/n\ (\text{mV}) \tag{63}$$

$$W_q{}^a / W_q{}^c > 1.00 \tag{64}$$

It should be noted again that these trends are the opposite of those from CE_r type.

3.1.4.3 For the second derivative

Following relationships have been observed from the analysis of the second derivative.

1. Peak-Potentials and Peak Separation
   As shown in **Figure 6**, both peak potentials shift anodically, and the separations of the two peaks decrease from 68.0n (mV) and approach to 54 mV as the rate constant (k) increases.

$$54/n\ (\text{mV}) < E_p{}^{2a} - E_p{}^{2c} < 68.0/n\ (\text{mV}) \tag{65}$$

   This suggests that the homogeneous rate constant, k. can be directly determined from the measurements of the peak separation of the second derivative.

2. Peak-Currents and their Ratio
   As shown in **Figure 7**, normalized values of both peak currents increase with inceasing k. However, the peak current ratios always degreases from 1.0 to about 0.71 as k increases.

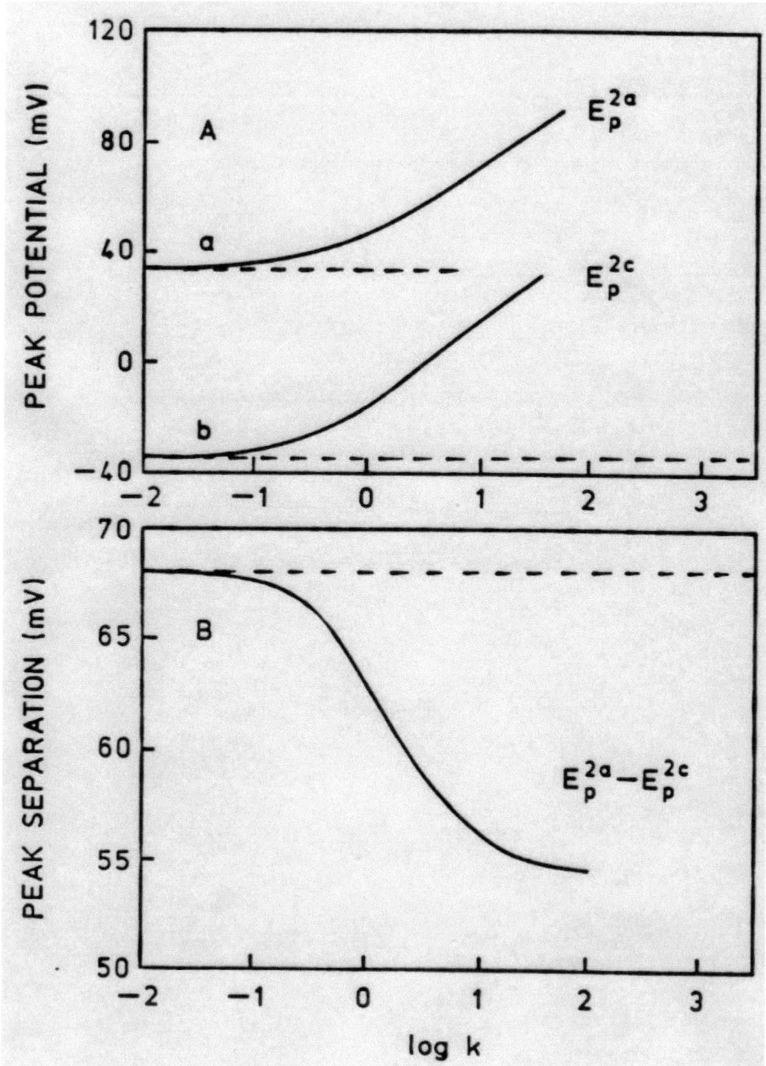

Figure 6.
Effect of k (A) on the (a) the anodic, (b) cathodic peak potentials of the second derivative, (B) effect of k on the peak separation ($E_p{}^{2a}$-$E_p{}^{2c}$).

$$0.71 < |i_p{}^{2a}/i_p{}^{2c}| < 1.00 \tag{66}$$

This implies that the rate constant, k. can be directly determined from the measurements of the peak currents ratio of the second derivatives.

3. Peak-Widths and their Ratio
As shown in **Figure 8**, as a follow-up chemical reaction occurs faster, both anodic and cathodic half-peak-width decreases (i. e., become sharper), but cathodic one become sharper than the anodic one at a higher rate, resulting in an increase in the half-peak ratio of larger than 1 as illustrated in the **Figure 8**.

$$W_{1/2}{}^{c} < W_{1/2}{}^{a} < W_{1/2}{}^{a}\ (\text{rev}) = W_{1/2}{}^{c}\ (\text{rev}) = 63/n\ (\text{mV}) \tag{67}$$

$$W_{1/2}{}^{a}/W_{1/2}{}^{c} > 1.00$$

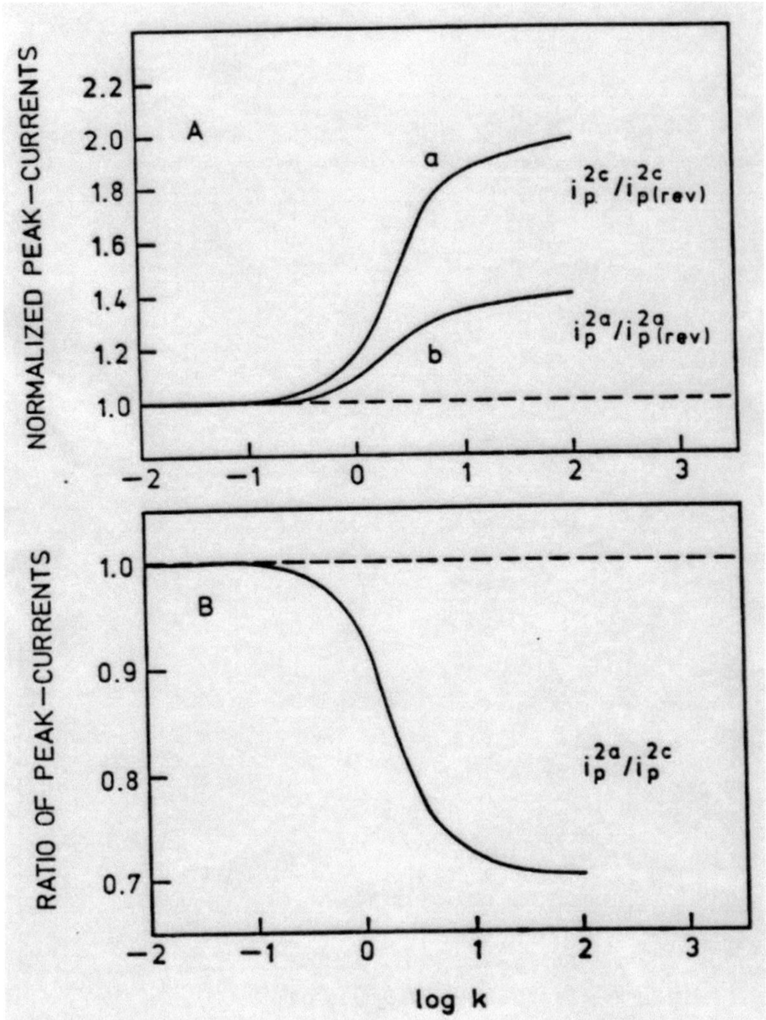

Figure 7.
Effect of k (A) on the (a) the anodic, (b) cathodic peak currents of the second derivative, which are normalized with respect to the values of reversible process. (B) Effect of k on the ratio of the anodic to cathodic peak current ratio, i_p^{2a}/i_p^{2c}.

This suggests that the homogeneous rate constant, k. can be directly determined from the measurements of a ratio of the two half-peak-width of the second derivatives.

3.1.4.4 For the third derivative

Third derivatives can be also analyzed graphically in the same fashion, and the details of the graphic analysis and descriptions can be found elsewhere [14]. Only key results are summarized in **Table 2** for the various symmetry parameters, and **Table 4** for key parameters of ratios (several peak-current ratios, half-peak ratios, peak separations or their ratios).

3.2 Advantages of derivative voltammetric method

The scheme of derivative voltammetric (DV) approach is simpler, more straight-forward and faster compared with other methods of studying electrode reaction mechanisms, namely, CV [1, 38–41] and CSWV [66–70]. The present DV

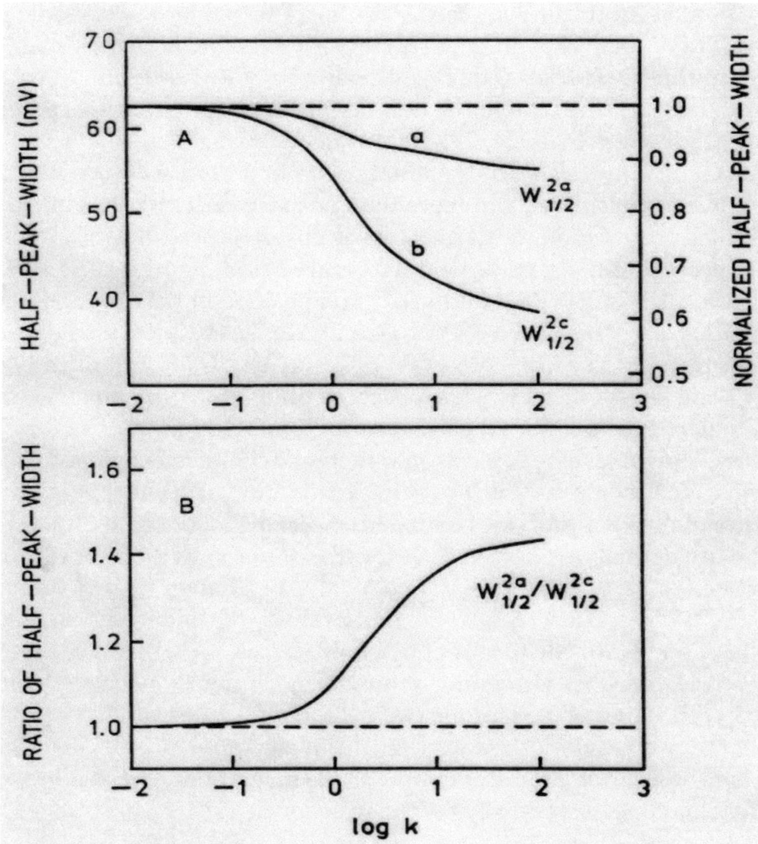

Figure 8.
Effect of k (A) on the (a) the anodic, (b) cathodic half peak-widths of the second derivative, an those values normalized to the values for the reversible case whose scales are given at the right side, (B) on the anodic to cathodic half-peak ratios, $W_{½}^{2a}/W_{½}^{2c}$.

only requires, in general, a single scan for a system under investigation: few experimental variables need to be changed although some experimental variables (such as duration of the pulse) can be optimized at an initial stage of measurement. Mostly, a single scan at a optimized condition will suffice while other methods such as CV and CSWV requires multiple scans (mostly six or more). In present DV method, analysis of the second derivative can be enough for most cases: however, for a system requires a better resolution and a higher sensitivity, one may resort to the third derivatives for more diagnostic parameters.

4. Conclusions

A quantitative measure of symmetry *in the original current are not readily available*, can be found indirectly from a ratio of difference in several potentials defined from a voltammogram: namely, a quarter-wave potentials ($E_{1/4}$), half-wave potentials ($E_{1/2}$) and a three-quarter-wave potentials); then a ratio of an anodic to cathodic quarter-wave potential differences (i.e., $\mathbf{\Delta E_q^a/\Delta E_q^c}$ = $(E_{1/2}-E_{1/4})/(E_{1/2}-E_{3/4})$ is very limited. Nevertheless, the ratio is unity for a symmetrical curve (for a reversible case), but it deviates from one if the curve becomes asymmetrical, yielding 1.14 for E_{irr} (**Table 2**). On the other hand, asymmetry *in first derivatives is more readily found* in terms of a ratio of anodic-to-cathodic quarter-peak ratio ($\mathbf{W_q^a/W_q^c}$): this ratio increases from 1.00 (for E_{rev}) to 1.27 for E_{irr}. The anodic and cathodic parts of the peak-width (W_q and W_q^c) in the first derivative can be more conveniently measured than three potentials (i.e., $E_{1/4}$, $E_{1/2}$ and $E_{3/4}$) in the zeroth derivative; in addition, the change of the ratio in the first derivative is more sensitive than the ratio in the zeroth derivatives (namely, 27% vs. 14%).

In the second and third derivatives, compared to first derivatives, *asymmetry in the peak shape is more pronounced*, and more than one symmetry parameters are readily obtained, which are given in terms of ratios of different peak-heights, and peak-widths, and peak-potentials; these parameters are based on peak heights (i_p^a and i_p^c) as well as peak widths ($W_{½}^a$ and $W_{½}^c$) and peak potentials (E_p and E_p). Thus, symmetry ratios of $\mathbf{i_p^{2a}/i_p^{2c}}$, $\mathbf{W_{1/2}^{2a}/W_{1/2}^{2c}}$ and $\mathbf{\Delta E_p^{2a}/\Delta E_p^{2c}}$ are introduced for second derivatives, and $\mathbf{i_p^{3a}/i_p^{3c}}$, $\mathbf{W_{½}^{3a}/W_{½}^{3c}}$, and $\mathbf{\Delta E_p^{3a}/\Delta E_p^{3c}}$ for third derivative (in **Tables 2** and **4**). All of these values become unity (1.00) for the reversible processes, indicating the symmetry in the curves; but they become no longer unity for quasi-reversible/irreversible electron transfer or chemically coupled processes as the shapes become asymmetrical. The results are summarized in **Tables 2** and **4** with all types of types of processes included. Basically, all of the symmetry parameters for the 0th derivative ($\Delta E^a_q/\Delta E^c_q$), for the 1st derivative (W_q^{1a}/W_q^{1c}), and for 2nd derivative (i_p^{2a}/i_p^{2c}, $W_{1/2}^{2a}/W_{1/2}^{2c}$, & $\Delta E_p^{2a}/\Delta E_p^{2c}$) and for third derivative (i_p^{3a}/i_p^{3c}, $W_{½}^{3a}/W_{½}^{3c}$, $\Delta E_p^{3a}/\Delta E_p^{3c}$) are unity for E_{rev}, they become no longer unity for E_{irr}, $E_{quasi\text{-}rev}$ and two chemically coupled processes of CE_{rev} and $E_{rev}C$. With the third derivatives, there are also minor symmetry parameters associated with the middle peak with values of 0.32 for the two peak-height ratios (i_p^{3a}/i_p^{3m} and i_p^{3c}/i_p^{3m}) and 1.32 for the two peak-width ratios ($W_{1/2}^{3a}/W_{1/2}^{2m}$, and $W_{1/2}^{3c}/W_{1/2}^{3m}$) for reversible but these ratios are not as simple as the major ones, but changes to 0.22, 0.52, and 1.62 and 0.98 respectively for the irreversible case.

5. Summary

From careful analysis of derivatives of voltammetric current-potential curves, it is possible to extract various parameter which are associated with asymmetry of the derivatives associated with various types of electrode mehanisms. Among those parameters, some of them are strongly influenced by electron transfer kinetics and electrode reaction mechanisms associated with a system, and these parameters are particularly useful in elucidating the reaction mechaisms. In particular, the ratios of the anodic to cathodic peak-currents (i_p^a/i_p^c), and the ratios of the anodic to cathodic peak-widths ($W_{½}^a/W_{½}^c$) and the ratio of the difference in the anodic and cathodic peak potentials ($\Delta E^a/\Delta E^c$) or the anodic and cathodic peak potential difference (ΔE^a-ΔE^c) are most sensitive and useful when a simple reversible electron system is disturbed by other kinetics, breaking the symmetry in the derivatives. The parameters of each particular systems respond differently to each type of electrode process; thus analysis of symmetry parameters can provide much insight to the mechanistic nature of electrochemical systems. A useful master table for such a diagnostic criteria for differentiating several common types of electrode mechanisms is devised and presented.

Acknowledgements

Several parts of the work have been presented at various conferences including the 3rd Chemical Congress of North America (Toronto, Canada, 1988), the 57th Georgia Academy of Science Meeting (Lawrenceville, GA 1999), the 221st National

Meeting of American Chemical Society (ACS, San Diego, 2001), and the 65th South East Regional Meeting (SERM of ACS, Atlanta, GA 2013). This work was partially supported by Academic Computing Services of Old Dominion University, and with Computer Services through Cornell National Super Computer Center (NSF supported), and by Faculty Fellowship Awards (2008-2009) from the Writers Institute of Georgia Perimeter College (GPC). The author acknowledges the work of all collaborators, Prof. Tae-Kee Hong (Haseo University, Korea) in particular. GPC has been consolidated to Georgia State University (GSU) as Perimeter College in 2016.

Author details

Myung-Hoon Kim
Department of Physical Sciences, Perimeter College, Georgia State University, Dunwoody, USA

*Address all correspondence to: mkim124@gsu.edu

References

[1] A. M. Bond, *Modern Polarographic Methods in Analytical Chemistry*, Marcel Dekker, New York (1980) 145–157, 212–216, 288–389.

[2] T. C. O'Haver, and G. L. Green, *Anal. Chem.,* **48** (1976) 312.

[3] A. F. Fell, *Trends in Analytical Chemistry,* **2** (1983) 63–66.

[4] S. V. Romanenko, *J. Anal. Chem*, **52** (1997) 822–896.

[5] K. Nagashima, M. Matsumoto, and S, Suzuki, *Anal. Chem*, **57** (1985) 2065–2067.

[6] D. L. Roelke, C. D. Kennedy, and A. D. Weidemann, *Gulf Mex. Sci.*, **2** (1999) 75–86.

[7] R. Mehrotra, G Tyagi, D.K. Jangir, R. Dawar, and N. Gupta, J. Ovarian Cancer, **3** (2010) 27.

[8] M. I. Toral, *N. Lara*, J. Gomez, and P. Richter, *Anal. Lett.*, **35** (2002) 153–166.

[9] A. P. Kumar, P.R. Reddy, and V. K. Reddy, *J. Anal. Chem.*, **63** (2008) 26–29.

[10] V. D. Parker, *Electroanalytical Chemistry,* A.J. Bard, Editor, Vol. 14, Marcel Dekker, New York (1986) p28.

[11] M.-H. Kim, *Anal. Chem.,* **59** (1987) 2136–2144.

[12] M.-H. Kim, *Anal. Sci. Tech. (J. Korean Soc. Anal. Sci.),* **2** (1989) 225–236.

[13] M.-H. Kim, V. P. Smith and T.-K. Hong, *Bull. Korean Chem. Soc.,* **11** (1990) 497–505.

[14] M.-H. Kim, *J. Electrochem. Soc.,* **137** (1990) 3815–3825.

[15] M.-H. Kim, V. P. Smith and T.-K. Hong, *J. Electrochem. Soc.* **140** (1993) 712–7215

[16] J. Gonzales. Molina, M. Lopez-Tenes, and C. Serna, *J. Electrochem. Soc.* **147** (2000) 3429–3435.

[17] A. Molina, J. Gonzales, and M. M. Moreno, *Electroanalysis*, **14** (2002) 281–291.

[18] A. Molina, and I. Morales, *Int. J. Electrochem. Sci.*, **2** (2007) 386–405.

[19] Y. P. Ding, W. L. Liu, Q. S. Wu, and X. G. Wang, *J. Electroanal. Chem.*, **575** (2005) 275–280

[20] J. Mbindyo, L. Zhou, Z. Zhang, J. D. Stuart, and J. E. Rusling, *Anal. Chem.*, **72** (2000) 2059–2065

[21] A. Murthy, and A Manthyram, *J. Phys. Chem. C*, **116** (2012) 3827–3832.

[22] M. Lovric, J. J. O'Dea, and J. Osetryoung, *Anal. Chem.,* **55** (1983) 704–708.

[23] C. G. Chan, and J. Kelly, *Earthquake Engineering & Structural Dynamics*, **19** (1990) 220–241

[24] C. L. Vaughn, *Int. J. Biomedical Computing*, 13 (1980), 375–386

[25] H. J. Woltring, *Adv. Eng. Software*, **8** (1986), 104–113.

[26] G. Ferrigno, and M. D'Amico, *J. Biomechanics*, **22** (1989), 1010.

[27] G. Giakas, and V. Baltzopoulos, *J. Biomechanics*, **30** (1997), 851–855.

[28] A. M. Zoubiv, M. Viberg, R. Chellappa, and S. Theodoridis, in *Academic Press Library in Signal Processing, Vol. 3 Array and Statistical Signal Processing*, Amsterdam (2014).

[29] M. Jakubowska, *Electroanalysis,* **23** (2011), 553–572.

[30] L. Meitis, *Polaographic Techniques,* 2nd Ed., John Wiley & Sons, New York (1965), p 114, 418, 457.

[31] M.-H. Kim, and T.-K. Hong, *Georgia J. Science,* **57**(1999), 46.

[32] T-K Hong, I. Rusodimos, and M.-H. Kim, *J. Electroanal. Chem.,* **785**(2017), 255–264.

[33] G. Giakas, and L. Stergiolas, and A. Vourdas, *J. Biomechanics*, 33 (2000), 567–574.

[34] T. C. O'Haver and T. Begley, *Anal. Chem.,* **53** (1981) 1876–1878.

[35] A. Cobelo-Garcia, J. Santos-Echeandia, D. E. Lopez-Sanchez, C. Amecija, and D. Omanovic, *Anal. Chem.,* **86** (2014) 2306–2313.

[36] A. P. Murthy, K. Duraimurugan, J. Sridar, and J. Madhavan, *Electrochim. Acta*, **317** (2019), 182–190.

[37] G. Ziyatidinova, E. Yacupova, E. Ziganshina, and H. Budnikov, *Electroanalysis*, **31** (2019) 2130–2137.

[38] A. J. Bard, and L. Faulkner, *Electrochemical Methods and Applications,* 2nd Ed. John Wiley & Sons, New York (2000).

[39] N. Elgrishi, K. J. Rountree, B. D. McCarthy, E. S. Rountree, T. T. Eisenhart, and J. L. Dempsey, *J. Chem. Educ.* , **95** (2018) 197–206.

[40] P. T, Kissinger, and W. R. Heineman, *Laboratory Techniques in Electroanalitical Chemistry,* 2nd Ed., Dekker, Monticello, NY (1996).

[41] D. K. Gosser, Cyclic Voltammetry: Simulation and Analysis of Reaction Mechanisms, VCH, New York, NY (1993).

[42] M.-H. Kim and V.P. Smith, *Anal. Sci. Tech. (J. Korean Soc. Anal. Sci.)* **2** (1989) 237–245.

[43] K. B. Oldham, and J. Myland, *Fundamentals of Electrochemical Sciences,* Academic Press, New York (2012), p 434–435.

[44] D. D. Macdonald, *Transient Techniques in Electrochemistry* Plenum Press, New York (1977) p76.

[45] M. L. Olmstead ans R.S. Nicholsen, *J. Electroanal. Chem.,* **14** (1967) 133–141.

[46] J. Galvez, A. Molina, and T. Fuente, *J. Electroanal. Chem.,* **107** (1980) 217–231.

[47] J. Galvez, M. L. Alkaraz, T. Perez, M. L. Cordova, *Anal. Chem.,* **57** (1985) 2116–2120.

[48] J. Galvez, *Anal. Chem.,* **57** (1985) 585–591.

[49] J. Galvez, A. Molina, C. Serna, and R. Saura, *J. Electroanal. Chem.,* **119** (1986) 37–45.

[50] J. Goodisman, , *J. Electroanal. Chem.,* **114** (1983) 33–43.

[51] F. Martinez-Oritz, M. L. Alkaraz, I. Roca, and M. Lopez-Tenes, *J. Electroanal. Chem.,* **443** (1998) 243–252.

[52] A. Molina, C. Serna, and F. Martinez-Oritz, *J. Electroanal. Chem.,* **486** (2000) 9–15.

[53] C. Serna, A. Molina, M. M. Moreno, and M. Lopez-Tenes, *J. Electroanal. Chem,* **546** (2003) 97–108.

[54] M. Lovric, and Y.I. Tur'yan, *Croatia Chem. Acta,* **76** (2003) 189–197.

[55] F. Martinez-Oritz, M. L. Alkaraz, and I. Roca, *J. Electroanal. Chem.,* **568** (2004) 79–86.

[56] A. Molina, F. Martinez-Oritz, E. Laborda, R. G. Compton, *Electrochim. Acta*, **55** (2010) 5163–5172.

[57] R. L. Birke, M.-H. Kim and M. Strassfeld, *Anal. Chem.,* **53** (1981) 852–856.

[58] M.-H. Kim and R. L. Birke, *Anal. Chem.,* **55** (1983) 522–527.

[59] M.-H. Kim and R. L. Birke, *Anal. Chem.,* **55** (1983) 1735–1741.

[60] M.-H. Kim, L. Yan, M.-Z. Czae, and R. L. Birke, *Electroanalysis,* **15** (2003) 1541–1553.

[61] M.-H. Kim and R. L. Birke, *Proceedings of the Korean Federation of Science and Technology: 11th Biennial Meetings, Physical Sciences Section, Korea University,* (1990) 606–610.

[62] J. J. O'Dea, J. Osteryoung, and R. A. Osteryoung, *Anal. Chem.*, **53** (1981) 695–701.

[63] J. G. Osteryoung and R. A. Osteryoung, *Anal. Chem.*, **57** (1985) 101A-110A.

[64] J. J. O'Dea, and J. Osteryoung, *Anal. Chem.*, **85** (1993) 3090–3097.

[65] L. Ramaley, and M. S. Klause, *Anal. Chem.*, **41** (2002) 1632–1635

[66] J. C. Helfrick, and L. A. Bottomley, *Anal. Chem.*, **81** (2009) 9041–9047.

[67] M. A. Mann, C. Helfrick, and L. A. Bottomley, *Anal. Chem.*, **86** (2014) 8183–8191.

[68] J. C. Helfrick, M. A. Mann, and L. A. Bottomley, *Electrochim. Acta*, **205** (2016) 20–28.

[69] M. A. Mann, J. C. Helfrick, and L. A. Bottomley, *J. Electrochem. Soc.*, **163** (2016) H3101-H3109.

[70] J. C. Helfrick, M. A. Mann, and L. A. Bottomley, *ChemPhysChem*, **17** (2016) 2596–606.

[71] Eduardo Laborda, José María Gómez-Gil, Manuela López-Tenés, and Angela Molina, *J. Electroanal. Chem.*, **873** (2020), 114421 (https://doi.org/10.1016/j.jelechem.2020.114421).

[72] J. R. Delmastro and D. E. Smith, Anal. Chem., 38 (1966) 169–179.

[73] A. Molina and J. Gonzalez, Pulse Voltammetry and Physical Chemistry and Electroanalysis, Springer, Heidelberg, (2016) p155.

Chapter 2

UPLC-MS: An Emerging Novel Technology and Its Application in Food Safety

Syed Amir Ashraf, Sadaf Nazir, Mohd Adnan and Zulfiqarur Rashid Azaz Ahmad Azad

Abstract

Over the past decade, food safety has become an important issue worldwide due to higher incidences of food contamination. Currently, one of the great challenges in food safety is the analysis of emerging food contaminants. Moreover, the scope, relevance, and level of food safety and testing have never been in such complexity than in today's global marketplace. In recent years, a novel technology ultra performance liquid chromatography (UPLC) coupled with mass spectroscopy (MS) has been developed to estimate the food contaminants, as well as food components with better accuracy, sensitivity, precision, and high throughput. UPLC–MS works on van Deemter principle, which states that, the flow rate of smaller particles are much faster in compare with large particles as well as unfolding the correlation of flow rate and plate height. Additionally, various food components as well as food contaminants such as vitamins, amino acid, metabolite identification, adulteration, forensic testing, toxicity studies, phytoconstituents, pesticide in agriculture, antibiotic residue, hormones, dyes and pigment analysis can be performed using UPLC–MS. Moreover, uniqueness of UPLC-MS and its wide range of application makes it an important tool for food safety laboratory around the world.

Keywords: food safety, food contaminants, liquid chromatography, van Deemter, UPLC-MS

1. Introduction

Food safety has become an important key issue worldwide, because of the emergence of several new chemical hazards present in food [1]. In addition to that, maintaining food safety has become very challenging at the operational level, as production of food and their consumptions are currently involved in a series of events that must be adequately accomplished to ensure the safety of food [2]. Therefore, food safety has become an increasingly important public health issue all over the world and due to which governments are escalating their efforts to improve and ensure food safety. These efforts can also be recognized in response to a growing number of food safety problems and increasing consumer health safety concerns [3]. A very well-know proverb from nutritionists or dietitians is "we are what we eat". Definitely, it does not mean that if we eat apple we become apple, but for good or for ill, the components we eat must be incorporated, transformed, and/or excreted by our bodies. Because, food is an indispensable ingredient of life, and access to food is often the limiting factor in the size of a given populace [4]. There are several incidents of food safety outbreak, which has received major attention from all parts of the world such as occurrence of benzene in carbonated drinks (UK), foods contaminated with pesticides (Japan), presence of dioxins in milk products and pork sample (Belgium), incidence of pesticides in soft drinks (India) and occurrence of melamine in dairy products (China). Such incidents have made people distressful of their food consumption worldwide [5]. In addition to that, such contemporaneous incidents are growing concerns, mainly because of mass production of agronomic products and industrialization at a very fast pace to meet the requirement of current population. Moreover, it has been considered that mainly increasing worldwide population is making farming people to force mass production of agronomic products without giving ample consideration to the safety and quality of food produce. In addition to that, changes in life style patterns of consumers have been called responsible for food safety hazards [6]. Due to fast-paced urbanization, food products such as ready-to-eat, processed food and junk foods has increased, but due to rise in application of chemicals usage, such processed food has also come under the scanner of food safety professionals [1, 7].

Moreover, the scope, relevance, and level of food safety and testing have never been in such complexity than in today's global marketplace. In recent years, a novel technology UPLC-MS has been developed to estimate the food contaminants as well as food components with better accuracy, sensitivity, precision, and high throughput. In addition to that, this advanced novel technique provided the platform to estimate different analytes at very lower levels, with better accuracy, and more importantly in less time. Moreover, the uniqueness of UPLC-MS has marked several applications to food safety. Various food safety parameters such as residual analysis, vitamins, amino acid, metabolite identification, adulteration, forensic testing, toxicity studies, phytoconstituents analysis, pesticide in agriculture, antibiotic residue, hormones, dyes and pigment analysis can be performed by using UPLC-MS [8, 9]. In addition to that, wide range of analysis makes UPLC-MS as an integral part of food safety laboratory around the globe. Moreover, in this chapter a detailed study and exploration has been made for better understanding of principles and applicability of UPLC-MS in food safety.

2. Chromatography and food safety

Today, our food supply is more diverse and highly processed than ever before. However, to ensure the nutritive value and to improve the food safety several states have disseminated regulations that states the acceptable limit for each components likewise, food additives, food residues and contaminants in food or food products. Consequently, a better and safe food can only be ensured when we have good approach to analyze such food components, contaminants, or chemical contaminants. In past few decades, chromatography has been recognized as one of important tool to identify and quantify food contaminants to ensure food safety. This novel technique allows the separation, purification, and identification from a mixture of the components for both qualitative and quantitative analysis. In current years, a unique technology UPLC-MS has been developed to estimate the food contaminants as well as food components for improving food safety. Therefore to obtain such targets, in 2004 Waters launched a brand of liquid chromatography (LC) called UPLC having a significant advancement in column particle size and column dimension having a small and porous particle (sub 2 μm) [10, 11].

3. Ultra performance liquid chromatography

UPLC is a novel technique that offers a new pathway for LC. UPLC enhances the capability of LC in four main areas like increasing speed, sensitivity, resolution and accuracy. UPLC is also known as ultra high-performance liquid chromatography (UHPLC). In comparison to high-performance liquid chromatography (HPLC), UPLC has been upgraded with column packing materials of less than 2 μm in diameter, which increases the speed, accuracy, resolution and sensitivity. Moreover, particle size used in HPLC, UPLC column ranges from 3 to 5 μm and < 2 respectively as well as mobile phase flow rate in HPLC is usually 3.0 ml/min compared to UPLC flow rate 0.6 ml/min. The basic difference in the principle of UPLC and HPLC is the column packing material, which makes a huge difference over the sensitivity and accuracy of the novel techniques. Apart from the principle involved in the LC, there is not much change in basic principle except the pressure generated or created in the instruments make it a more efficient technology. The development of UPLC techniques has urged the scientists to improve the prevailing instrumentation capability for LC, which has the advantage of improved parting performance and constant pressure. Efficiency of this technique is equivalent to the dimension of the column and inversely proportional to the radius of the atoms. As the name suggest ultra performance or ultra-pressure, UPLC works under very high pressure up to 1000 bars, however for HPLC, pump pressure not go more than 300–400 bars. A schematic diagram of UPLC and its internal diagram are presented here in **Figure 1**. In recent years, UPLC has become an integral part of any food safety laboratories, as it reduces the time of run as well as cost of analysis for any analysis [9, 12, 13].

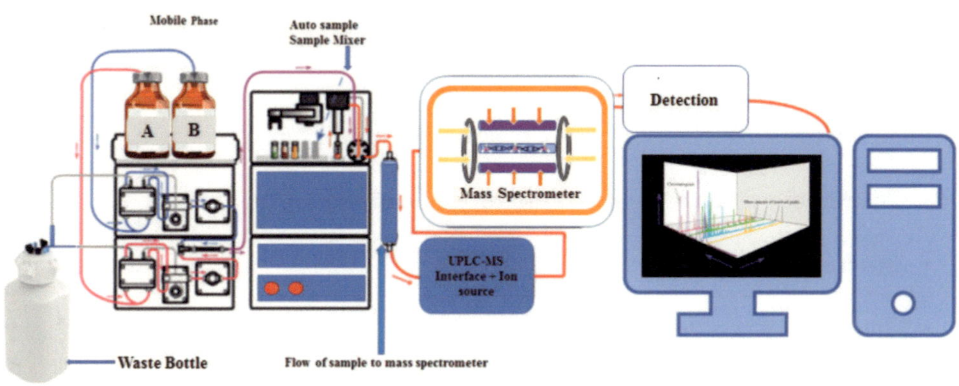

Figure 1.
Flow diagram of ultra performance liquid chromatography-mass spectrometry.

3.1 Principle

UPLC works on the van Deemter principle, which describes the correlation between the flow rate and height of chromatogram. The van Deemter states that, "the flow rate of smaller particles are much faster in compare with large particles as well as unfolding the correlation of flow rate and plate height". According to van Deemter equation, when the porous particle size reduced to less than 2.5 μm, there will be increase in efficiency; however, the efficiency does not weaken at increased flow rates or linear velocities.

The following equation describes the relationship between linear velocity (flow rate) and plate height [13, 14].

$$H = A + B / v + C_v \quad (1)$$

where,

A, B and C = Constants.

v = Linear velocity of carrier gas flow rate.

A = It is independent of velocity and represents "eddy" mixing. This is smallest when the packed column particles are small and uniform.

B = It stands for axial diffusion or the natural diffusion tendency of molecules. This effect is diminished at high flow rates and so this term is divided by v.

C = It represent kinetic resistance to equilibrium in the separation process.

According to van Deemter equation, resistance of kinetics is the time lag involved in traveling from the gas phase to the packing stationary phase and back again. Moreover, higher the gas flow, greater will be a molecule to lag behind in the mobile phase on packed stationary phase. Therefore, the term is proportional to v. Moreover, there will be a chance to surge throughput, and thus the rapidity of analysis without affecting the chromatographic performance [15]. However, UPLC performance is not much efficient until unless it is coupled with tandem mass spectrometry or other spectrometry techniques as it helps in molecular analysis by using mass-by-charge ratio [16, 17].

3.2 Mass spectrometry

Spectrometry method for the molecular analysis of any compound requires mass spectrometry (MS). The principle of MS was first proposed by Dr. Wien, which suggests that, refraction of charged particle in electric or magnetic field can analyzed by using MS. Mass spectrometer is an important tool to for the molecular mass analysis [18]. MS methods identifies the ionized molecules in gaseous phase in different ways

- Qualitative analysis of unknown compounds or mixture
- Quantitative estimation of any mixture or solution
- Structure characterization
- Molecular weight determination

MS works on the principle of fragmentation of molecule and separation or filtration of ions on the basis of their mass-to-charge (m/z) ratio. The molecular mass resulting from mass spectrum and produced ions are a function of mass by charge ratio [19]. Consequently, fragmentation of molecular mass in MS make it principally a very important technique over any other traditional chromatographic techniques. Notwithstanding that, on account of the capacity of MS to create m/z proportion, it considered as an exceptionally novel, straightforward, sensitive, accurate, and particular for the quantitative investigation of any mixture or blend [20, 21].

3.3 Tandem mass spectrometry (MS/MS)

There are mainly five techniques for analyzing mass of any compound by using MS like, quadrupole mass filter (single and triple), time of flight, quadrupole ion trap and Fourier transform ion-cyclotron resonance instruments. Furthermore, MS gave a thought of molecular mass, however on the other hand it does not give authentication of molecular structure. In this way, to conquer the restriction of past mass spectrometry, improvement of couple mass spectroscopy (MS/MS) rises. This MS/MS system work into two stages, first to choose parent ions generated from parent ion cells and to disintegrate into daughter ions after the collision of parent ion into at least one daughter ions. In mass spectrometry parent ions and daughter ions gets isolated, divided, and distinguished into single ion cell. In addition to that, fast collisions of compounds performed in argon cell, where translational energy gets transformed into ion internal energy to make ions in excited state and unimolecular decay progresses [22]. The breaking of compound in ion cell of MS/MS spectrum is selected based upon parent and daughter ions. Collision of compound can be performed in in single ionization cell or triple quadrupole system (TQS). TQS is the most frequently used now a day MS/MS techniques as compared to other mass analyzer [23].

3.4 Small-size particles and their chemistry

Small-size particles not only enhance proficiency, nonetheless it also increases the flexibility to enhance linear velocity without losing efficiency of the column. Moreover, efficiency is the essential separation factor in UPLC, as it depends on the selectivity and retention activity as in HPLC. Below equation shows that: (Rs) resolution is directly proportional to the square root of N.

$$\mathrm{Rs} = \frac{\sqrt{N}}{4}\left(\frac{\alpha - 1}{\alpha}\right)\left(\frac{k}{k+1}\right) \tag{2}$$

However, N is inversely proportional to particle size (dp): as the particle size is lowered by a factor of three, from, for example, 5 μm (HPLC scale) to 1.7 μm (UPLC-scale), N is increased by three and resolution by the square root of three or 1.7. N is also inversely proportional to the square of the peak width:

$$N \alpha \frac{1}{w^2} \tag{3}$$

This demonstrates that the narrower the peaks are, the easier would be to separate from each other. Moreover, peak width height is inversely proportional to the peak height:

$$H \alpha \frac{1}{w} \tag{4}$$

Therefore, decrease in particle size increases N and subsequently Rs, and by virtue of which sensitivity increased, taller peak as well as narrower peak mean more peak capacity per unit time in gradient separations, as per the requirement in several food safety application notes. Moreover, another equation comes into play when migrating toward smaller particles:

$$F_{xxx}\alpha = \frac{1}{\text{dpc}} \quad (5)$$

Van Deemter equation revealed that, as particle size decreases, the optimum flow *F* opt to reach maximum *N* increases. However, flow rate is directly proportional to back pressure as smaller particle sizes needed much higher operating pressures. Efficiency is inversely proportional to the particle size however proportional to column length.

$$N\alpha \frac{L}{dp} \quad (6)$$

Moreover, the column can be shortened by the same factor as the particle size without loss of resolution. Although non-porous, high-efficiency 1.5-μ particles are easily available in market, but these non-porous particles suffer poor loading capacity as well as poor retention because of low surface area. However, silica-based column have good mechanical strength nonetheless, it can undergo to a number of disadvantages, such as limited pH range and tailing of basic analytes. In addition to that, polymeric columns can overcome pH limitations. Moreover, packed column bed and their uniformity are also important, mainly if shorter columns have to uphold resolution while achieving the objective of faster separations [9, 13, 15].

4. Application of ultra performance liquid chromatography in food safety

In recent years, the demand of UPLC-MS/MS in food analysis has increased, because of the novel characteristics of UPLC with good resolution, better accuracy and sensitivity and reproducibility. Since its inception, it has reduces the time of food scientists as well as cost of the analysis because of its capability of producing more valuable, reliable, and reproducible data. The UPLC sensitivity has reached to ppb and ppt levels by virtue of which a food analyst would be more confident in ensuring safe food for consumption. Analysis of several food components as well as food contaminants has been performed using UPLC-MS/MS technique. By using this technique, below-mentioned food matrices can be tested for ensuring better food safety and we can also get more accurate qualitative and quantitative data of samples with high standards [11].

- Determination of antibiotic residue in food matrices
- Quantification of pesticides residues in food [24–26]
- Amino acid profiling [9, 27]
- Multi-drug residue quantitation in food matrix [28]
- Metabolomics study in food safety [9]
- Analysis of food contaminants in food matrices

- Determination of phytoconstituents
- Analysis of natural medicine and herbal medicine [9]
- Determination of acrylamide in food matrix [29]
- Analysis of mycotoxin in food [30, 31]
- Determination of bromate in drinking water [32]
- Pesticide in fruit and vegetables [33]
- Determination of food-borne carcinogens heterocyclic amines [34]
- Capsaicinoids analysis in capsicum species [35]
- Analysis of vitamin in food
- Determination of alkaloids in cocoa
- Lactose content determination in milk
- Phenolic content determination in fruits and vegetables
- Analysis of food based coloring agent [36]

4.1 Determination of antibiotic residue in honey

Several antibiotic residues such as streptomycin (**Figure 2**), chloramphenicol, tetracycline etc. has been identified and quantified in honey by using UPLC-MS coupled along with electron spray ionization [37, 38].

4.2 Multi pesticide residue analysis in cereal grains

Pesticides are chemicals widely used against plant pests in agriculture and farming to increase crop production, either against plant diseases or prophylactic usage. Currently, more than 350 pesticides are known, which are used to protect plants or plant products; however these pesticide are not allowed more than the permitted level. In addition to that, these chemicals could be dangerous to human health. The function of full scan UHPLC-Orbitrap-MS/UPLC-MS is adequate enough to enable detection and accurate analysis of mass measurement of a broad range pesticides residue at very lowest concentration in complex sample matrices [24–26].

4.3 Amino acid profiling

Amino acid profiling is one of the important proximate analyses parameter in food safety, as it contributes major portion of protein and an essential component of human diet. However, among the several protein food resources mammalian milk is purest food available over the globe. However, free amino acids are calculated from total nitrogen present in milk. UPLC coupled to electrospray ionization tandem mass spectrometry (ESI-MS/MS) system has been estimated for free amino acid analysis in milks of human, rat, and cow as presented in **Figure 3**. Moreover, UPLC-ESI-MS/MS allowed the quantitation of 21 free amino acids in 10-minute run time using labeled amino acids as internal standard in mammalian milk [27].

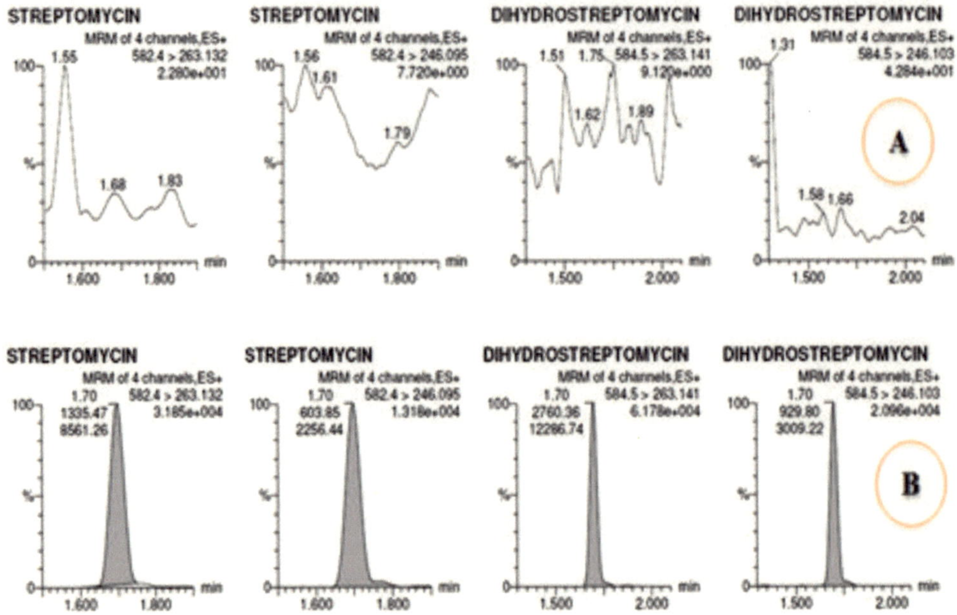

Figure 2.
Chromatogram showing blank honey sample (A) vs. spiked honey sample (B).

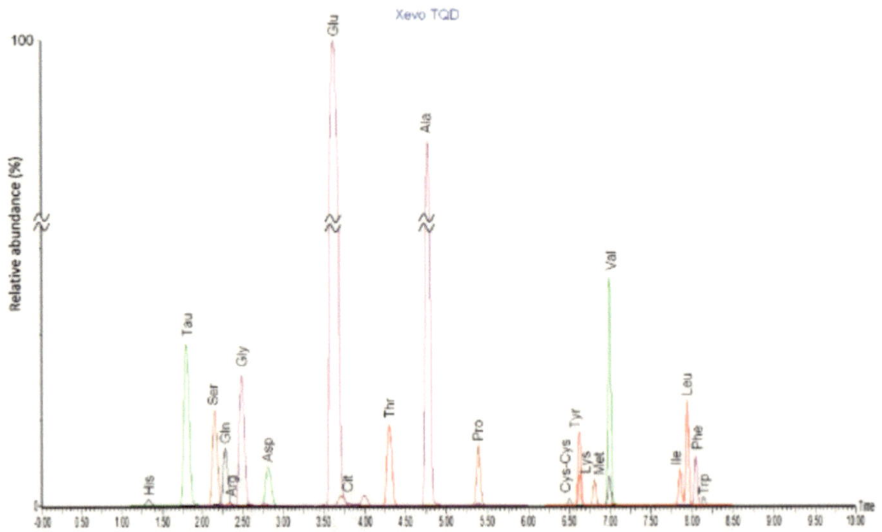

Figure 3.
Free amino acid ion chromatogram obtained in human milk (cumulative).

4.4 Metabolomics study in food safety

In recent years, the performance of UPLC has set the stage for a myriad of metabolomics analysis in plants and plant products. UPLC along with qTOF (quadrupole time of flight) system has been applied for semi-polar metabolite analysis in tomato fruit model. Moreover, UPLC coupled with qTOF mass spectrometer produces high-resolution and mass accuracy, good dynamic range, and a fast spectral acquisition capacity, which makes UPLC one of the most appropriate techniques for extensive profiling of many plant metabolites. In addition to that UPLC-MS

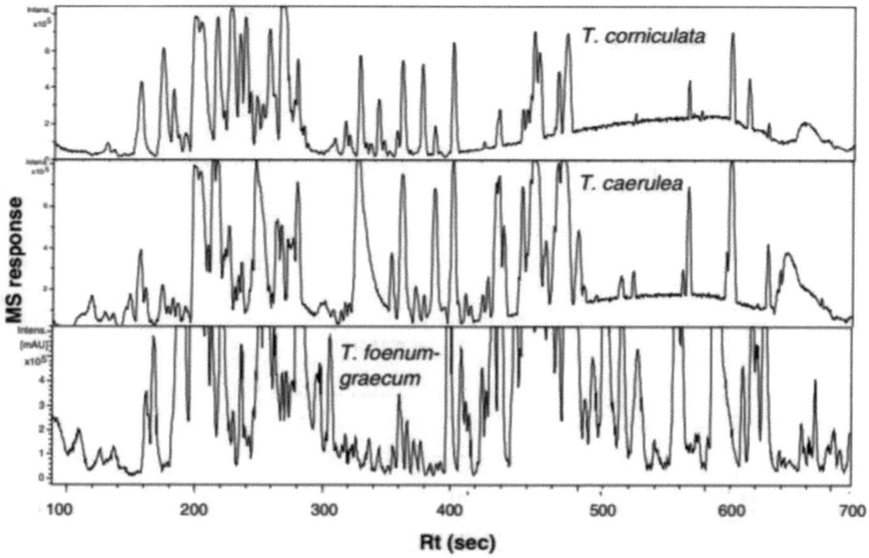

Figure 4.
Metabolomic analysis of T. caerulea, T. corniculata, *and* T. foenum-graecum *using UPLC-qTOF-MS.*

along with multivariate data analysis has been used for metabolomics profiling of Trignella seed. Metabolomic study of all the three Trigonella species *T. caerulea*, *T. corniculata*, and *T. foenum-graecum* identified 93 metabolites including 26 saponins, 5 peptides, 22 C/O-flavonoid conjugates, and 9 fatty acids as determined in **Figure 4**. Out of which, various novel compounds such as dipeptides, flavonoids were reported for first time [39–41].

4.5 Multi-drug residue quantitation in poultry muscle

In recent year, poultry industries have become million dollar industries due to higher consumption among the world population. However, multi-drug residue is very common in poultry muscles as poultry husbandry people illegally feed several drugs such as quinolones, amantadine, sulfonamides, tetracycline, amoxicillin, lincomycin, and so on. UHPLC-ESI-MS/MS has been used to analyses such veterinary drug residues in poultry muscle ranging from very polar to nonpolar compounds. UHPLC-ESI-MS/MS operating in positive multiple reactions monitoring (MRM) has been operated to quantify most of the multi-drug residue in sample [28].

4.6 Method development and validation

Method development plays a great role in concluding for any analytical method. In quantitative evaluation, development of method can roughly divided into three parts

- Optimization of chromatography conditions
- Mass spectrometry parameters
- Preparation of sample

Depending upon physical or chemical characteristics of analyzing components method development could be easily performed considering the following factors like selection of column, mobile phase, pH, and particle size and flow rate in any chromatographic setting.

The benefits of using UPLC-MS method over others were better recovery, good repeatability, and amount of extraction solvent volume. The selection of ionization techniques is depending on analytical results with pretreated samples. UPLC-MS/MS tuning parameters and scan modes are decided by uninterrupted infusion of standard solution, depending on the sensitivity and specificity needed. Few key elements for method development are sample pre-treatment, chromatography, internal standard, choice between electrospray ionization (ESI) and APCI, and mass spectrometry [42]. On the other hand, method validation results support for new analytical procedures or new drug development such as Carnosol, Carnosic acid, and Rosmarinic acid in food matrices. Validation required defining performance of developed method and reliability of obtained results. The analytical developed method could be utilized for quantitation application then it would be better to be validated to ensure minimum requirement of validation experiments along with satisfactory results [43].

4.7 Determination of acrylamide in food matrices

Acrylamide as a risk factor come to scientists attention recently, as its discovery in food was accidental. Formation of acrylamide in different types of cooked food or processed food at high temperatures reported recently. Several researchers have validated an analytical method for the analysis of acrylamide in food by UPLC-MS/MS as determined in **Figure 5**. Various reports suggests that processed food such as potato, coffee, bakery and other human dietary products contain acrylamide. One of the study carried out in Cyprus found that potato crisp had highest amount of acrylamide (642 ppb), followed by French fries and biscuits. Concurrently, regular consumption of such food products may lead to carcinogenicity [29].

4.8 Determination of phytoconstituents

Determination of phytoconstituents analysis involves usage of several analytical techniques for the isolation and characterization of phytoconstituents. Primitive techniques basically involved usage of UPLC-MS for the isolation and determination of phytoconstituents. Analysis and identification of chemical constituents of fenugreek by UPLC-MS and UPLC-Q-TOF-MS revealed that, 57 saponins and 19 flavonoid components. In addition to that, characterizations and quantitation of phytoconstituents has been reported in *Piper betle*. Moreover, quantitative data revealed significant variances in the contents of the major bioactive components in *Piper betle* species [44, 45].

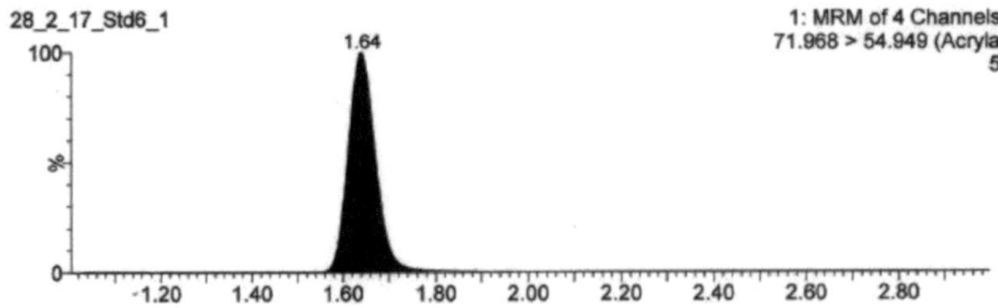

Figure 5.
UPLC-MS chromatogram for acrylamide standard solution at 500 ng ml^{-1}.

4.9 Determination of food contaminants in food matrices

In current years, various food such as legumes, cereals, potatoes, eggs, aquatic foods, dairy products, vegetables, fruits, and beverages reported to have several mycotoxins such as beauvericin, enniatin A, enniatin B, alternariol, tentoxin, and tenuazonic acid (**Figure 6**). These mycotoxins have been considered as a major food contaminates. In recent years, UPLC-MS has emerged as one of the most suitable method for the determination of these food contaminants. UPLC-MS has advantages over other instruments because of having better detection level, fast and accurate. UPLC-MS has emerged as a powerful tool for monitoring and measuring dietary exposure assessment of such mycotoxins [30].

4.10 Analysis of antioxidant and phenolic compound using UPLC-MS

It's been well-know that antioxidant has ability to fight against free radicals since free radicals are considered as a causative agent for several diseases. However, use of antioxidant has increased in food industry due to its antimicrobial property. Nowadays, natural as well as synthetic antioxidant such as butylated hydroxyanisole (BHA), butylated hydroxytoluene has been extensively used in food industry. However, the safety and toxicity of synthetic antioxidant is still a matter of concern for human health. On the other hand, several phenolic compounds have been well known for human nutrition. Moreover, these components are used for retarding microbial growth, increasing shelf life, reducing undesirable fragrances, enhancing nutritional value as well as delaying the formation of toxic oxidation. Phenolic profiling as well as antioxidant activities can be analyzed UPLC-ESI-MS/MS in Salvia species in some of the medicinal plants from South West Anatolia, Turkey. Moreover, it is assumed that, it was first reported for the analysis of individual phenolic profiles of *S. potentillifolia, S. albimaculata,* and *S. nydeggeri* [46].

4.11 Bromate in drinking water

Most of the drinking water contains bromide, as the primary source of bromide is soils containing bromide or sea water containing excess amount of

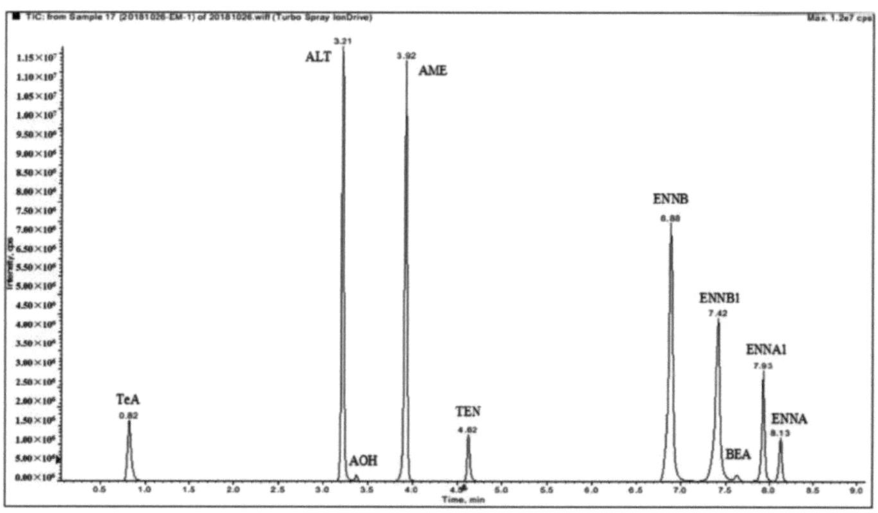

Figure 6.
Chromatogram (100 ng/mL) showing complete separation of a mixture of 10 mycotoxin standards at (100 ng/mL) using UPLC-MS method.

bromide. During the ozonation process bromide gets converted into carcinogenic bromate [47]. International Cancer research agency has found that, bromate has carcinogenic property in human beings. UPLC-MS techniques have been reported to quantify bromate at very low detection levels, that is, 0.01 ng/mL as found in **Figure 7**. UPLC-MS method is found to be rapid, selective, and sensitive for routine analysis of bromate at very low level in drinking water as well as sea water [32].

4.12 Analysis of capsaicinoids in capsicum species

Capsaicinoids are the pungent metabolites of the fruit capsicum. Capsaicinoids are a group of more than 13 alkaloids having structure of vanillylamide with branched fatty acid in the 9–11 carbons. Moreover, the most predominant capsaicinoids are capsaicin and dihydrocapsaicin. These two major capsaicinoids are responsible for the spiciness of capsicum (**Figure 8**). UPLC-MS is used to analyze capsaicinoids in various capsicum species. Analysis is carried out to measure the amount of all the capsaicinoids such as capsaicin, dihydrocapsaicin, nordihydrocapsaicin, homocapsaicin, and homodihydrocapsaicin in different species of capsicum. Based upon the UPLC-MS analysis limit of detection is calculated 0.05, 0.06, 0.15, 0.2, and 0.1 g/g for capsaicin, dihydrocapsaicin, nordihydrocapsaicin, homocapsaicin, and homodihydrocapsaicin, respectively [35].

4.13 Foodborne carcinogens

Foodborne carcinogens are a metabolic product of food after food processing (e.g., heating, curing, smoking) and during food preparation (e.g., baking, frying, grilling). Sometimes, fungi and plant-derived products also tend to produce foodborne carcinogens. Dietary carcinogens produced by chemical and physical food processing are N-nitroso compounds, heterocyclic aromatic amines, polycyclic aromatic hydrocarbons, and acrylamide. However, infected grains and peanuts have been reported to contain mold *Aspergillus flavus* and *Aspergillus parasiticus*, which is considered for producing secondary metabolite such as aflatoxins (carcinogenic potential) [48].

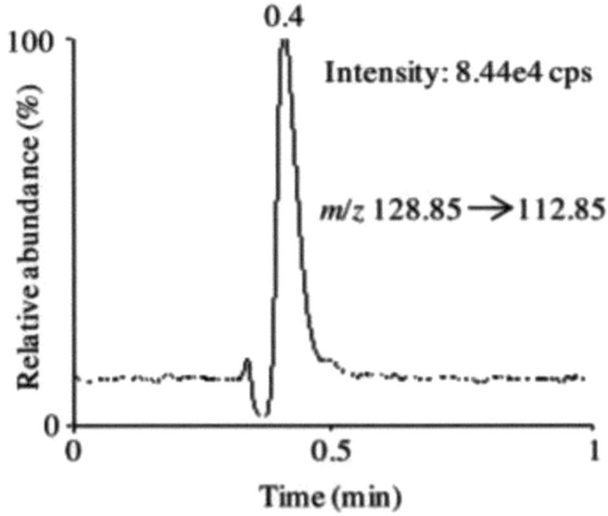

Figure 7.
Chromatograms showing a UPLC-MS/MS peak of bromate in drinking water.

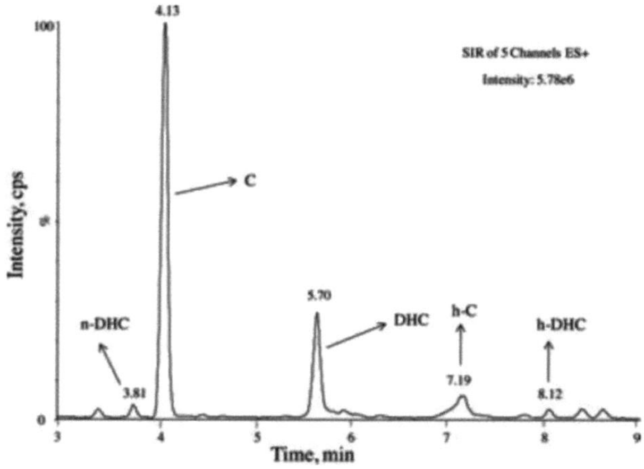

Figure 8.
UPLC-MS chromatogram showing different capsaicinoids extracted from red chili.

Researchers have developed an analytical method for biomonitoring of cooked meat carcinogens and their metabolites in human urine [34].

4.14 Vitamin analysis using UPLC-MS

Vitamins can be defined as biologically active organic compounds that have a relatively low molecular weight. Vitamins are present in minute quantity; however it is very important for human health and overall growth. Vitamin can be fulfilled only from regular diet or nutrition supplement, because these nutrients help in the metabolism of carbohydrates, fat, and proteins. In addition to that, it is also reported that, it reduces damage from free radicals. On the other hand deficiency in vitamin may lead to various diseases. UPLC-MS is very well known for the analysis of vitamins. Several UPLC-MS methods have been reported for the analysis of vitamin B complex (thiamin, riboflavin, biotin, nicotinic acid, pyridoxine, pyridoxamine, pyridoxal, pantothenic acid, FAD, and nicotinamide) analysis in human milk. UPLC-MS coupled with ESI techniques is used to analyze vitamin B from milk sample [49].

4.15 Determination of alkaloids, theobromine, and caffeine in cocoa

Ortega et al. [50] reported identification and quantification of alkaloids, theobromine, and caffeine in cocoa sample using UPLC-MS/MS. UPLC instrumentations are the most common techniques for routine analysis of such components in field of trace analysis. On the other hand, UPLC-MS has also been reported for alkaloid profiling of medicinal plants having cytotoxic properties. It is used for analysis of various alkaloids such as sanguinarine, berberine, protopine, and chelidonine [50].

4.16 Analysis of lactose in human and cow's milk

Sugars are found in a variety of food matrices as either naturally or artificially added. Fructose, glucose, and sucrose are important constituents of various fruit juices. Maltose is found in products derived from corn and grain products. Lactose, also known as milk sugar, exists in dairy products. This set of sugars is known as the five food sugars. Analysis of these sugars is important for quality control purposes, or to determine authenticity or adulteration of food products (**Figure 9**).

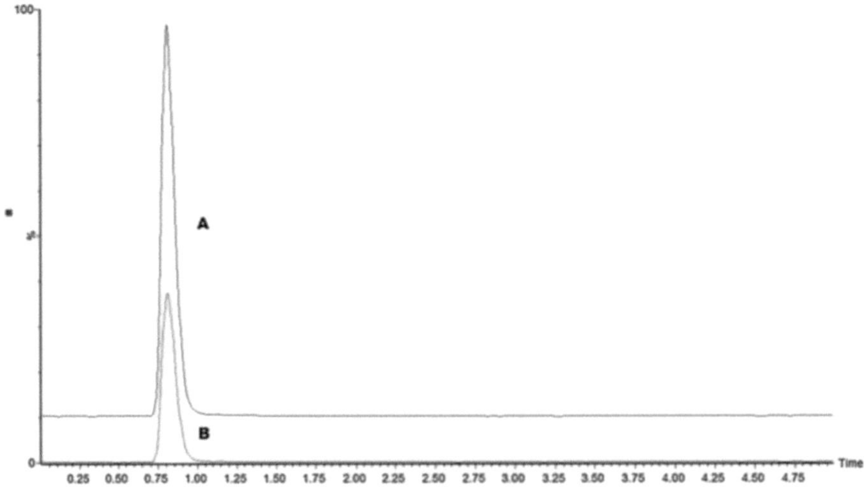

Figure 9.
Chromatogram showing standard lactose (A) vs. milk sample (B).

In addition to that, lactose is most important source of sugar for infant, kids as well as adult. UPLC-MS/MS can be easily used for determination of lactose in cow's or human milk as well as other food products [51].

4.17 Pesticides in fruit and vegetables

Fruit and vegetables are important crops of horticulture, as they are an integral part of the human diet. Fruit and vegetables provide carbohydrate, protein, vitamins, minerals, fiber and help in the maintenance of a healthy life style. However, in current years demand of fruit and vegetables has increased tremendously, because of high consumption and population demand. Therefore to boost the production, farmers are using so many chemicals in terms insecticides, fungicides, herbicides, acaricides, and rodenticides for prophylactic use or in diseased condition. However, it has been reported that, these chemical has very harmful effect on human health [52]. Savini et al. [33] reported a quick and sensitive UPLC method coupled with Orbitrap for determining highly polar pesticides and contaminants in processed fruits and vegetables.

4.18 Analysis of food-based dyes using UPLC-MS

Synthetic oil-soluble mono-azo coloring agents such as Sudan dyes and Para Red are very common in food industries. Due to minimal expense and high intensity color it is very commonly used as food additives particularly in chili. However, International Agency for Research on Cancer (IARC 1975) categorized these dyes a potential cancer-causing agent. Moreover, illegal use of these dye such as Sudan Red 7B, Sudan I–IV and Para Red have been still found in food impacting consumer health. UPLC-MS has been reported as one of best choice of instrument analysis of such dyes due to their highest sensitivity.

Author details

Syed Amir Ashraf[1], Sadaf Nazir[2], Mohd Adnan[3] and Zulfiqarur Rashid Azaz Ahmad Azad[4]*

1 Department of Clinical Nutrition, College of Applied Medical Science, University of Hail, Hail, Kingdom of Saudi Arabia

2 Department of Food Technology, Bundelkhand University, Jhansi, Uttar Pradesh, India

3 Department of Biology, College of Science, University of Hail, Hail, Saudi Arabia

4 Department of Post Harvest Engineering and Technology, Aligarh Muslim University, Aligarh, Uttar Pradesh, India

*Address all correspondence to: zrazad@gmail.com

References

[1] Pinu FR. Metabolomics: Applications to Food Safety and Quality Research. Microbial Metabolomics. Switzerland: Springer International Publishing; 2016. DOI: 10.1007/978-3-319-46326-1_8

[2] Chaves RD, Alvarenga VO, Campagnollo FB, Caturla MYR, Oteiza JM, Sant'Ana AS. Food safety. In: Current Developments in Biotechnology and Bioengineering. Food and Beverages Industry. 2017. pp. 245-259. DOI: 10.1016/B978-0-444-63666-9.00009-1

[3] Farhang B. Nanotechnology and applications in food safety. IUFoST World Congress Book: Global Issues in Food Science and Technology. Elsevier Inc.; 2009. ISBN: 9780123741240

[4] Lehotay SJ. Application of gas chromatography in food analysis. Trends in Analytical Chemistry. 2002;**21-9**(10):686-696

[5] Malik AK, Cristina B, Yolanda P. Liquid chromatography–mass spectrometry in food safety. Journal of Chromatography A. 2010;**1217**:4018-4040

[6] Motarjemi Y, Stadler RH, Studer A, Damiano V. Application of the Haccp Approach for the Management of Processing Contaminants. Process-induced Food Toxicants: Occurrence, Formation, Mitigation, and Health Risks. New York: Wiley; 2008. pp. 563-620

[7] Motarjemi Y, Lelieveld H. Food Safety Management: A Practical Guide for the Food Industry. Amsterdam: Elsevier; 2014

[8] Naushad M, Khan MR, Alothman ZA. History and Introduction of UPLC/MS. Ultra Performance Liquid Chromatography Mass Spectrometry: Evaluation and Applications in Food Analysis. 1st ed. CRC Press; 2014. ISBN: 9781466591547; Available from: https://www.researchgate.net/deref/http%3A%2F%2Fdx.doi.org%2F10.1201%2Fb16670-2?_sg%5B0%5D=k2YFvHkkxbDjrB8CsA8Z_7lD6IQH3mgcmaI-rQqwNTPs9IJGWaoRFJ69LjSBb8NfI1Db8KNJv5gToJKuVK74jGUKWw.08ofxad9RNGM3nRhQtsyvVLVvrjaVdEb70azZMyzB4kiU5ReVYQBnsvYmyFeEaBgO6cu8ZgWJl2iDlzZTnkHvg" 10.1201/b16670-2

[9] Taleuzzaman M, Ali S, Gilani SJ, Imam SS, Hafeez A. Ultra performance liquid chromatography (UPLC)—A review. Austin Journal of Analytical and Pharmaceutical Chemistry. 2015;**2**(6):1056

[10] Tamosiunas V, Padaraukas A. Comparison of LC and UPLC coupled to MS–MS for the determination of sulfonamides in egg and honey. Chromatographia. 2008;**67**(9/10):783-788. DOI: 10.1365/s10337-008-5-0009-5893/08/05

[11] Steiner WE, English WA. Emerging trends in liquid chromatography and mass spectrometry instrumentation for analytical & bioanalytical techniques. Journal of Analytical and Bioanalytical Techniques. 2012;**3**:e106. DOI: 10.4172/2155-9872.1000e106

[12] Available from: https://upload.wikimedia.org/wikipedia/commons/7/70/Liquid_Chromatography_Mass_Spectrometer.png

[13] Chawla G, Chanda R. Principle, instrumentation, and applications of UPLC: A novel technique of liquid chromatography. Open Chemistry Journal. 2016;**2016**(3):1-16. DOI: 10.2174/1874842201603010001

[14] Van Deemter JJ, Zuiderweg EJ, Klinkenberg A. Longitudinal diffusion

and resistance to mass transfer as causes of non ideality in chromatography. Chemical Engineering Science. 1956;**1956**(5):271-289

[15] Naresh K, Bhawani S, Kumar TM. Ultra performance liquid chromatography. International Journal of Pharma and Bio Sciences. 2014;**3**(3):84-94

[16] Jerkovich AD, Mellors JS, Jorgenson JW. The use of micrometer-sized particles in ultrahigh pressure liquid chromatography. LCGC North America. 2003;**21**(7):600-610

[17] Swartz ME. Ultra pe1rformance liquid chromatography (UPLC): An introduction, separation science re-defined. LCGC Supplement. 2005;**8**:8-14

[18] Munzenberg G. Development of mass spectrometers from Thomson and Aston to present. International Journal of Mass Spectrometry. 2013;**349-350**:9-18. DOI: 10.1016/j.ijms.2013.03.009

[19] Urban PL. Quantitative mass spectrometry: An overview. Philosphical Transactions A. 2016;**374**:1-5. DOI: 10.1098/rsta.2015.0382

[20] Hoffmann E, Stoobant V. Mass Spectrometry—Principles and Applications. 3rd ed. John Wiley & Sons, Ltd; 2007. ISBN: 978-0-470-03310-4

[21] Parasuraman S, Anish R, Balamurugan S, Vijayan V. An overview of liquid chromatography-mass spectroscopy instrumentation. Pharmaceutical Methods. 2014;**5**(2):49-55, Part A. DOI: 10.1080/19440049.2018.1508893

[22] Kang J. Principles and applications of LC-MS/MS for the quantitative bioanalysis of compounds in various biological samples. Tandem Mass Spectrometry-Applications and Principles. 2012:441-492. DOI: 10.5772/32085

[23] Madeira PJA, Florencio MH. Applications of tandem mass spectrometry: From structural analysis to fundamental studies. Tandem Mass Spectrometry—Applications and Principles. 2012:1-32. DOI: 10.5772/31736

[24] Mastovska K, Kelly JD, Steven JL, Jennifer SW, Kelli AS. Pesticide multiresidue analysis in cereal grains using modified QuEChERS method combined with automated direct sample introduction GC-TOFMS and UPLC-MS/MS techniques. Journal of Agricultural and Food Chemistry. 2010;**2010**(58):5959-5972. DOI: 10.1021/jf9029892

[25] Gomez-Pereza ML, Patricia PN, Roberto RG, José LM, Antonia G. Comprehensive qualitative and quantitative determination of pesticides and veterinary drugs in honey using liquid chromatography–Orbitrap high resolution mass spectrometry. Journal of Chromatography A. 2012;**1248**:130-138

[26] Zhang X, Yue S, Qi J, Lin Z, Wei Z, Pengqian M, et al. Simultaneous determination of 58 pesticides and relevant metabolites in eggs with a multi-functional filter by ultra-high performance liquid chromatography-tandem mass spectrometry. Journal of Chromatography A. 2019;(1593):81-90. DOI: 10.1016/j.chroma.2019.01.074

[27] Roucher VF, Emmanuelle D, Charlotte N, Aurore MA, Marie-Cécile A, Dominique D, et al. Use of UPLC-ESI-MS/MS to quantitate free amino acid concentrations in micro-samples of mammalian milk. Springerplus. 2013;**2**:622

[28] Cao G, Zhan J, Shi X, Deng X, Zhu J, Wu W, et al. Analysis of 140 veterinary drugs and other other contaminants in poultry muscle by ultrahigh performance liquid

hromatography-tendem mass spectrometry. Chromatographia. 2018;(81):707-718. DOI: 10.1007/s10337-018-3475-7.r

[29] Kafouris D, Georgios S, Maria C, Xenia I, Eftychia C, Lefkios P, et al. Determination of acrylamide in food using a UPLC–MS/MS method: Results of the official control and dietary exposure assessment in Cyprus. Food Additives & Contaminants. 2018;**35**(10):1928-1939. DOI: 10.1080/19440049.2018.1508893

[30] Sun D, Nannan Q, Shuang Z, Bing L, Shuo Z, Jingguang L, et al. Development of sensitive and reliable UPLC-MS/MS methods for food analysis of emerging mycotoxins in China Total Diet Study. Toxins (Basel). 2019;**11**:166. DOI: 10.3390/toxins11030166

[31] Malachova A, Milena S, Marta V, Christopher TE, Connor B, Julie M, Jana H, Chibundu NE. Advanced LC–MS-based methods to study the co-occurrence and metabolization of multiple mycotoxins in cereals and cereal-based food. Analytical and Bioanalytical Chemistry. 2018;(410):801-825. DOI: 10.1007/s00216-017-0750-7

[32] Alsohaimi IH, Zeid AA, Mohammad RK, Mohammad AA, Rosa B, Ahmad KA. Determination of bromate in drinking water by ultraperformance liquid chromatography–tandem mass spectrometry. Journal of Separation Sciences. 2012;**35**(19):2538-2543. DOI: 10.1002/jssc.201200312

[33] Savini S, Mirella B, Anna S. An improved, rapid and sensitive UHPLC-MS/MS analysis for the determination of highly polar pesticides and contaminants in processed fruits and vegetables. Journal of Agricultural and Food Chemistry. 2019;(67):2716-2722. DOI: 10.1021/acs.jafc.8b06483

[34] Gu D, Melissa MR, Fred FK, Robert JT. An ultra performance liquid chromatography—tandem mass spectrometry method for biomonitoring cooked meat carcinogens and their metabolites in human urine. Analytical Chemistry. 2011;**83**(3):1093-1101. DOI: 10.1021/ac102918b

[35] Alothman ZA, Saikh MW, Mohammad RK, Ayman AG, Mohamed AH, Yacine BHA. Determination of capsaicinoids in capsicum species using ultra performance liquid chromatography-mass spectrometry. Journal of Separation Science. 2012;**35**:2892-2896

[36] Li C, Wu YL, Shen JZ. UPLC-ESI-MS/MS analysis of Sudan dyes and Para Red in food. Food Additives and Contaminants. 2010;**27**(9):1215-1221. DOI: 10.1080/1'9440049.2010.483600

[37] Ashraf SA, Azad ZRAA. Development and validation of an UPLC-ESI-MS/MS analytical method for the determination of streptomycin and dihydrostreptomycin residues in honey. Biomedical & Pharmacology Journal. 2017;**10**(4):1983-1992

[38] Ashraf SA, Azad ZRAA. Validation of an analytical methodology for the determination of chloramphenicol residues in honey using UPLC-MS/MS. Oriental Journal of Chemistry. 2018;**34**(2):723-729

[39] Rogachev I, Asaph A. UPLC-MS-based metabolite analysis in tomato. In: Hardy NW, Hall RD, editors. Plant Metabolomics: Methods and Protocols, Methods in Molecular Biology. Vol. 860. Springer Science+Business Media, LLC; 2012. DOI: 10.1007/978-1-61779-594-7_9

[40] Vaclavik L, Jaroslava O, Ladislav K, Jan H, Katerina D, Jana H. Application of ultra-high performance liquid chromatography-mass spectrometry (UHPLC-MS) metabolomic fingerprinting to characterise GM

and conventional maize varieties. Czech Journal of Food Sciences. 2013;**31**(4):368-375

[41] Farag MA, Dalia MR, Matthias K, Andreas GH. Metabolite profiling in Trigonella seeds via UPLC-MS and GC-MS analyzed using multivariate data analyses. Analytical and Bioanalytical Chemistry. 2016;**408**(28):8065-8078. DOI: 10.1007/s00216-016-9910-4

[42] Shabir GA. Systematic strategies in high performace liquid chromatography method development and validation. Separation Science and Technology. 2010;**45**(5):670-680

[43] Choi S, Gill-Woong J, Sun-Il C, Tae-Dong J, Bong-Yeon C, Wan-Sup S, et al. Development and validation of an analytical method for carnosol, carnosic acid and rosmarinic acid in food matrices and evaluation of the antioxidant activity of rosemary extract as a food additive. Antioxidants. 2019;**8**(76):1-12. DOI: 10.3390/antiox8030076

[44] Pandey R, Preeti C, Mukesh S, Arya KR, Praveen KS, Brijesh K. A rapid analytical method for characterization and simultaneous quantitative determination of phytoconstituents in Piper betle landraces using UPLC-ESI-MS/MS. Analytical Methods. 2014;**6**(18):7349-7360. DOI: 10.1039/C4AY00975D

[45] Wang J, Jiang W, Liu Z, Wang J, Fu T, Wang Y. Analysis and identification of chemical constituents of fenugreek by UPLC-IT-MS*n* and UPLC-Q-TOF-MS. Chemical Research in Chinese Universities. 2017;**33**(5):721-730. DOI: 10.1007/s40242-017-7136-4

[46] Kivrak S, Tolga G, Ibrahim K, Ergun K, Erşan K. Investigation of phenolic profiles and antioxidant activities of some *Salvia* species commonly grown in Southwest Anatolia using UPLC-ESI-MS/MS. Food Science and Technology. 2019;**39**(2):423-431. DOI: 10.1590/fst.32017

[47] Yassin KE. Bromate in sudanese bottled water. Annual Conference of Postgraduate Studies and Scientific Research. 2012;**1**:7-8

[48] Fahrer J. Food-borne carcinogens. In: Schwab M, editor. Encyclopedia of Cancer. Springer-Verlag Berlin Heidelberg; 2016. DOI: 10.1007/978-3-642-27841-9_7235-1

[49] Nan RX, Yin SA, Yang ZY, Yang XG, Shao B, Ren YP, et al. Application of UPLC-MS/MS method for analyzing B-vitamins in human milk. Biomedical and Environmental Sciences. 2015;**28**(10):738-750

[50] Ortega N, Maria-Paz R, Alba M, Jordi R, Neus A, Maria-Jose M. Comparative study of UPLC–MS/MS and HPLC–MS/MS to determine procyanidins and alkaloids in cocoa samples. Journal of Food Composition and Analysis. 2010;**23**:298-305. DOI: 10.1016/j.jfca.2009.10.005

[51] Fusch G, Arum C, Niels R, Christoph F. Quantification of lactose content in human and cow's milk using UPLC–tandem mass spectrometry. Journal of Chromatography B. 2011;**879**(2011):3759-3762

[52] Syed JII, Ambreen A, Ashiq M, Karam A, Zunera S, Haroon A, et al. Pesticide residues in fruits and vegetables from Pakistan: A review of the occurrence and associated human health risks. Environmental Science and Pollution Research. 2014;(21):13367-13393. DOI: 10.1007/s11356-014-3117-z

Chapter 3

Analysis of the Electrochemical Transport Properties of Doped Barium Cerate for Proton Conductivity in Low Humidity Conditions: A Review

Laura I.V. Holz, Vanessa C.D. Graça, Francisco J.A. Loureiro and Duncan P. Fagg

Abstract

Proton-conducting perovskites are among the most promising electrolytes for Proton Ceramic Fuel Cells (PCFCs), electrolysers and separation membranes. Particularly, yttrium-doped barium cerate, $BaCe_{1-x}Y_xO_{3-\delta}$ (BCY), shows one of the highest protonic conductivities at intermediate temperatures ($\sigma \sim 10^{-3}$ S cm^{-1} at 400°C); values that are typically achieved under humidified atmospheres ($p_{H2O} \sim 10^{-2}$ atm). However, BCY has commonly been discarded for such applications due to its instability in the presence of water vapour and carbonaceous atmospheres. A recent discovery has shown that BCY10 exhibits pure protonic conductivity under very low humidity contents ($\sim 10^{-5}$–10^{-4} atm), owing to its very high equilibrium constant for hydration. This peculiar characteristic allows this material to retain its functionally as a proton conductor in such conditions, while preventing its decomposition. Hence, this chapter explores the electrochemical properties of the $BaCe_{0.9}Y_{0.1}O_{3-\delta}$ (BCY10) composition, comprehensively establishing its limiting operation conditions through defect chemistry and thermodynamic analyses. Moreover, the importance of such conditions is highlighted with respect to potential industrially relevant hydrogenation/de-hydrogenation reactions at low temperatures under low humidity.

Keywords: perovskite, barium cerate, protonic conductivity, transport number, nominally dry conditions

1. Introduction

Ceramic proton conductors have been highlighted for electrochemical synthesis, as potential membranes in hydrogenation and dehydrogenation reactions [1]. One of the best compositions for this role is that of the doped barium cerate, *e.g.* $BaCe_{1-x}M_xO_{3-\delta}$ (M = Y^{3+}, In^{3+}, Gd^{3+}, etc.), which can show very high levels of proton conductivity at intermediate temperatures (*i.e.* $\sigma \sim 10^{-3}$ S cm $^{-1}$ at 400°C) [2–9]. This material belongs to the perovskite family with ABO_3 ceramic oxide structure, including a divalent alkaline earth element, such as Ba^{2+} (also, Sr^{2+} or Ca^{2+}), in the A-cation site, while a tetravalent rare-earth element, Ce^{4+}, is present in the B-cation site. The introduction of dopants in the B-site with suitable acceptor elements, such as Y^{3+}, In^{+3} or Gd^{3+} trivalent cations, leads to the formation of charge compensating oxygen vacancies [9]:

$$BaO + (1-y)CeO_2 + \left(\frac{y}{2}\right)M_2O_3 \Longrightarrow Ba_{Ba}^{x} + (1-y)Ce_{Ce}^{x} + (y)\,M'_{Ce} + \left(3-\frac{y}{2}\right)O_O^{x} + \left(\frac{y}{2}\right)V_O^{\bullet\bullet} \quad (1)$$

In addition to potential oxide-ion conductivity, these acceptor-substituted materials are also capable of offering both protonic and electronic conductivity, depending on the temperature and atmospheric conditions. The protonic conductivity is the most significant characteristic of these materials that is usually associated with the existence of protonic defects ($OH_O^{\bullet}$), upon filling of these oxygen vacancies in the presence of water vapour, as expressed by Eq. (2) [10–13]:

$$H_2O + V_O^{\bullet\bullet} + O_O^{x} \Leftrightarrow 2OH_O^{\bullet} \quad (2)$$

Accordingly, the equilibrium constant for hydration, K_w, is given by the following equation:

$$K_w \approx \frac{\left[OH_O^{\bullet}\right]^2}{p_{H2O}\left[V_O^{\bullet\bullet}\right]\left[O_O^{x}\right]} \quad (3)$$

Due to the significant importance of humidity to promote protonic conductivity, most of the reported studies of barium cerate based materials have focused on highly wetted atmospheres with typical water vapour partial pressure $p_{H2O} \sim 3 \times 10^{-2}$ atm [14–18]. Unfortunately, these works also underline the tendency of this material for reacting with acidic gases, *viz.* carbon dioxide (CO_2) and water vapour (H_2O), leading to the formation of insulating carbonate or hydroxide phases, respectively, on the surface of the material. This complication impedes the ability of this material to be used in highly humidified and carbon-based fuels, thus, limiting its potential application range [3, 14–20]. The typical degradation reactions in such atmospheres include:

$$BaCeO_{3(s)} + CO_{2(g)} \rightarrow BaCO_{3(s)} + CeO_{2(s)} \quad (4)$$

$$BaCeO_{3(s)} + H_2O_{(g)} \rightarrow Ba(OH)_{2(g)} + CeO_{2(s)} \quad (5)$$

The chemical stability of doped barium cerates is well documented in the literature and huge efforts have been made to explore the reasons behind its chemical instability, using both conventional and non-conventional techniques [21–25]. For instance, Matsumoto *et al.* [22] studied the effect of dopant M in $BaCe_{0.9}M_{0.1}O_{3\text{-}\delta}$ (M = Y, Tm, Yb, Lu, In, or Sc) on the electrical conductivity in the temperature range 400–900°C and on the chemical stability with respect to CO_2 by thermogravimetry (TG). Both the electrical conductivity (moistened H_2 or O_2, p_{H2O} = 1.9×10^{-2} atm) and the stability against carbonate formation were shown to decrease with increasing ionic radius (**Figure 1**), corresponding to an increase in basicity. Nonetheless, all compounds were found to interact with pure CO_2 at temperatures below 900°C, failing to succeed in the mitigation of the chemical instability in the doped barium cerate.

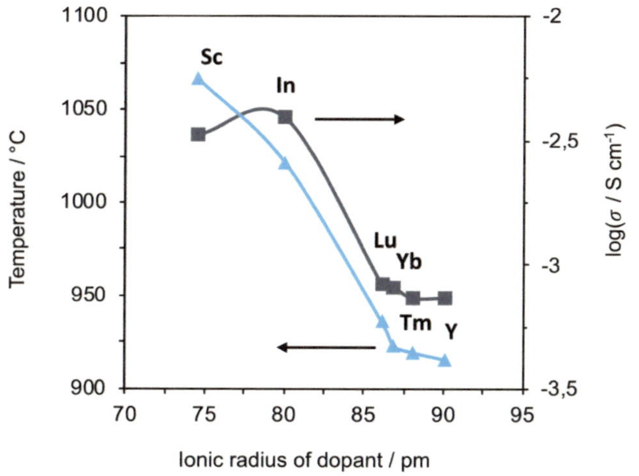

Figure 1.
Carbonate formation temperature (blue) and the conductivity isotherm at 400°C of $BaCe_{0.9}M_{0.1}O_{3-\delta}$ *(M = Y, Tm, Yb, Lu, In or Sc) in moist* H_2 *as a function of the ionic radius of the dopant. Adapted from [22].*

Against this scenario, one common alternative is the use of the barium zirconates or compounds containing both Ce and Zr elements, where the introduction of Zr can significantly increase their chemical stability. Nonetheless, it has also been demonstrated that increased amounts of Zr negatively impact the total conductivity of these materials, due to an increase in their refractive nature and in their grain growth, which aggravate the problem of resistive grain boundaries. As such, much lower values of total conductivity are, typically, reported for the zirconate materials than for their cerate analogues, even though their bulk protonic conductivities are actually greater [9, 26–33].

More recently, the work of Kim *et al.* [34] reported that the chemical instability of the barium cerates is due to the presence of a nanometre-thick amorphous phase found at the grain boundaries in proton-conducting $BaCeO_3$ polycrystals, which not only leads to a reduced proton mobility, but also can act as a penetration path for H_2O and CO_2 gas molecules, facilitating chemical decomposition and collapse of the microstructure (**Figure 2a**). Furthermore, this effect could be minimised by controlling the composition to obtain Ba-deficient samples in which the intergranular amorphous layer could be minimised, leading to a mitigation of the reactivity with such gases (**Figure 2b**). The presence of an amorphous layer on the interfaces between grains has also been documented in barium zirconate-based compositions [26, 35], where this feature can exert significant complications during fabrication of complete electrochemical cells [19, 36].

In summary, the high electrical conductivity and the facile processing of the doped barium cerates demands further investigation to succeed to overcome their limited stabilities. In fact, it is only very recently that research in these materials has moved towards a more fundamental and, yet, critical aspect, concerning a deeper understanding of the limiting atmospheric conditions that are necessary to retain their functionality. Taking this into account, Loureiro *et al.* [37] reanalysed the barium cerate stability limits by thermodynamic calculations, considering its decomposition products in the presence of water vapour and CO_2 (**Figure 3**). According to this theoretical study, no degradation would be expected for humidity values of $\sim 3 \times 10^{-2}$ atm and temperatures higher than ~500°C. However, when considering the formation of barium carbonate (**Figure 3**), the thermodynamics predict that much stricter conditions need to be applied, where only very low partial

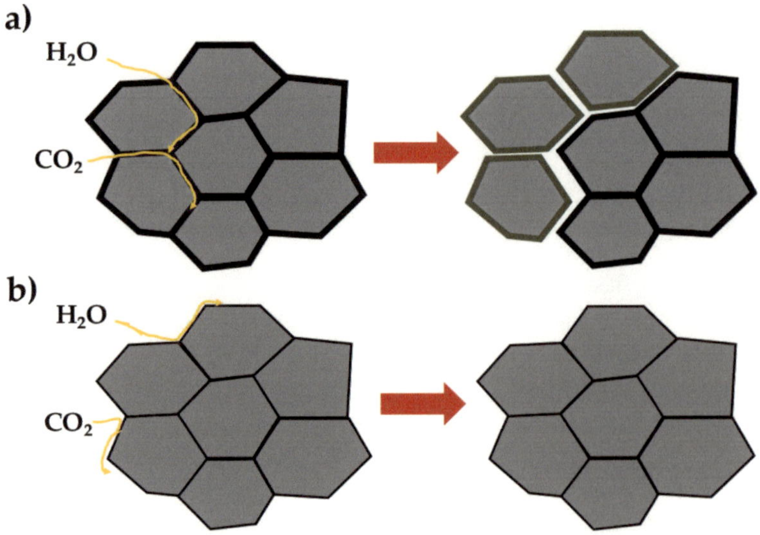

Figure 2.
Schematic representation of microstructural changes upon reaction with water and carbon dioxide: (a) Ba-stoichiometric compositions (thick amorphous intergranular phase); (b) Ba-deficient compositions (thin amorphous intergranular phase).

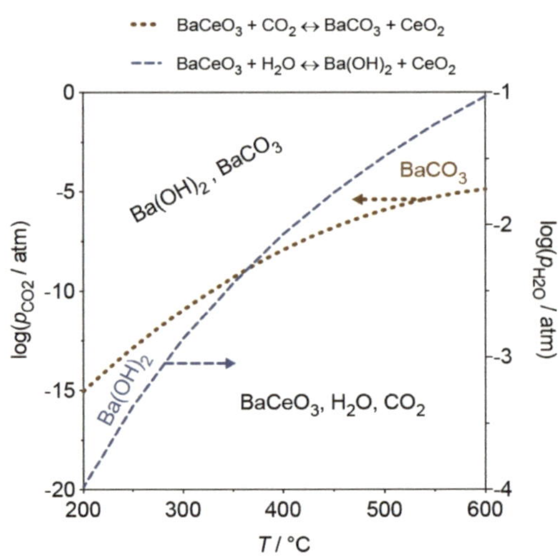

Figure 3.
Thermodynamic stability of carbon dioxide partial pressure (p_{CO_2}) and water vapour partial pressure (p_{H_2O}) as function of temperature considering the equilibrium of $BaCeO_3$ and its decomposition products (i.e. $BaCO_3$ and $Ba(OH)_2$) [38] (reproduced by permission of The Royal Society of Chemistry).

pressures of CO_2 (*e.g.* $p_{CO2} < \sim 10^{-8}$ atm at 400°C) are able to avoid barium cerate degradation.

For this reason, only very few reports can be found on successful applications of BCY membranes for chemical reactions. Most of these have concerned, ammonia synthesis [39–41], or the conversion of propane to propylene [42]. In these cases, no chemical instability has been reported and the survival of the BCY material is likely to be related to the effective absence of CO_2 or significant water vapour in these operations. To understand this further, **Figure 4** presents the maximum water vapour partial pressure (p_{H2O}) that could be tolerated in different carbonaceous atmospheres to provide an equilibrium partial pressure of CO_2 that remains below

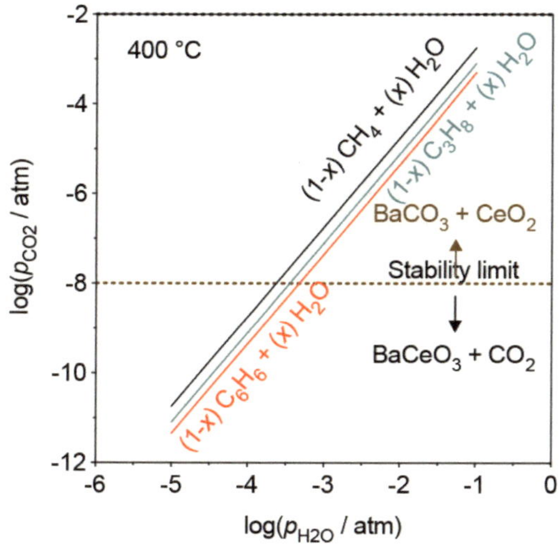

Figure 4.
Thermodynamic equilibrium for the formation of carbon dioxide from a hydrocarbon-based mixture and water at 400°C [38] (reproduced by permission of The Royal Society of Chemistry).

that of the BCY stability limit. These results demonstrate that, for example, at 400°C, these values should range between the values of $10^{-3} < p_{H2O} < \sim 10^{-4}$ atm in order to avoid decomposition of the perovskite phase, for the potential hydrocarbon atmospheres of CH_4, C_3H_8 or C_6H_6 [43].

Nonetheless, one of the requirements for operating in such low water vapour partial pressures is that the protonic conductivity must be maintained in order to ensure the functionality of the electrolyte membrane in these applications. In this respect, protonic conductors are complex materials as they are capable to offer mixed conductivity (protonic, oxide-ion and electronic), depending on the temperature and on the nature of the surrounding atmosphere [37, 38]. One of the most promising compositions for this type of application is that of the yttrium-doped barium cerate, $BaCe_{1-x}Y_xO_{3-\delta}$ (BCY), which has very high protonic conductivity at lower temperatures under humidified atmospheres (*e.g.* $\sim 10^{-3}$ S cm^{-1} at 400°C, $p_{H2O} \sim 10^{-2}$ atm) [1, 38].

Therefore, the current chapter will be focus on the electrochemical transport properties of the $BaCe_{0.9}Y_{0.1}O_{3-d}$ (BCY10) in reducing and oxidising conditions when operating in very low humidity levels. The aim of this chapter is to comprehensively explain the working limits of BCY10 and to assess its applicability as an electrolyte membrane for fuel cell, electrolysers and other electrochemical-based applications, with special focus on operation under low water vapour partial pressures.

2. Electrochemical properties of BCY10 in nominally dry reducing conditions

Figure 5 depicts the total conductivity of BCY10 analysed by impedance spectroscopy between 100 and 500°C in H_2, 10%H_2-N_2 and N_2, highlighting that no significant differences can be observed in the conductivity measured under these atmospheres. In addition, at the higher temperature range, a notable decrease of the activation energy is observed in all cases, as a result of the exsolution of protons from the structure of BCY10, and the concomitant decrease of the protonic contribution to the electrical transport [37]. Interestingly, and also surprisingly at first

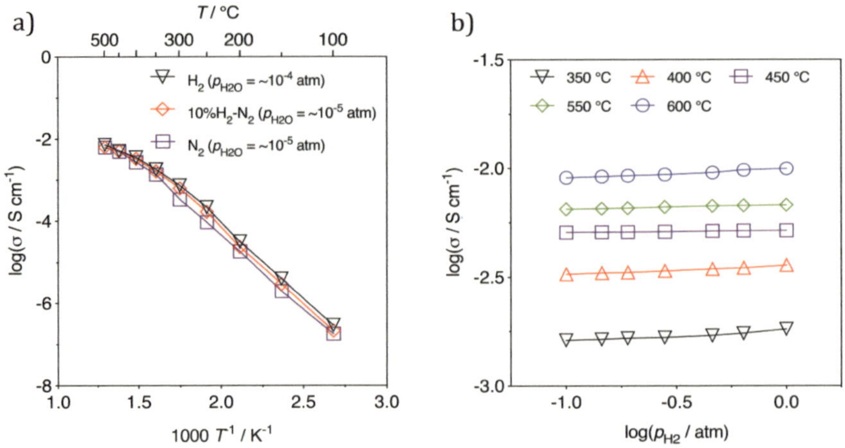

Figure 5.
(a) Temperature dependency of the total conductivity of BCY10 obtained in the temperature range 100–500°C in nominally dry conditions for H_2, 10% H_2-N_2 and N_2; (b) BCY10 total conductivity as function of hydrogen partial pressure (p_{H2}) under nominally dry conditions in the temperature range of 350–600°C [37] (reproduced by permission of The Royal Society of Chemistry).

Conductivity (S cm^{-1})	p_{H2O} (atm)	Reference
3.59×10^{-3}	~10^{-5} atm (dry H_2)	[37]
2.67×10^{-3}	~10^{-2} atm (wet H_2)	[14]
1.85×10^{-3}		[17]
1.96×10^{-3}		[15]
2.60×10^{-3}		[16]
8.48×10^{-4}		[18]

Table 1.
Comparison of literature studies of total conductivity of BCY10 in nominally dry and wet H_2 at 400°C [37] (reproduced by permission of The Royal Society of Chemistry).

sight, current results of total conductivity in nominally dry H_2 are close to those corresponding to the available data in literature for humidified H_2 (**Table 1**).

To analyse the contribution of electronic conductivity to this material in nominally dry conditions, the total conductivity was also analysed as a function of hydrogen partial pressure (p_{H2}) [37], as shown in **Figure 5b**. A slight increase in total conductivity can be observed towards higher p_{H2} values.

To be able to understand this behaviour, firstly the potential for an electronic component to conductivity must be assessed. In reducing conditions (*e.g.* H_2-containing atmospheres), the cerium cations from the B-site of the perovskite structure of BCY10 can reduce from a higher oxidation state, Ce^{4+}, to a lower one, Ce^{3+}, altering the contribution of the concentration of the electronic charge carriers. This phenomenon is well documented in the literature for various cerium-based compositions [37, 38, 44–47], being described as small-polaron electronic conductivity (*i.e.*, a localised, mobile electron, Ce'_{Ce}). Due to the high mobility of electronic conductors, such electronic contribution can exceed that of the ionic, under very reducing conditions and high temperatures [37, 45–47]. However, in the case of BCY10, the extent of cerium reduction has been assessed by Loureiro *et al.* [37], who performed coulometric titration measurements to study the potential role of electronic contribution in $BaCe_{0.9}Y_{0.1}O_{3-\delta}$ in reducing conditions as a function of temperature. This technique has been widely adopted to quantify the changes in the oxygen non-stoichiometry ($\Delta\delta$), which can be associated with the reduction of Ce^{4+} to Ce^{3+}, following the equation:

$$2Ce_{Ce}^{x} + O_O^{x} \Leftrightarrow V_O^{\bullet\bullet} + 2Ce_{Ce}' + \frac{1}{2}O_2 \tag{6}$$

with the equilibrium constant for reduction reaction given by:

$$K_R = \frac{[V_O^{\bullet\bullet}][Ce_{Ce}']^2 p(O_2)^{\frac{1}{2}}}{[O_O^{x}][Ce_{Ce}^{x}]^2} \tag{7}$$

The results of coulometric titration (**Figure 6**) [37] show considerable variations of $\Delta\delta$ with oxygen partial pressure only at very high temperature, with a lower impact as temperature decreases. Thus, **Figure 6** demonstrates that very extreme reducing conditions and very high temperatures are required to produce appreciable increase in the oxygen-vacancy and electronic concentrations in BCY [47, 48]. These results contrast with those of fluorite-ceria-based materials which usually show high reducibility under milder conditions [46, 49].

Thus, to take the possibility of reduction into account, the methodology applied by Loureiro *et al.* [37] for the determination of reduction equilibrium follows the method reported elsewhere [50], as described below.

The corresponding mass action constant (Eq. (7)) can be combined with the electroneutrality condition:

$$2[V_O^{\bullet\bullet}] + [OH_O^{\bullet}] \approx [Y_{Ce}'] + [Ce_{Ce}'] \tag{8}$$

and other mass and lattice position restrictions, on neglecting defect interactions and assuming nearly ideal behaviour, with the following relations between the concentrations of relevant species, stoichiometric changes ($\Delta\delta$), and fraction of trivalent additive (x):

$$[Ce_{Ce}'] = \frac{Z}{v_0}(2\Delta\delta) \tag{9}$$

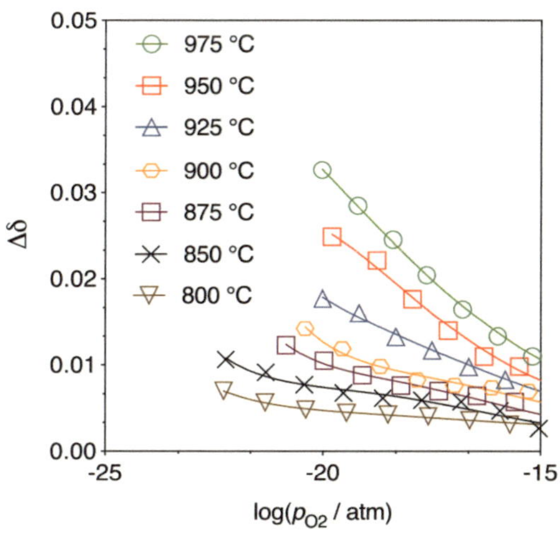

Figure 6.
Oxygen non-stoichiometry as function of oxygen partial pressure (p_{O_2}) [37] (reproduced by permission of The Royal Society of Chemistry).

$$[V_O^{\bullet\bullet}] = \frac{Z}{v_0}\left(\Delta\delta + \frac{x}{2}\right) \tag{10}$$

$$[Ce_{Ce}^{x}] = \frac{Z}{v_0}(1 - x - 2\Delta\delta) \tag{11}$$

$$[O_O^{x}] = \frac{Z}{v_0}\left(3 - \frac{x}{2} - \Delta\delta\right) \tag{12}$$

where Z is the number of atoms per unit cell and n_0, the unit cell volume. Substitution in Eq. (7) leads to the values of the equilibrium constant for reduction (K_R) from the entire range of values of $\Delta\delta$ *versus* p_{O2} at a given temperature *T*:

$$K_R(T) = \frac{4\Delta\delta^2\left(\Delta\delta + \frac{x}{2}\right)p_{O2}^{1/2}}{\left(3 - \frac{x}{2} - \Delta\delta\right)(1 - x - 2\Delta\delta)^2} \tag{13}$$

The following equation was then determined to describe the temperature dependence of K_R, from the results of oxygen-nonstoichiometry shown in **Figure 6**:

$$K_R(T) = 4.47 \cdot 10^{14} \exp\left(-7.85 \cdot 10^4/T\right) atm^{1/2} \tag{14}$$

with an enthalpy for reduction, ΔH_R = 804.99 kJ mol^{-1}. This value is significantly higher than those obtained by other authors for fluorite ceria-based materials (**Table 2**) [46, 49], underscoring the low reducibility of BCY10 in such conditions from intermediate to low temperatures.

On the basis of these results, the potential rehydration of the BCY10 material was then assessed by thermogravimetric experiments [37]. **Figure 7** depicts the concentration of protonic charge carriers as a function of temperature, calculated from the following methodology.

By expressing the equilibrium constant for water incorporation reaction (Eq. (3)) in terms of entropy, ΔS_w, and enthalpy, ΔH_w:

$$K_w = \exp\left(\frac{\Delta S_w}{R}\right) . \exp\left(-\frac{\Delta H_w}{RT}\right) \tag{15}$$

where *T* and R have usual meanings. Given Eq. (1) and knowing that the number of oxygen sites per formula unit of barium cerate is restricted to 3, implying the site restriction relationship:

Compound	δh_r (kJ mol^{-1})	Reference
$BaCe_{0.9}Y_{0.1}O_{3-x/2-\Delta\delta}$	805	[37]
$Ce_{0.9}Gd_{0.1}O_{2-x/2-\Delta\delta}$	410-420	[46]
	438	[51]
$Ce_{0.8}Gd_{0.2}O_{2-x/2-\Delta\delta}$	430	[46]
	385	[51]
$Ce_{0.9}Sm_{0.1}O_{2-x/2-\Delta\delta}$	400	[52]
$Ce_{0.8}Sm_{0.2}O_{2-x/2-\Delta\delta}$	385	[52]
	375	[49]

Table 2.
Enthalpy (ΔH_R) for reduction of different ceria-based based solid solutions materials.

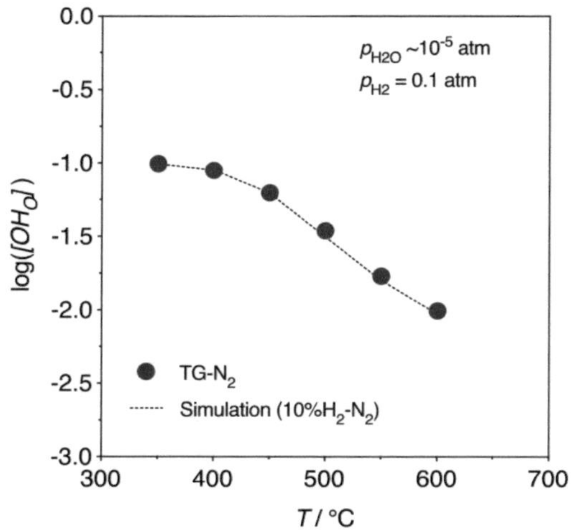

Figure 7.
Concentration protonic defects obtained from TG in N_2 and from the simulation performed in [37] (reproduced by permission of The Royal Society of Chemistry).

$$2\left[V_O^{\bullet\bullet}\right] + \left[OH_O^{\bullet}\right] + \left[O_O^{x}\right] = 3 \quad (16)$$

with Eqs. (3), (15), (16), K_w can be reformulated as

$$K_w = \exp\left(\frac{4\left[OH_O^{\bullet}\right]^2}{p_{H2O}\left(S - \left[OH_O^{\bullet}\right]\right)\left(6 - S - \left[OH_O^{\bullet}\right]\right)}\right) \quad (17)$$

and then, the concentration of protonic defects is given by

$$\left[OH_O^{\bullet}\right] = \frac{3.K' - \sqrt{9K' - 6K'.S + K'.S^2 + 24S - 4S^2}}{K' - 4} \quad (18)$$

where $K' = K_w.p_{H2O}$ and S = [Y_{Ce}']. Because the formation of protonic defects is usually accompanied by a significant weight increase, the concentration of protonic defects as a function of temperature and water vapour partial pressure is generally measured by thermogravimetric analysis (TG). From **Figure 7**, one can observe an increase in the concentration of protonic species as a function of decreasing temperature, even in nominally dry 10%H_2/N_2. This factor is most likely related to the intrinsic formation of water vapour under the presence of hydrogen and oxygen impurities in the feed stream:

$$H_{2(g)} + \frac{1}{2}O_{2(g)} \leftrightarrow H_2O_{(g)} \quad (19)$$

This result emphasises the existence of protonic conductivity in nominally dry hydrogen-containing atmospheres, as even trace amounts of oxygen can form water vapour, potentially contributing to the hydration of the BCY10 material. Hence, the partial conductivities can be obtained by combining the results from both coulombic titration and TG experiments using a defect chemistry methodology [37].

Figure 8 shows the partial conductivities of all species (protons, oxide-ions and electrons) obtained at the temperature range (350–600°C) in nominally dry H_2 ($p_{H2O} \sim 10^{-4}$ atm). One can observe a dominance of the ionic charge carriers over the electronic carriers in the whole temperature range, corroborating the negligible reducibility of cerium cations measured by coulombic titration (**Figure 6**).

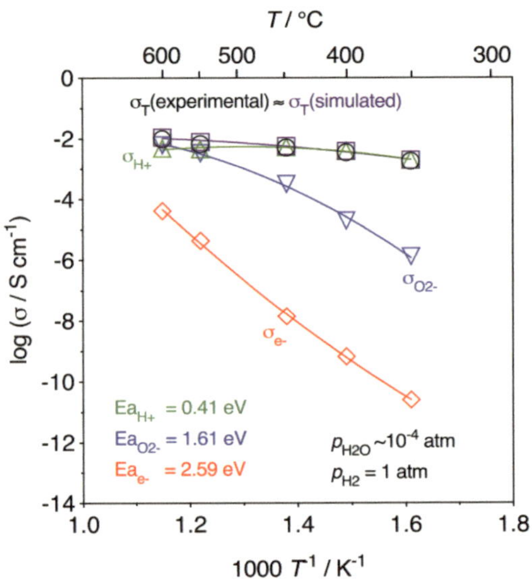

Figure 8.
Total (experimental and calculated) and partial conductivities vs. *temperature. Data obtained in the temperature range 350–600°C in nominally dry conditions [37] (reproduced by permission of The Royal Society of Chemistry).*

Furthermore, at the low temperature range (350–400°C), the dominance of protonic conductivity is related to the high equilibrium constant for water incorporation in BCY10, allowing a significant hydration even at p_{H2O} values as low as $\sim 10^{-4}$ atm [53], as confirmed by TG (**Figure** 7). This behaviour also explains the slight p_{H2} dependence of conductivity shown in **Figure 5b** that is due, not to electronic behaviour, but to changes in the effective water vapour partial pressure arising from Eq. (19) and subsequent slight increase in ionic conductivity due to a higher level of hydration Eq. (18). In contrast at higher temperatures in the (550–600°C) range, oxide-ion conductivity starts to become dominant at due to the loss of protons from the structure (**Figure** 7).

3. Electrochemical properties of BCY10 in low humidity oxidising conditions

The transport numbers of BCY10 in oxidising atmospheres were firstly studied by Oishi *et al.* [54] and by Grimaud *et al.* [55]. Later, Lim *et al.* [56] determined the concentration of charge carriers in BCY10 by thermogravimetric analysis (TGA) under two different humidity conditions (dry and wet, $p_{H2O} \sim 10^{-5}$ and 10^{-3} atm, respectively). More recently, Loureiro *et al.* [38] focused on the determination of the transport properties of this composition at temperatures below 600°C and under very low humidity levels ($p_{H2O} \leq 10^{-4}$ atm).

In oxidising conditions, the absence of hydrogen species, shifts the water formation reaction, Eq. (19), away from the water product, leading to a lower intrinsic water vapour partial pressure that can, in turn, decrease the protonic transport number [38]. Therefore, at the intermediate temperature range, 350–600°C, it is necessary to externally add humidity to guarantee a sufficient level of protonic conductivity. Moreover, BCY10 is known to possess p-type electronic conductivity in oxidising atmospheres, which can importantly impact the total conductivity in these conditions [38], as expressed by

$$\frac{1}{2}O_{2(g)} + V_O^{\bullet\bullet} \longleftrightarrow O_O^x + 2h^{\bullet} \tag{20}$$

with the following mass action constant

$$K_O \approx \frac{[h^{\bullet}]^2}{[V_O^{\bullet\bullet}] \cdot p_{O2}^{1/2}} \tag{21}$$

Figure 9 shows the total conductivity of BCY10 measured in the temperature range 350–600°C in wet and low humidity O_2 and N_2. From **Figure 9**, this expected decrease in the concentration of protonic species is corroborated, as in both, N_2 and O_2, total conductivity is shown to be higher in wet conditions ($p_{H2O} \sim 10^{-3}$ atm) than in low humidity conditions ($p_{H2O} \sim 10^{-7}$ atm). It is also possible to observe that low humidity N_2 ($p_{H2O} \sim 10^{-7}$ atm) the total conductivity is lower in the whole measured temperature range in comparison to wet N_2 ($p_{H2O} \sim 10^{-3}$ atm), as a result of dehydration of the sample according to Eq. (22). In contrast, in O_2, the total conductivity in low humidity and wet conditions are similar, particularly at higher temperatures, a factor that can be explained due to the presence and dominance of p-type electronic conductivity [57, 58] (see Eq. (20)):

$$2OH_O^{\bullet} \Leftrightarrow H_2O + V_O^{\bullet\bullet} + O_O^x \tag{22}$$

In agreement, the presence of p-type electronic conductivity can explain the slightly higher activation energy registered in low humidity O_2, 0.49 eV, in comparison to the other studied atmospheres.

Figure 10 illustrates the partial conductivities obtained in wet ($p_{H2O} \sim 10^{-3}$ atm) and low humidity ($p_{H2O} \sim 10^{-7}$ atm) conditions in N_2 and O_2. **Figure 10a** and **b** show that in moderate wet conditions ($p_{H2O} \sim 10^{-3}$ atm) the protonic conductivity is dominating in both atmospheres with activation energies similar to that obtained for the protonic conduction (~0.4–0.5 eV) [16, 17]. In contrast, in low humidity conditions (**Figure 10c** and **d**) a drop on protonic conductivity with increasing temperature is observed, due to predominant oxide-ion conductivity in both

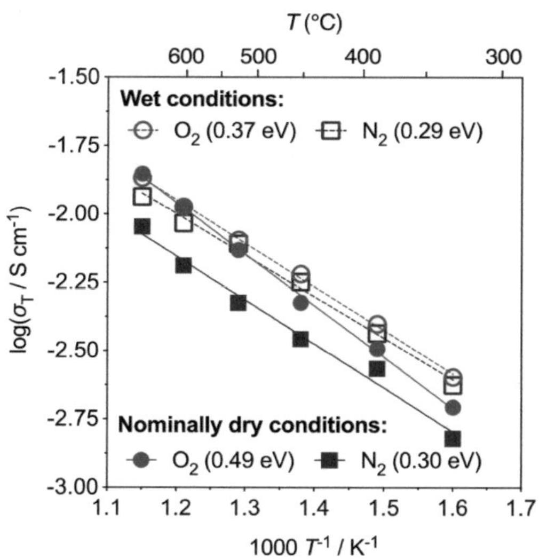

Figure 9.
Total conductivity of BCY10 in wet ($p_{H2O} \sim 10^{-3}$ atm) and low humidity ($p_{H2O} \sim 10^{-7}$ atm) N_2 and O_2. Reproduced from [38] with permission from Elsevier.

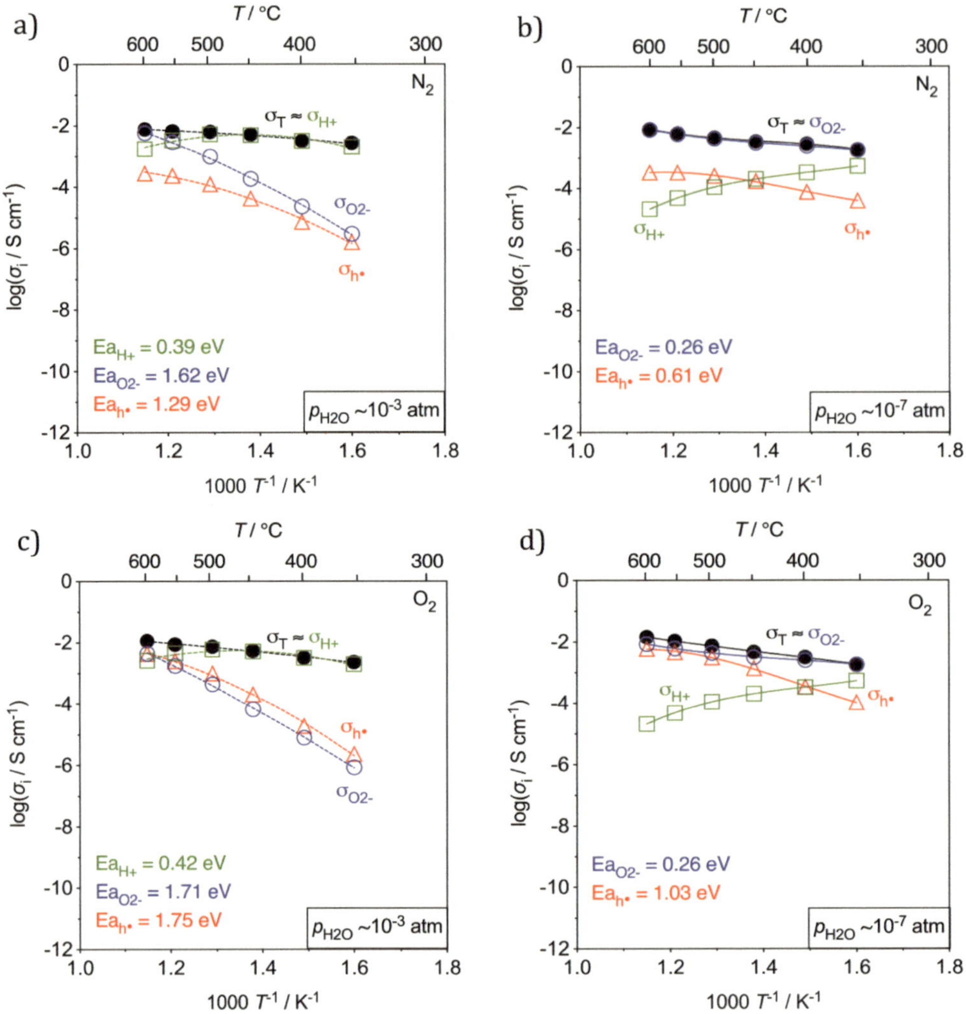

Figure 10.
Partial conductivities obtained in wet and low humidity conditions in (a) and (b) N_2, and (c) and (d) O_2. The activation energy values, E_a, were calculated in the temperature range 350–500°C. Reproduced from [38] with permission from Elsevier.

atmospheres. In the case of hole conductivity, the activation energies obtained were found to be lower at low humidity conditions (0.61–1.03 eV, T = 350–500°C) in comparison with those obtained in wet conditions (1.29–1.75 eV, T = 350–500°C). This can be explained by the creation of electronic defects (Eq. (20)), upon filling the oxygen vacancies.

4. Comparison between reducing and oxidising conditions under low humidity

As discussed previously, to maximise the protonic conductivity is necessary to maintain a minimum level of humidity in the order of 10^{-4}–10^{-5} atm. It is also important to emphasise that, while this level of humidity is intrinsically formed in nominally dry hydrogen-containing atmospheres, in the case of oxidising atmospheres this level must be externally supplied. A comparison of the partial conductivities in all cases is shown in **Figure 11**, for $p_{H2O} \sim 10^{-4}$ atm. At temperatures

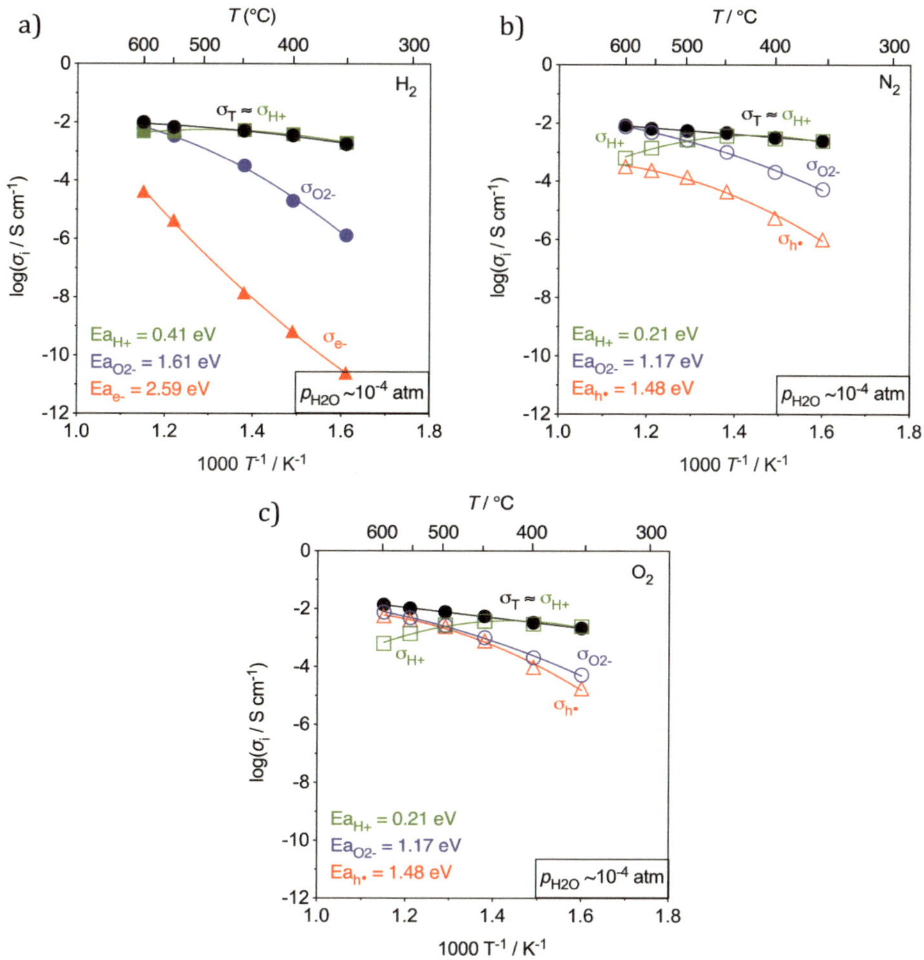

Figure 11.
Temperature dependence of partial conductivities in at $p_{H2O} \sim 10^{-4}$ atm: (a) H_2, (b) N_2 and (c) O_2. Activation energy values, E_a, calculated in the temperature range 350-500°C. Reproduced from [38] with permission from Elsevier.

below 450°C, the total conductivity is dominated by protonic conductivity, with the oxide-ion conductivity taking a negligible role. In contrast, at higher temperatures ($T > 450$°C), the oxide- ion conductivity dominates the total conductivity with a simultaneous decrease of protonic conductivity. With respect to the electronic conductivity, this term increases as p_{O2} increases, being only relevant in oxidising conditions and/or high temperatures. This can be explained due to the creation of electronic holes, which become more relevant with increasing p_{O2} and temperature (Eq. (20)).

Overall, BCY10 is shown to be a predominant protonic conductor in both reducing and oxidising atmospheres at sufficiently low temperatures ≤500°C, even under relatively low water vapour partial pressures ($p_{H2O} \sim 10^{-4}$–10^{-5} atm). Moreover, the level of conductivity measured at 400°C in these conditions is high, *e.g.* $\sim 10^{-3}$ S cm^{-1}. The origin of protonic conductivity is due to a high equilibrium constant for water absorption that allows this material to offer high bulk protonic conductivity at intermediate temperatures in these very low humidity conditions. From **Figure 12**, one can immediately envisage that this is a particular behaviour of BCY10 that cannot be obtained in other competing proton-conducting perovskites, due to their much lower values of K_w.

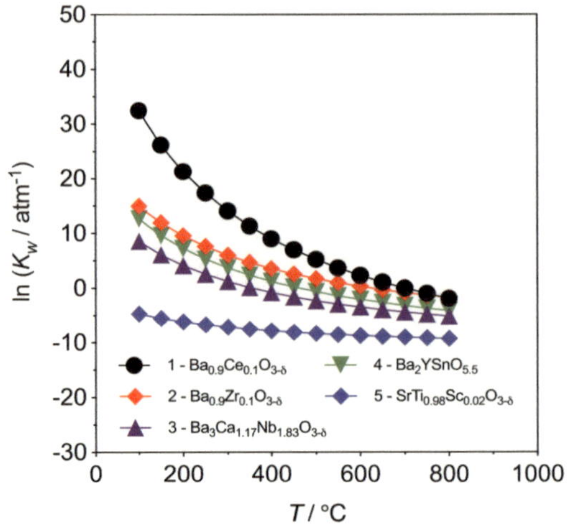

Figure 12.
Equilibrium constant for hydration of several perovskite proton conductors. Adapted from [9].

Type	Reaction	ΔH^0_{298}/kJ mol^{-1}
Dehydrogenations	$2CH_{4(g)} \rightleftharpoons C_2H_{4(g)} + 2H_{2(g)}$	202
	$6CH_{4(g)} \rightleftharpoons C_6H_{6(g)} + 9H_{2(g)}$	89
	$C_3H_{8(g)} \rightleftharpoons C_3H_{6(g)} + H_{2(g)}$	124.3
	$iC_4H_{10(g)} \rightleftharpoons iC_4H_{8(g)} + H_{2(g)}$	122
	$C_8H_{10(g)} \rightleftharpoons C_8H_{8(g)} + H_{2(g)}$	117.6
Hydrogenations	$C_{10}H_{8(g)} + 2H_{2(g)} \rightleftharpoons C_{10}H_{12(g)}$	−134
	$C_6H_{10(g)} + H_{2(g)} \rightleftharpoons C_6H_{12(g)}$	−120
	$C_6H_{6(g)} + 3H_{2(g)} \rightleftharpoons C_6H_{12(g)}$	−207
	$N_{2(g)} + 3H_{2(g)} \rightleftharpoons 2NH_{3(g)}$	−109

Reproduced from [38] with permission from Elsevier.

Table 3.
Examples of dehydrogenation/hydrogenation reactions that can occur at very low humidity conditions.

This is a very exciting result since it opens a wide range of possibilities for using the BCY material, in particular, in different applications that involve very low humidity levels and low temperatures of operation. The most well-known is that of ammonia electrochemical synthesis [39–41], although many other processes concerning hydrogenation and de-hydrogenation reactions can also be considered that agree with these operating conditions (**Table 3**).

5. Conclusions

The current chapter highlights that the transport properties of $BaCe_{0.9}Y_{0.1}O_{3-\delta}$ (BCY10) in very low humidity conditions are dependent on the nature of the surrounding atmosphere and on the temperature, being significantly different in reducing and oxidising conditions and at high and low temperatures. In reducing conditions, BCY10 shows a very high protonic conductivity (*e.g.* ~ 10^{-3} S cm^{-1}) at low temperatures *i.e.* < 400°C, even in nominally dry atmospheres with negligible oxide-ion/electronic influence.

In the other hand, in oxidising conditions, the same behaviour can only be obtained by externally supplying humidity in the range ($p_{H2O} \sim 10^{-4}$–10^{-5} atm) at low temperatures ≤500°C. At higher temperatures, at this low humidity, the onset of hole conductivity can be noted at higher oxygen partial pressures due partial material dehydration.

The present discussion shows the importance of controlling the humidity levels in order to maximise the protonic conductivity of BCY under operation. The very low levels of humidity required ($p_{H2O} \sim 10^{-4}$–10^{-5} atm), to ensure predominant proton conductivity in both reducing and oxidising atmospheres at low temperatures ≤500°C, are highly interesting as they highlight the possibility of using this composition in applications where low humidity levels and temperatures are required, such as the suggested de-hydrogenation/hydrogenation chemical reactions, while maintaining its stability against decomposition.

Acknowledgements

The authors acknowledge Fundação para a Ciência e Tecnologia (FCT) for the PhD grants – PD/BDE/142837/2018, SFRH/BD/130218/2017, and PD/BDE/114353/2016. The authors also acknowledge the projects UID/EMS/00481/2019-FCT and CENTRO-01-0145-FEDER-022083 - Centro Portugal Regional Operational Programme (Centro2020), under the PORTUGAL 2020 Partnership Agreement, through the European Regional Development Fund (ERDF).

Author details

Laura I.V. Holz, Vanessa C.D. Graça, Francisco J.A. Loureiro* and Duncan P. Fagg
Mechanical Engineering Department, Centre for Mechanical Technology and Automation, University of Aveiro, Aveiro, Portugal

*Address all correspondence to: francisco.loureiro@ua.pt

References

[1] Morejudo SH, Zanón R, Escolástico S, Yuste-Tirados I, Malerød-Fjeld H, Vestre PK, et al. Direct conversion of methane to aromatics in a catalytic co-ionic membrane reactor. Science (80-) [Internet]. 2016 Aug 5;353(6299):563 LP – 566. Available from: http://science.sciencemag.org/content/353/6299/563.abstract

[2] Bi L, Boulfrad S, Traversa E. Steam electrolysis by solid oxide electrolysis cells (SOECs) with proton-conducting oxides. Chem Soc Rev [Internet]. 2014; 43(24):8255–70. Available from: http://dx.doi.org/10.1039/C4CS00194J

[3] Fabbri E, Bi L, Pergolesi D, Traversa E. Towards the Next Generation of Solid Oxide Fuel Cells Operating Below 600 °C with Chemically Stable Proton-Conducting Electrolytes. Adv Mater [Internet]. 2012;24(2):195–208. Available from: http://dx.doi.org/10.1002/adma.201103102

[4] Bannykh A V, Kuzin BL. Electrical conductivity of BaCe0.9Nd0.1O3−α in H2+H2O+Ar gas mixture. Ionics (Kiel) [Internet]. 2003;9(1):134–9. Available from: https://doi.org/10.1007/BF02376550

[5] Kuzin BL, Beresnev SM, Bannykh A V, Perfil'yev M V. Transport numbers of H+ and O2- in the electrochemical system (H2 + H2O),Me/BaCe0.9Nd0.1O3-α/Me,(H2 + H2O). Russ J Electrochem [Internet]. 2000;36(4):424–30. Available from: https://doi.org/10.1007/BF02756951

[6] Qiu L-G, Ma G-L, Wen D-J. Properties and Application of Ceramic BaCe0.8Ho0.2O3−α. Chinese J Chem [Internet]. 2005 Dec 22;23(12):1641–5. Available from: https://doi.org/10.1002/cjoc.200591641

[7] Virkar AN, Maiti HS. Oxygen ion conduction in pure and yttria-doped barium cerate. J Power Sources [Internet]. 1985;14(4):295–303. Available from: http://www.sciencedirect.com/science/article/pii/0378775385800458

[8] Suksamai W, Metcalfe IS. Measurement of proton and oxide ion fluxes in a working Y-doped BaCeO3 SOFC. Solid State Ionics [Internet]. 2007;178(7):627–34. Available from: http://www.sciencedirect.com/science/article/pii/S0167273807000549

[9] Kreuer KD. Proton conducting oxides. Annu Rev Mater Res. 2003 Aug; 33(1):333–59.

[10] Malavasi L, Fisher CAJ, Islam MS. Oxide-ion and proton conducting electrolyte materials for clean energy applications: structural and mechanistic features. Chem Soc Rev [Internet]. 2010;39(11):4370–87. Available from: http://dx.doi.org/10.1039/B915141A

[11] Fabbri E, D'Epifanio A, Di Bartolomeo E, Licoccia S, Traversa E. Tailoring the chemical stability of Ba(Ce0.8-xZrx)Y0.2O3-d protonic conductors for Intermediate Temperature Solid Oxide Fuel Cells (IT-SOFCs). Solid State Ionics. 2008;179(15–16):558–64.

[12] Brett DJL, Atkinson A, Brandon NP, Skinner SJ. Intermediate temperature solid oxide fuel cells. Chem Soc Rev [Internet]. 2008;37(8):1568–78. Available from: http://dx.doi.org/10.1039/B612060C

[13] Liu M, Lynch ME, Blinn K, Alamgir FM, Choi Y. Rational SOFC material design: new advances and tools. Mater Today [Internet]. 2011;14(11):534–46. Available from: http://www.sciencedirect.com/science/article/pii/S1369702111702796

[14] Coors WG, Readey DW. Proton Conductivity Measurements in Yttrium Barium Cerate by Impedance Spectroscopy. J Am Ceram Soc [Internet]. 2004 Dec 20;85(11):2637–40. Available from: http://www3.interscience.wiley.com/journal/118935979/abstract

[15] Ma G, Shimura T, Iwahara H. Simultaneous doping with La3+ and Y3+ for Ba2+ − and Ce4+ −sites in BaCeO3 and the ionic conduction. Solid State Ionics. 1999;120(1):51–60.

[16] Kreuer KD, Dippel T, Baikov YM, Maier J. Water solubility, proton and oxygen diffusion in acceptor doped BaCeO3: A single crystal analysis. Solid State Ionics [Internet]. 1996;86–88, Par: 613–20. Available from: http://www.sciencedirect.com/science/article/pii/0167273896002214

[17] Bonanos N, Ellis B, Knight KS, Mahmood MN. Ionic conductivity of gadolinium-doped barium cerate perovskites. Solid State Ionics [Internet]. 1989;35(1–2):179–88. Available from: http://www.sciencedirect.com/science/article/pii/0167273889900283

[18] Slade RCT, Singh N. The perovskite-type proton-conducting solid electrolyte BaCe0.90Y0.10O3−α in high temperature electrochemical cells. Solid State Ionics [Internet]. 1993;61(1–3): 111–4. Available from: http://www.sciencedirect.com/science/article/pii/016727389390342Z

[19] Fabbri E, Pergolesi D, Traversa E. Materials challenges toward proton-conducting oxide fuel cells: a critical review. Chem Soc Rev [Internet]. 2010; 39(11):4355–69. Available from: http://dx.doi.org/10.1039/B902343G

[20] Fabbri E, Pergolesi D, Traversa E. Electrode materials: a challenge for the exploitation of protonic solid oxide fuel cells. Sci Technol Adv Mater [Internet]. 2010 Aug 10;11(4):44301. Available from: http://www.ncbi.nlm.nih.gov/pmc/articles/PMC5090333/

[21] Kruth A, Irvine JTS. Water incorporation studies on doped barium cerate perovskites. Solid State Ionics. 2003;162–163:83–91.

[22] Matsumoto H, Kawasaki Y, Ito N, Enoki M, Ishihara T. Relation Between Electrical Conductivity and Chemical Stability of BaCeO3-Based Proton Conductors with Different Trivalent Dopants. Electrochem Solid-State Lett [Internet]. 2007 Apr 1;10(4):B77–80. Available from: http://esl.ecsdl.org/content/10/4/B77.abstract

[23] Eriksson Andersson AK, Selbach SM, Grande T, Knee CS. Thermal evolution of the crystal structure of proton conducting BaCe0.8Y0.2O3-[small delta] from high-resolution neutron diffraction in dry and humid atmosphere. Dalt Trans [Internet]. 2015;44(23):10834–46. Available from: http://dx.doi.org/10.1039/C4DT03948C

[24] Lacz A. Effect of microstructure on chemical stability and electrical properties of BaCe0.9Y0.1O3 {\thinspace}−{\thinspace}$δ$. Ionics (Kiel) [Internet]. 2016;22(8):1405–14. Available from: http://dx.doi.org/10.1007/s11581-016-1665-6

[25] Li Y, Su P-C, Wong LM, Wang S. Chemical stability study of nanoscale thin film yttria-doped barium cerate electrolyte for micro solid oxide fuel cells. J Power Sources [Internet]. 2014; 268:804–9. Available from: http://www.sciencedirect.com/science/article/pii/S0378775314009926

[26] Sun Z, Fabbri E, Bi L, Traversa E. Lowering grain boundary resistance of BaZr0.8Y0.2O3-[small delta] with LiNO3 sintering-aid improves proton conductivity for fuel cell operation. Phys Chem Chem Phys [Internet]. 2011;

13(17):7692–700. Available from: http://dx.doi.org/10.1039/C0CP01470B

[27] Donglin H, Naoyuki H, Tetsuya U. Chemical Expansion of Yttrium-Doped Barium Zirconate and Correlation with Proton Concentration and Conductivity. J Am Ceram Soc [Internet]. 2016 Jun 30; 99(11):3745–53. Available from: https://doi.org/10.1111/jace.14377

[28] Muccillo R, Muccillo ENS, Andrade TF, Oliveira OR. Thermal analyses of yttrium-doped barium zirconate with phosphor pentoxide, boron oxide and zinc oxide addition. J Therm Anal Calorim [Internet]. 2017; 130(3):1791–9. Available from: https://doi.org/10.1007/s10973-017-6523-x

[29] Narendar N, Mather GC, Dias PAN, Fagg DP. The importance of phase purity in Ni–BaZr0.85Y0.15O3−δ cermet anodes – novel nitrate-free combustion route and electrochemical study. RSC Adv [Internet]. 2013;3(3): 859–69. Available from: http://dx.doi.org/10.1039/C2RA22301E

[30] Soares HS, Zhang X, Antunes I, Frade JR, Mather GC, Fagg DP. Effect of phosphorus additions on the sintering and transport properties of proton conducting BaZr0.85Y0.15O3−δ. J Solid State Chem [Internet]. 2012;191:27–32. Available from: http://www.sciencedirect.com/science/article/pii/S0022459612001557

[31] Yang T, Loureiro FJAFJA, Queirós RP, Pukazhselvan D, Antunes I, Saraiva JAJA, et al. A detailed study of hydrostatic press, sintering aids and temperature on the densification behavior of $Ba(Zr,Y)O_{3-d}$electrolyte. Int J Hydrogen Energy. 2016;41(27): 1–10.

[32] Bozza F, Bator K, Kubiak WW, Graule T. Effects of Ni doping on the sintering and electrical properties of BaZr0.8Y0.2O3−δ proton conducting electrolyte prepared by Flame Spray Synthesis. J Eur Ceram Soc [Internet]. 2016;36(1):101–7. Available from: http://www.sciencedirect.com/science/article/pii/S0955221915301291

[33] Grant H, Anthony M, G. CW, Sandrine R. Chemical expansion in BaZr0.9−xCexY0.1O3−δ (x = 0 and 0.2) upon hydration determined by high-temperature X-ray diffraction. J Am Ceram Soc [Internet]. 2017 Oct 7; 101(3):1298–309. Available from: https://doi.org/10.1111/jace.15275

[34] Kim H-S, Bae H Bin, Jung W, Chung S-Y. Manipulation of Nanoscale Intergranular Phases for High Proton Conduction and Decomposition Tolerance in BaCeO3 Polycrystals. Nano Lett [Internet]. 2018 Feb 14;18(2):1110–7. Available from: https://doi.org/10.1021/acs.nanolett.7b04655

[35] Park J-S, Lee J-H, Lee H-W, Kim B-K. Low temperature sintering of BaZrO3-based proton conductors for intermediate temperature solid oxide fuel cells. Solid State Ionics [Internet]. 2010;181(3):163–7. Available from: http://www.sciencedirect.com/science/article/pii/S0167273809002549

[36] Loureiro FJA, Nasani N, Reddy GS, Munirathnam NR, Fagg DP. A review on sintering technology of proton conducting BaCeO3-BaZrO3 perovskite oxide materials for Protonic Ceramic Fuel Cells. J Power Sources [Internet]. 2019;438:226991. Available from: http://www.sciencedirect.com/science/article/pii/S037877531930984X

[37] Loureiro FJA, Pérez-Coll D, Graça VCD, Mikhalev SM, Ribeiro AFG, Mendes A, et al. Proton conductivity in yttrium-doped barium cerate in nominally dry reducing conditions for application in chemical synthesis. J Mater Chem A [Internet]. 2019;7: 18135–42. Available from: submitted

[38] Loureiro FJA, Ramasamy D, Ribeiro AFG, Mendes A, Fagg DP.

Underscoring the transport properties of yttrium-doped barium cerate in nominally dry oxidising conditions. Electrochim Acta [Internet]. 2019;334: 135625. Available from: https://doi.org/10.1016/j.electacta.2020.135625

[39] Otomo J, Noda N, Kosaka F. Electrochemical Synthesis of Ammonia with Proton Conducting Solid Electrolyte Fuel Cells at Intermediate Temperatures. ECS Trans [Internet]. 2015 Jun 2;68(1): 2663–70. Available from: http://ecst.ecsdl.org/content/68/1/2663.abstract

[40] Li Z, Liu R, Wang J, Xu Z, Xie Y, Wang B. Preparation of double-doped BaCeO3 and its application in the synthesis of ammonia at atmospheric pressure. Sci Technol Adv Mater [Internet]. 2007;8(7–8):566–70. Available from: http://www.sciencedirect.com/science/article/pii/S1468699607001404

[41] Marnellos G, Stoukides M. Ammonia Synthesis at Atmospheric Pressure. Science (80-) [Internet]. 1998 Oct 2;282(5386):98 LP – 100. Available from: http://science.sciencemag.org/content/282/5386/98.abstract

[42] Feng Y, Luo J, Chuang KT. Propane Dehydrogenation in a Proton-conducting Fuel Cell. J Phys Chem C [Internet]. 2008 Jul 1;112(26):9943–9. Available from: http://dx.doi.org/10.1021/jp710141c

[43] Bale CW, Bélisle E, Chartrand P, Decterov SA, Eriksson G, Gheribi AE, et al. FactSage thermochemical software and databases, 2010–2016. Calphad [Internet]. 2016;54:35–53. Available from: http://www.sciencedirect.com/science/article/pii/S0364591616300694

[44] Bonanos N, Willy Poulsen F. Considerations of defect equilibria in high temperature proton-conducting cerates. J Mater Chem [Internet]. 1999;9(2):431–4. Available from: http://dx.doi.org/10.1039/A805150J

[45] Pérez-Coll D, Aguadero A, Núñez P, Frade JR. Mixed transport properties of Ce1−xSmxO2−x/2 system under fuel cell operating conditions. Int J Hydrogen Energy [Internet]. 2010;35(20):11448–55. Available from: http://www.sciencedirect.com/science/article/pii/S0360319910009900

[46] Pérez-Coll D, Marrero-López D, Ruiz-Morales JC, Núñez P, Abrantes JCC, Frade JR. Reducibility of Ce1−xGdxO2−δ in prospective working conditions. J Power Sources [Internet]. 2007;173(1):291–7. Available from: http://www.sciencedirect.com/science/article/pii/S0378775307009044

[47] Bonanos N. Transport study of the solid electrolyte BaCe0.9Gd0.1O2.95 at high temperatures. J Phys Chem Solids [Internet]. 1993;54(7):867–70. Available from: http://www.sciencedirect.com/science/article/pii/002236979390258S

[48] Chen W, Nijmeijer A, Winnubst L. Oxygen non-stoichiometry determination of perovskite materials by a carbonation process. Solid State Ionics [Internet]. 2012 Dec 14;229:54–8. Available from: http://www.sciencedirect.com/science/article/pii/S0167273812005723

[49] Abrantes JCC, Pérez-Coll D, Núñez P, Frade JR. Electronic transport in Ce0.8Sm0.2O1.9−δ ceramics under reducing conditions. Electrochim Acta [Internet]. 2003 Aug 15;48(19):2761–6. Available from: http://www.sciencedirect.com/science/article/pii/S0013468603003955

[50] Pérez-Coll D, Marrero-López D, Ruiz-Morales JC, Núñez P, Abrantes JCC, Frade JR. Reducibility of Ce1-xGdxO2-δ in prospective working conditions. J Power Sources. 2007;173(1):291–7.

[51] Wang S, Inaba H, Tagawa H, Hashimoto T. Nonstoichiometry of Ce0.8Gd0.2O1.9 − x . J Electrochem Soc

[Internet]. 1997 Nov 1;144(11):4076–80. Available from: http://jes.ecsdl.org/content/144/11/4076.abstract

[52] Kobayashi T, Wang S, Dokiya M, Tagawa H, Hashimoto T. Oxygen nonstoichiometry of Ce1−ySmyO2−0.5y−x (y=0.1, 0.2). Solid State Ionics [Internet]. 1999;126(3):349–57. Available from: http://www.sciencedirect.com/science/article/pii/S0167273899002593

[53] Kreuer KD. Aspects of the formation and mobility of protonic charge carriers and the stability of perovskite-type oxides. Solid State Ionics [Internet]. 1999 Oct;125(1–4):285–302. Available from: http://www.sciencedirect.com/science/article/pii/S0167273899001885

[54] Oishi M, Akoshima S, Yashiro K, Sato K, Mizusaki J, Kawada T. Defect structure analysis of B-site doped perovskite-type proton conducting oxide BaCeO3: Part 2: The electrical conductivity and diffusion coefficient of BaCe0.9Y0.1O3−δ. Solid State Ionics [Internet]. 2008;179(39):2240–7. Available from: http://www.sciencedirect.com/science/article/pii/S016727380800564X

[55] Grimaud A, Bassat JM, Mauvy F, Simon P, Canizares A, Rousseau B, et al. Transport properties and in-situ Raman spectroscopy study of BaCe0.9Y0.1O3−δ as a function of water partial pressures. Solid State Ionics [Internet]. 2011;191(1):24–31. Available from: http://www.sciencedirect.com/science/article/pii/S0167273811001767

[56] Lim D-K, Im H-N, Song S-J, Yoo H-I. Hydration of Proton-conducting BaCe0.9Y0.1O3−δ by Decoupled Mass Transport. Sci Rep [Internet]. 2017;7(1):486. Available from: http:https://dx.doi.org/10.1038/s41598-017-00595-w

[57] Heras-Juaristi G, Pérez-Coll D, Mather GC. Temperature dependence of partial conductivities of the BaZr0.7Ce0.2Y0.1O3-δ proton conductor. J Power Sources [Internet]. 2017;364:52–60. Available from: http://www.sciencedirect.com/science/article/pii/S0378775317310352

[58] Triviño-Peláez Á, Pérez-Coll D, Mather GC. Electrical properties of proton-conducting BaCe0.8Y0.2O3-δ and the effects of bromine addition. Acta Mater [Internet]. 2019;167:12–22. Available from: http://www.sciencedirect.com/science/article/pii/S1359645419300436

Chapter 4

Application of Carbon Nanomaterials Decorated Electrochemical Sensor for Analysis of Environmental Pollutants

Sunil Kumar and Abhay Nanda Srivastva

Abstract

Carbon nanomaterials (CNMs), especially carbon nanotubes and graphene, have been attracting tremendous attention in environmental analysis for rapid and cost effective detection of various analytes by electrochemical sensing. CNMs can increase the electrode effective area, enhance the electron transfer rate between the electrode and analytes, and/or act as catalysts to increase the efficiency of electrochemical reaction, detection, adsorption and removal are of great significance. Various carbon nanomaterials including carbon nanotubes, graphene, mesoporous carbon, carbon dots exhibited high adsorption and detection capacity. Carbon and its derivatives possess excellent electro catalytic properties for the modified sensors, electrochemical methods usually based on anodic stripping voltammetry at some modified carbon electrodes. Metal electrode detection sensitivity is enhanced through surface modification of working electrode (GCE). Heavy metals have the defined redox potential. A remarkable deal of efficiency with the electrochemical sensors can be succeeded by layering the surface of the working electrode with film of active electro-catalytic species. Usually, electro catalysts used for fabrication of sensors are surfactants, nano-materials, polymers, carbon-based materials, organic ligands and biomaterials.

Keywords: nanomaterials, redox active site, carbon nanotube, surface functionalization, organic pollutants, electrochemical reaction

1. Introduction

Sustainable environment requires development of portable sensors for monitoring heavy and toxic metallic pollutants. Nanomaterials and nanostructures play a vital role as an adsorption sites into sensors [1] that leads to shift sensitivity, selectivity, multiplexed detecting ability towards high performance in terms of capability and portability [2]. Nanomaterials-based sensors exhibit an extremely high surface area, which can increase the number of binding sites [2] available for the adsorption of metal ions. Heavy metal pollution becomes a concern for global sustainability. Carbon nanomaterials [3] act as electrochemical sensors because they have higher sensitivities, lower limit detection, and faster electron transfer kinetics than traditional detection electrodes [4]. An electrochemical sensor is an analytical device in which a recognition element is integrated within or intimately associated with a physical transducer [5] (an electrode) that transfers the analytical signal to an electronic circuit for the purpose of detecting a target analyte. The development of active electro catalysts plays a key role in the design of efficient, reliable, stable, and innovative sensing devices. Electrochemical detection is highly favored by the characteristics of rapid detection, high sensitivity and selectivity, high adsorption capability and large surface area [6]. Functionalized CNTs are good electrochemical sensing materials and can impart strong electro catalytic activity [7] to electrochemical reaction for most environmental pollutants such as heavy metal ions, organic pollutants containing electro active group. Environmental pollution is considered as a worldwide public problem, including heavy metals, inorganic/organic compounds, toxic gases, pesticides, antibiotics [8], bacteria, etc., which becomes a serious issues to human health and smooth environment [9]. The catechol (1, 2-dihydroxybenzene) is a phenolic compound which is extensively used in dye, petroleum refinery, plastic, antioxidant, cosmetics, medicines. The high toxicity and low degradability cause eczematous dermatitis, depression of the central nervous system (CNS) and a prolonged rise of blood pressure. With industrial development, many metal ions have discharged into natural environment. Unfortunately, metal ions, especial heavy metal ions, are easily caused soil and water polluted. Ordered mesoporous carbon have well-ordered and tunable porous structures and surface which have pore sizes in the range of 2–50 nm, Porosity offers high specific surface areas (more than 2000 m^2g^{-1}). However, the grafting of organic, inorganic or biomaterials into mesoporous carbon produces different functional groups and binding capacity which further improving their analytical performances. CNMs have received significant attention as candidate materials for detecting [10, 11] NO_x, NH_3, CO, SO_2 etc. For example, sensing of nitrogen oxide (NO_x), a major air pollutant emitted from power plants, which causes neurodegenerative diseases. The interfacial interaction can be enhanced by the surface-functionalization of nanotubes. The polar groups [12] on the nanotube surface increase the adsorption affinity of the electron-donor or acceptor pollutants and consequently offer better response. The detection of mercury ion at the Au-NPs interface is more sensitive and selective because they can form amalgam only with Hg compared to other metal ions. The electrochemical sensing performance had a relationship with the adsorption capacity, which excites the design of new sensing materials. The amino group on the surface of functionalized CMS [13] is bringing increased attractive force in adsorption of heavy metal ions. Though increasing the deposition time improves the sensitivity, it also lowers the detection limit because of the surface saturation at high metal ions concentrations. Carbon nanomaterials endowed with unique physiochemical properties were found to be most suitable for electrochemical detection of heavy metal due to their ease to modify, high sensitivity, good selectivity and high reproducibility. Unmodified CNTs are unable to chelate metal ions in aqueous solutions and cannot work as good electrode materials for the ASV analyses. The hydrophilic hybrid nanocomposites are able to adsorb heavy metal ions from aqueous solution due to the rich chelating groups. Carbon nano tubes (CNTs) exhibited effective adsorbent as well as sorbents for heavy metal ions. Therefore, it is reasonable to construct electrochemical sensors using the CNT or graphene-functionalized redox electrodes entity for detection of heavy metal because they are capable to detect simultaneously a majority of heavy metal ions with high resolution for defined and measured concentrations. The stripping techniques and particularly square wave and differential pulse anodic stripping voltammetry ensured alternative and extensive explored sensitive electrochemical methods for heavy metal ions detection. The carbon bound Fluorine [14] exhibited both ionic as well as covalent interface and significantly enhanced the capacitive performance of fluorinated GO compared to pristine GO. Further, the fluorinated GO has high affinity for the simultaneous detection of heavy metal ions Cd^{2+}, Pb^{2+}, Cu^{2+} and Hg^{2+} using square wave anodic stripping voltammetry (SWASV) as electrochemical tool. Fluorinated-graphene has gained great attention because of unique properties such as its high temperature resistance and enhanced electro catalytic activity. Electron withdrawing nature is arising from the strong electronegativity of F and electron donating nature from the lone-pair electrons. Graphene or reduced graphene oxide (rGO) are used as an working electrode material; however, low sensitivity and potential interferences lowers their sensing capacity due to inherent irreversible agglomeration of graphene particle which excites researcher to develop green idea in the designing of native grapheme [15] based detecting electrodes as electrochemical sensor. Low sensitivity and poor selectivity related with the large over potential and the interference from the reduced substance, such as oxygen, H_2O, encountered in the nitro aromatic compound (organic pollutants). Nitrophenols readily accumulate in organisms and are difficult to naturally degrade because of the high structural stability. The sensitivity can be improved by incorporating metal nano-particle over the surface of functional sites [16]. Owing to the high specific surface area, chemical stability, high p-conjugation and hydrophilic properties, GO can offer an excellent electrode platform for adsorbing other molecules. High surface to volume ratio with active sites, controlled distribution of pore size, exceptional sorption capacity and high sorption proficiency make CNTs suitable material for the development of electro-analytical systems dedicated for the detection of heavy metal ions. Hence, ionated CNTs play important role in the metal ion sensing due to their better ion exchange capacity. Oxidized CNTs have a great potential for cation uptake compared to non oxidized CNTs. In other words, non oxidized CNTs have tendency towards uptake of anions compared to oxidized CNTs. The presence of an extended π-conjugation in organic conducting polymer (OCPs) confers the required mobility to charges that are present on polymer backbone and makes them electrically conducting [17]. The sensing intensifier played a facilitating role between the GCE surface and the target metal ions by bringing analytes closer to transducer surface resulting in appearance of intense electrochemical signals. CNFs with high length-to-diameter ratio are capable of offering additional active sites for nanoparticle loading or deposition. The carbon nanotubes alone as well as in their oxidized and in their composite forms have tremendous ability to adsorb the heavy metal ions. Unmodified CNTs are unable to chelate metal ions in aqueous solutions and cannot work as good electrode materials. This is due to deficiency of functional group and sufficiency of hydrophobic environment. The effective combination of two carbon nanostructures can not only improve solubility and conductivity but also make up functional deficiencies [18]. Functionalization could significantly assist in the improvement of surface capacitance. Thus, the modified GCE exhibits good electro oxidative activity towards pollutants.

2. Nanomaterials extended electrochemical sensing platforms

Electrochemical Carbon Nanotube Filter Oxidative Performance [19] as a Function of Surface adsorption. The presences of surface resident reactive groups, or edge-plane like sites that are situated at the ends of their structures, and at defect sites, are responsible for the excellent electro catalytic activity of carbon nanomaterials. Nanoparticles exhibited high surface to volume ratio with functional and highly redox active core center leading to increasing the sensitivity and selectivity of the sensor. Thus, a highly active site has great affinity towards molecules result in molecule gets adsorbed on the surface of electrode to undergo a redox reaction. The conducting and chelating group has marked effect on the designation of sensor. Nanomaterials provide a special platform for the purification of contaminated water due to the high surface area of nano-sorbents and their capability of chemical modification and easier regeneration. NPs, QDs with some functionalization are used as tools, immobilization platforms [20] or electro active labels to improve the sensing performance exhibiting higher sensitivity and stability. The nano-particles and quantum dots [20] structures from the electrodes have significantly made a contribution to increasing the electro-catalytic properties because the functionalization of the structures could improve the high surface area, conductivity, stability, porosity, and mechanical rigidity.

2.1 Nanomaterials and its chemical functionalization

Nanomaterials have one dimension <100 nm [1] and possess physico-chemical properties dictated by their unusually small size, large surface area, shape and chemical composition. Nanomaterials usually require the surface functionalization for specific detection of metal ions. The p-type (anion doped) CNTs can behave as an electron deficient surface which can easily adsorb reductive molecule (NO_2) on its surface. The electrochemical sensitivity can be enhanced through attachment of active redox center either via physical or chemical forces over the reactive surface of carbon nanotube. Non-covalent functionalization normally involves physical forces (ion dipole, dipole–dipole, electrostatic force) for the binding of CNTs with catalysts (e.g., metal nanoparticles and metal oxides). Covalent functionalization [21] involves chemical forces (chemical reaction) for tagging of functional group with CNTs. In other words, it is realized through covalent attachment of chemical groups on the conjugated surfaces (edge, plane core) of CNTs. The number of oxygenated functional groups (e.g., –OH, –CO, and –COOH) created during calcinations, purification and isolation processes. As a result, controlled functionalities are susceptible to determine the sorption capacity of CNTs. These chemical groups greatly reduce the hydrophilicity and improve the capacity of ion exchanging behavior, leading to strong interactions with pollutants (e.g., heavy metal ions and organic compounds). Especially, the hydrophilic –OH and –COOH groups on the surface of CNTs exhibit superior sorption phenomenon towards low molecular weights and polarity. Their large surface area as pore volumes, functional surface groups and two basal planes are quite useful for the adsorption of pollutants. CNTs have been exploited in multiple electrochemical sensors because of their ability to facilitate electron transfer reactions [22] with electroactive species in solution and the electrode interface. *Thiruppathi et al.* reported functionalities of a carbon surface may assist the heavy metal ion adsorption properties. To improve their conductivity, FGO and GO were electrochemically reduced at −1.2 V for 300 s in a 0.1 M acetate buffer (pH = 5.0).Fe_3O_4 possessed electrostatic adsorption interaction with lead, and amine [13] acted as a better ligand displaying good chelation with lead. *Xiong et al.* designed amine –Fe_3O_4 modified glassy carbon electrode [23] as electrochemical sensor for detection of Pb(II) with a detection limit of 0.15 μM and 10.07 μA/μM sensitivity. Graphene-based nano-adsorbents are excellent advanced materials for the removal of the organic contaminants [24] from the water because of their nano-scaled size, high surface area, and ability to interact via pi-pi stacking. F-doped carbon nanomaterials have gained great attention because of unique properties such as its high temperature resistance, capacitance and enhanced electrocatalytic activity. The cross linked and bridged group exhibited high affinity and attract environmental pollutants more efficiently. Strength of binding varies with functional group. The Hydrophilic HOOC-MWCNTs [25] can improve MWCNTs in synergistically electrocatalytic ability and adhesive ability. The introduction of organic, inorganic or biomaterials into ordered mesoporous carbon produces different functional groups which can tailor the sensing behavior.

3. Functionalized carbon nanomaterials and its sensing capacity

The functionalization of MWCNT [26] will give more active surface area and also the ionic interaction with anions would be more compared to the pristine MWCNT. The enhanced surface area and ionic interaction are very important for the real sample analysis at nanomolar concentrations, especially for the detection of harmful analytes. HOOC-MWCNTs [11] modified glassy carbon electrode (GCE) exhibited high sensing and adsorption capacity towards binary and ionic pollutants. The cyclic voltammetry resolve clear anodic peaks of SO_3^{2-} and NO_2^-. The anodic peak currents were gradually increases with concentration of ions. The peak separation between sulfite and nitrite are comparatively higher to probe the sensing of anions in nanomolar concentrations, it was found to be around 420 mV by cyclic voltammetry (CV) technique. This potential difference is highly attractive to determine the sulfite and nitrite simultaneously. The HOOC-MWCNTs decorated GCE had low limit of detection (LOD) of 215 nM and 565 nM for SO_3^{2-} and NO_2^-. The electrochemical sensing and detection was found to be two electron transfer oxidative reaction. The sulphate and nitrate ions were produced over the nano surface. *Sablok et al.* reported amine functionalized reduced grapheme oxide/carbon nanotube decorated novel electrochemical sensor for ultra-trace detection of Trinitrotoluene (TNT) up to 0.01 ppb with good reproducibility (n = 3). The sensing capacity was enhanced due to formation of charge transfer complex between electron rich surface of sensor and electron deficient ring of TNT. The binding of electron-deficient TNT to the amine [27] groups on the nano-sensor surface modulate electrical and optical properties of nano sensing elements. *Devi et al.* reported GCE/rGO-SH/Au-NPs [28] electrode as fascinating electrochemical sensor to analyze mercury ($Hg^{2+)}$ ions in the aqueous solution. The working electrode capture Hg^{2+} ions electrochemically and consequently get adsorbed on the redox active core surface followed by electrochemical oxidation by differential pulse voltammetry (DPV) with the increased oxidation current at +0.172 V. Moreover, this sensor platform revealed linear response for Hg(II) detection from 1–10 μM in phosphate buffer saline (PBS) solution and the detection limit was found to be 0.2 μM (S/N = 3). *Wang et al.* designed a GCE with MWCNT-CO-PANi [29] as a electrochemical sensor for detection of Pb^{2+} because the porous structure of conducting PANI surface can retard the bulk surface active compounds from reaching the sensing surface and thus minimizes the passivation of the working electrode. In addition, the PANI matrix offers binding capacity which can firmly hold the MWCNTs on the electrode surface. *Dai et al.* reported the improvement in stripping peak signals of heavy ions on PA/PPy/GO can be attributed to the high surface area of GO and the excellent electrical conductivity of PPy could enhance the electron transfer during the detection processes and peak intensity collaborated with number of functional groups with large negative charges on PA and GO is beneficial to improving the adsorption capacity of heavy metal ions. Phytic acid [30] consists of six membered rings with six phosphate group and two hydroxyl groups, could enhance complexation ability. **Figure 1(a)** shows strip peaks with resolved potential which demonstrates suitability of electrochemical sensor [30]. *Zhang et al.* reported size controlled AuNPs (5–15 nm)/CNFs/GCE electrochemical sensor for simultaneous tracing of

Cd(II), Pb(II) and Cu(II) with SWASV and detected three signal at −0.8, −0.5 and 0 V over a linear range of concentration 0.1 μM - 1 μM at a deposition potential of −1.8 V. *Mohamed Shaban* reported a porous Anodic Alumina (PAA) membrane was functionalized with $CoFe_2O_4$ nanoparticles and used as a substrate for the growing of very long helical-structured Carbon Nanotubes (CNTs) with a diameter less than 20 nm. The designed electrode was found to be suitable for detection of Hg^{2+}, Cd^{2+} and Pb^{2+} ions. **Figure 1(b)** indicates concentration dependent profile which demonstrates maximum range of concentration of detection for which sharp and intense peak appeared sensitively [30].

3.1 Design of selective electrochemical sensor

Ferrocene (Fc) functionalized MWCNTs works as a ratio metric electrochemical sensor. The Fc-MWCNTs/GCE modified sensor was used for detection of o-nitrophenol and p-nitrophenol present in water as toxic pollutants. When Fc-MWNTs/GCE [31] was dipped in 50 μM of o-NP and p-NP, the reduction peak of Fc remained fixed, but two well-separated peaks at about −0.66 V and-0.79 V could be detected which correspond to the reduced peaks of o-NP and p-NP, respectively. The process implies that the modified Fc can effectively separate the reduction peaks of o-NP and p-NP by about 0.13 V, which makes suitable it to detect o-NP and p-NP individually and simultaneously. **Figure 2(a)** demonstrates ferrocene functionalized MWCNTs as a ratio-metric and selective sensor [32]. **Figure 2(b)** indicates suitability of modified sensor and influence of metal nanoparticle on sensitivity [30].

3.2 Electro active carbon nanomaterials and its high surface density

Carbon nanomaterials mainly include zero-dimensional fullerene (C_{60}) and carbon dots (CDs), one-dimensional carbon nanotubes (CNTs) and carbon nanohorns (CNHs), Carbon nanofiber, two- dimensional graphene and its derivatives, and ordered mesoporous carbon (OMC). The hydrogen bonding interaction between the oxygenated groups of CNMs and hydroxyl groups has been utilized for the adsorption of pollutants containing functional groups (e.g., amine, hydroxyl and carboxyl groups. The CNF [33] is functionalized for improving its solubility and also remove the catalytic impurities for enhancing the electrochemical properties by

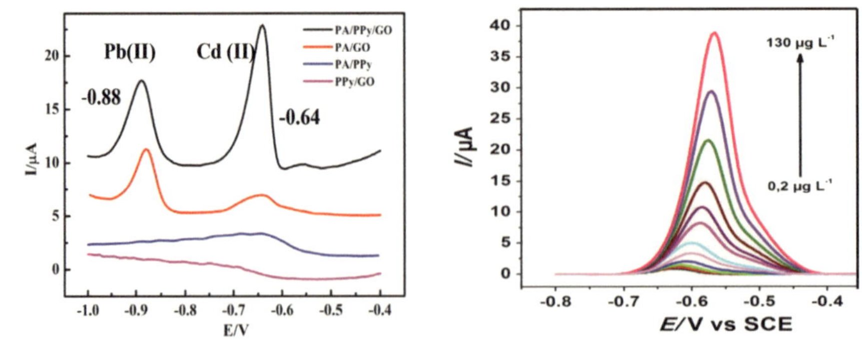

Figure 1.
(a) *DPV of PA/GO, PA/PPy, PPy/GO and PA/PPy/GO modified electrodes in 0.1 M acetate buffer solution (pH 4.5) containing 50 μg/L Cd(II) and Pb(II).* **(b)** *Calibration curve for Pb^{2+} determination, from down to up0.2, 0.5, 1, 5, 15, 20, 30, 40, 60,80, 100, and 130 μg/L Pb^{2+} in 0.1 M acetate buffer (pH 4.5) at PPy/CNFs/CPE under the optimized experimental.*

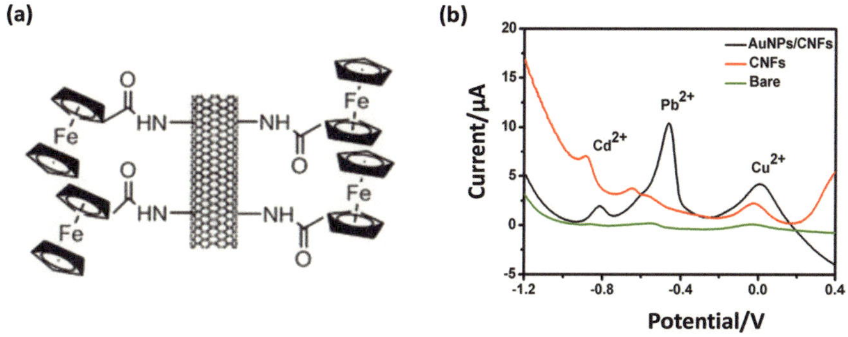

Figure 2.
(**a**) *Ferrocene functionalized MWCNT,* (**b**) *SWASVs for simultaneous detection of* Cd^{2+}*,* Pb^{2+} *and* Cu^{2+} *with the modified AuNPs/CNFs/GCE, CNFs/GCE and bare/GCE.*

the generation of more anchoring sites and surface reactive groups (carboxylic acid, hydroxyl, and carbonyl groups) on the open end and side walls of CNF. *Ramaraj et al.* reported $BiFeO_3$-F-CNF modified GCE is an effective electrode for electrochemical detection of catechol. The lowest value of ΔEp (0.106 V) and higher redox peak current response are indicating that $BiFeO_3$-CNF/GCE has faster electron transfer kinetics than that of other modified electrodes. The electrode is fabricated to anchor more recognition sites on the surface of the electrode and to achieve high affinity for the chemical adsorption of heavy metal ions. Polyaniline is combined with the rGO and glycine for strengthening the collective capacity for metal ions through nitrogen functionalities for example amine (-NH-) and imine (=N-) functional groups [13, 33]. Chitosan (CS) is a polysaccharide and its chemical modification can introduce new chelating groups along the CS chains, which can not only prevent its dissolution in acidic solutions but also improve the adsorption capacity and selectivity of an existing group for a specific metal ion. The MWCNT can be adhered through thiol functionalized Chitosan [21] which enhances the surface density to capture heavy metal ions. MWCNTs can lead to formation of good conduction pathway in the CS-SH film for better electro-analysis. *Li et al.* reported the simultaneous stripping analysis of Cd^{2+} and Pb^{2+} at the nitrogen doped carbon quantum dots modified grapheme oxide NCQDs-GO/GCE. The recorded ASV curves depict individual and highly resolved peaks at around −0.75 V for Cd^{2+} and − 0.50 V for Pb^{2+}, respectively. The peak to peak separation potential is about 0.250 V, which is large enough to recognize selective detection of Cd^{2+} and Pb^{2+} simultaneously. GO are served as a novel support to load nitrogen doped carbon quantum dots (NCQDs) and improve the conductivity and electron transfer rate of the hybrid. The AuNPs [34] modified working electrode can provide more electro active sites and faster electron transfer rate, all of which contribute to enhance the sensitivity of the simultaneous determination of Hydroquinone and catechol. It is difficult to simultaneously determine catechol and hydroquinone due to their overlapping peaks at ordinary electrode [35].

4. Analytical role of square wave anodic stripping voltammetry (SWASV)

Electrochemical techniques have the capability to maintain environmental interfacial processes at high rates and efficiencies by directionally and accurately controlling the electron transfer processes. An electrochemical technique where the analyte of interest is first electrodeposited onto the sensing electrode and removed

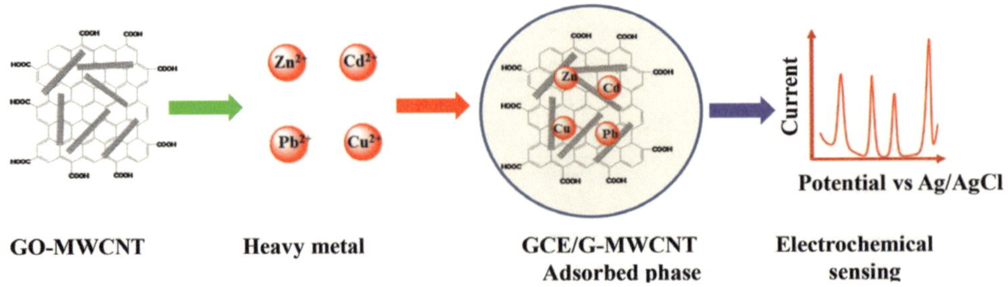

Figure 3.
General scheme of electrochemical sensing and detection of inorganic pollutants (heavy metal ions) through SWASV.

or 'stripped' with a sharp and intense peak by applying an oxidizing potential. During removal of pollutants, the peak current is measured as a function of time or function of the potential between the indicator (sensor) and reference electrodes. The redox probe is introduced as the inner reference to provide a built-in correction towards the signal transduction. The peak current ratio of analyte signal to probe signal is employed as the detected signal for analyte determination. The potential is varied as a square wave superimposed on a linear sweep. The potential separation between the stripping peaks can clear enough to distinguish the various heavy metal ions. The detection is expressed as sensing signals. The stripping peak currents are controlled by the amount of target metal ions adsorbed on the electrode surface. Striping peak current is directly proportionate to concentration of analyte. The SWASV is more prone over other voltammetry technique because of excellent sensitivity and unique ability to detect metals simultaneously. SWASV includes two independent procedures: deposition and stripping. First, in the deposition process (electrochemical reduction), metal ions can be reduced under a certain potential from the analyte solution to the working electrode. Inversely, when anodic potential is applied, the reduced metals are oxidized to their ions. Interference ions reduce the peak current for detected analyte during electrochemical analysis. Peak current varies with concentration of analyte and it increases linearly up to optimum concentration range which is also referred to as linear range concentration profile. Square-wave anodic stripping voltammetry is commonly used for metal detection due to its high sensitivity and low (nM–pM) detection limits. **Figure 3** indicates schematic sensing analysis and detected signal for pollutants [15].

5. Adsorption sites and its electrochemical sensitivity

The adsorption activity is related to the number of active functional groups on the surface of the carbon nanomaterials with highly oxidized surfaces showed a greater adsorption affinity for the stabilizers. Electro catalytic activity is related with hydrophobic or hydrophilic, positive or negative redox active groups of carbon nanomaterials. *Li et al.* reported MWCNTs are highly efficient to remove perfluorooctanoic acid (PFOA) and perfluorooctane sulfonate (PFOS) from aqueous solution in light of their environmental persistence [36]. Bismuth modified CNT polystyrene Sulfonate (PSS) composite electrode for simultaneous detection of Pb(II) and Cd(II) by anodic stripping voltammetry. The designed composite electrode shows synergistic effect of bismuth and Polystyrene sulfonate. Since, polymeric dopant acts as cation-exchanger and CNT as an efficient signal transducer for sensitive and simultaneous detection of lead and cadmium. The detection limits were estimated to be 0.04 ppb for Pb(II) and 0.02 ppb for Cd(II) at 2 min accumulation. The presence of MWCNTs can greatly enhance the conductivity of the hybrid nanocomposites and also make the GO plane unfold, whereas, the GO components can give the hybrid an important property to capture metal ions in aqueous. The CNT/rGO as nanohybrid materials exhibit strongly adsorption the organic components through the π-π interactions, a high electrical transport, and conductivities [37]. *Ruecha et al.* developed of an electrochemical sensor for simultaneous detection of Zn(II), Cd(II), and Pb(II) using a graphene–polyaniline (G/PANI) nanocomposite electrode in a linear working range of 1–300 μg/L. The anodic peak potential −1.31, −0.98 and − 0.75 V were recorded with well separation. **Table 1** shows different capacity for different electrochemical sensor towards inorganic /organic pollutants. The carbon nanomaterials are integral part of sensing materials. The CNMs are very prone to stabilize the structural integrity and reproducibility.

5.1 Influence of surface group (chromophores) and sensing sites

The electro catalytic properties of Carbon nanodots material depend on the presence of functional groups on the surface electrode because the material is a great electron acceptor and donor electron with the presence of some functional groups such as hydroxyl groups. Calixarenes [39] have three-dimensional spherical basket, cup or bucket shapes. **Figure 4(a)** depicts structural integrity of Calixarenes [39]. The spherical core volume is utilized in ion selective electrodes and membranes. It can capture stationary phases. The macrocyclic ring structure is efficient ionophores for metal ions viz. Na^+, Cd^{2+}, Pb^{2+} and Fe^{3+}. Coordination depends on macrocyclic ring size and ionic size of metal ion. Calixarenes can coordinate with the metal ions to increase the sensitivity of the electrochemical sensors. The metal ions, Fe(III), Cd(II), and Pb(II) gave a linear relationship with their concentrations at 1.0–10 nM on the CA/RGO/GCE.

Designed sensor/GCE	Pollutants	Sensitivity	LOD	Technique	Reference
GQDs-Au NPs	Hg(II)	2.47 μA/Nm	0.02 nM	SWASV	[38]
	Cu(II)	3.69 μA/nM	0.05 nM		
CA/RGO	Fe(III),Cd(II) Pb(II)	-	0.02 nM		[39]
Au NPs/CNF	Cd(II),Pb(II) Cu(II)	-	0.1 μM	SWASV	[32]
PAA-$CoFe_2O_4$/CNTs	Pb(II), Hg(II) Cd(II)	-	1 ppb	SWASV	[40]
CyS-MWCNT	Pb(II)	-	1 ppb	DPASV	[41]
	Cu(II)		15 ppb		
$BifeO_3$-CNF	Catechol	-	0.0015 μM	DPV	[42]
Gly/RGO/PANi	Cd(II)	15.20 μA/μM	0.07 nM	SWASV	[33]
	Pb(II)	41.3 μA/μM	0.02 nM		
CS-HS-MWCNTs	Hg(II)	36 μA/μM	3 nM	SWASV	[43]
CNPE-(CTS-ECH)	Cu(II)	212 μA/μM	10 nM	-	[44]
G/PANi	Zn(II)		1.0 μg/L		[45]
	Cd(II),Pb(II)	-	0.1 μg/L	SWASV	
PPy/CNFs	Pb(II)		0.05 μg/L	SWASV	[46]
NCQDs-GO	Cd(II)	-	7.45 μg/L	SWASV	[47]
	Pb(II)	-	1.17 μg/L	SWASV	
	Hydroquinone	-	0.17 μM	SWASV	[21]
AuNPs-HOOC-MWCNT	Catechol	45.53 μA/μM	0.89 μM	SWASV	
CNHs/GO	4-NCB	54.47 μA/μM	10 nM	SWASV	[48]

Table 1.
Different electrochemical sensor for detection of pollutants and analytical parameters.

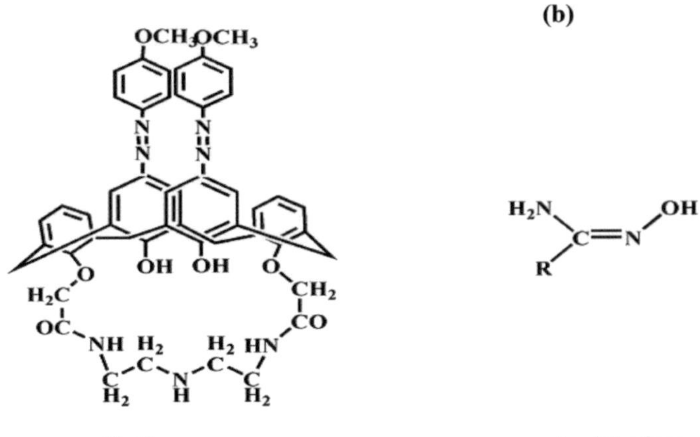

Figure 4.
Structure of organic compounds with core functional and sensing unit.

Different binding energies of functional groups have a great impact on absorption; weaker binding energies facilitate easier desorption [27]. As one of the best hydrophilic functional groups, the amidoxime groups modified on the electrode surface largely intensified the adsorption of heavy metals and lowered the impedance of the material when compared with an unmodified electrode. Amidoxime [49] group for functionalizing the carbon felt electrodes because of its superior adsorption ability for metal ions resulting from their coordination active sites. **Figure 4(b)** shows the structure of amidoxime [49]. The amidoxime group can coordinate with cations to form stable pentacyclic compounds, suggesting that this coordination bond should be stronger than other kinds of monodentate groups. The organic ligands having the amide functional moiety revealed strong and selective complexing ability towards metal ions when used to fabricate electrochemical sensor. The presence of carboxylate and pyridinium functional groups as negative and positive charge bearers over the surface of CNMs enhances the affinity of electrochemical sensing of cations and anions, respectively. The potential to modify carbon nanotubes with multiple chelating molecules with different selectivity towards various analytes attract designing of fancy sensor [12, 13].

5.2 Sensing reaction over electrochemical integrity

Composite phase provides more recognition sites on the surface of the electrode to achieve high affinity for the binding of inorganic pollutants. Conducting polymer enhances the collective capacity of carbon nanomaterials towards metal ions [33].

a. Accumulation/adsorption at working electrode

$$\text{Gly/RGO/PANi} + M^{2+}\ \text{(solution phase)} + 2e^{-} \rightarrow M^{0}\text{----Gly/RGO/PANi}$$

b. Anodic striping (electro analytical operation)

$$M^{0}\text{----Gly/RGO/PANi} \rightarrow \text{Gly/RGO/PANi} + M^{2+}\ \text{(solution phase)} + 2e^{-}$$

Highly efficient ionophores was developed for removing Cd^{2+} and Pb^{2+} using (PyTS-CNTs) [8]. The small conjugative surface is recalled 1, 3, 6, 8-pyrenetetrasulfonicacid sodium salt as the sensing material. The working window was from 1.0 μg/L

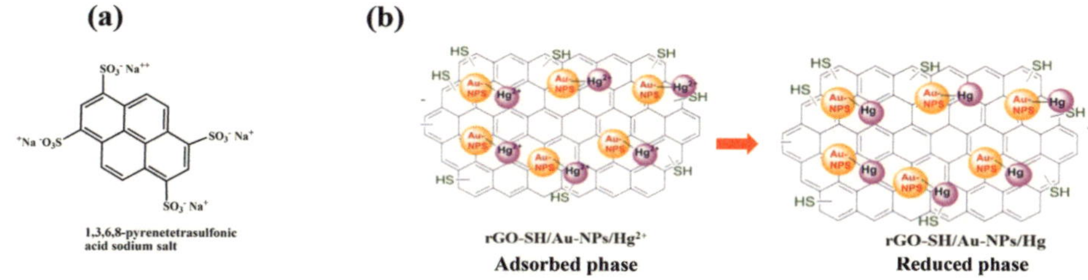

Figure 5.
(a) *Structure of functionalized pyrene.* **(b)** *Schematic representation for adsorption and electro-reduction of Hg(II) at rGO-SH/Au-NPs functionalized sensor.*

to 110 μg/L for both Pb^{2+} and Cd^{2+} ions. The limit of detection (LOD) was 0.02 and 0.08 μg/L, respectively. Functional groups greatly improve the hydrophobicity and ion exchange capacity, leading to strong interactions with pollutants. **Figure 5(a)** depicts the structural entity as a sulfonated salt of pyrene which can detect the pollutants through adsorption and ion exchange phenomenon [9]. **Figure 5(b)** demonstrates thiol and Au NPs anchor the adsorption and electrochemical reduction through enhanced charge transport [30].

5.3 Influence of cross linked structure and core integrity

Analytical performances of sensors are proportional to the surface concentration of the receptors. Cross-linking can enhance the electrochemical properties of electrochemical sensor. The CNT possess –COOH group. The cross linked and grafted CNT improve adsorption, adhesion, completion, chelation and ion exchange phenomenon along with fast charge transport [6]. This enhances selectivity and sensitivity. Cross linking increases the capturing integrity over the surface of carbon nanotube [43]. The chemical modifications of the chitosan by covalently attaching of selected molecules to the amino or hydroxyls groups can improve the ion-transport and ion-exchange proprieties of the biopolymer. Strong electron transmission from substituent (bridge) to carbon nanomaterial enhances the kinetics of sensing phenomenon [34]. *Janegitz et al.* developed functionalized carbon nanotubes paste electrode (CNPE) modified with cross-linked chitosan [crosslinked with glutaraldehyde (CTS-GA)] for determination of Cu(II) in industrial wastewater [44]. Chitosan is a cross linked [50] biopolymer having enriched $–NH_2$ and –OH functionality.

6. Factors influencing electrochemical detection

6.1 Nature and structure of sensor

The pore radius and pore volume decides enhanced redox peaks with much higher current and the kinetics of electro analytical process otherwise; signal distortion appears during detection of environmental pollutants. Different structures have different activation energy of adsorption and binding.

6.2 Deposition potential

The decrease in stripping current is attributed to the inadequate accumulation of the metal ions at lower negative potential and the initiation of hydrogen evolution reaction at a higher negative potential that may damage the surface of the electrode. The optimum deposition potential will resolve the ions efficiently and selectively.

6.3 Deposition time

The stripping peak currents show a linear increase with a prolonged period holding the maximum peak current which indicates the saturation of all the possible attachment sites on the functionalized electrode by the adsorption of the heavy metal contaminants.

6.4 pH (buffer capacity) of medium

The pH-value have significant influence onto the size of the square wave voltammetric peaks and also assists in the hydrolysis of metal ions, therefore, it is crucial to choose a suitable pH-value for the sensing of metal ions. At the high concentration of hydrogen ion, the intensity of peak current is reduced due to protonation of hydrophilic groups on the surface of sensing material which leads to the decrease in the attachment sites for the adsorption of the heavy metal ions. Proper pH maintains originality of electron rich functional group over the sensing integrity.

6.5 Supporting electrolyte

Supporting electrolytes are introduced to purge off the electro-migration effect. Therefore, the stripping voltammetric response of the peaks of current for the metal ions determination was also assessed by varying the nature of the stripping medium.

7. Conclusion and perspectives

In summary, non modified electrochemical sensor exhibited week binding and week adsorption capacity. The introduction and modification of surface functional groups was explored to improve the chemical selectivity and charge density at the active surface. The detection capacity can be improved through attachment of functional group having high affinity towards environmental pollutants. The electrochemical detection depends on the nature and structure of sensing electrode. The sensitivity and selectivity are critical core sites which enhances electrochemical analysis. CNTs enable faster transfer of electrical signal due to its high conductivity and conjugated polymers provide advanced affinity towards metal ions. Functionalized CNPs can result in highly sensitive redox sensors for a number of analytes. It can be demonstrated that the modified electrode showed excellent electro catalytic activity, increase the rate of electron transfer electron and the adsorption of the pollutants (inorganic/organic)molecules on the surface of the electrode.

Acknowledgements

The authors are thankful to the authorities of L.N.T. College and R.B.B.M. College Muzaffarpur (S.K.) and Nitishwar Mahavidyalaya, Muzaffarpur (A.N.S.) for providing necessary facilities related to proceed the work. We are also grateful to the authorities of B.R.A. Bihar University, Muzaffarpur as well as Higher education department, Govt. of Bihar, Patna for their kind support in terms of academic and research development.

Conflict of interest

The authors have declared no conflict of financial interest.

Author details

Sunil Kumar[1] and Abhay Nanda Srivastva[2]*

1 Department of Chemistry, L.N.T. College, B.R.A. Bihar University, Muzaffarpur, India

2 Department of Chemistry, Niteshwar Mahavidyalaya, B.R.A. Bihar University, Muzaffarpur, India

*Address all correspondence to: drabhaysss@gmail.com

References

[1] M. Farré, J. Sanchís, D. Barceló, Analysis and assessment of the occurrence, the fate and the behavior of nanomaterials in the environment, TrAC - Trends Anal. Chem. 30 (2011) 517-527. doi:10.1016/j.trac.2010.11.014.

[2] X. Lu, H. Qi, X. Zhang, Z. Xue, J. Jin, X. Zhou, X. Liu, Highly dispersive Ag nanoparticles on functionalized graphene for an excellent electrochemical sensor of nitroaromatic compounds, Chem. Commun. 47 (2011) 12494-12496. doi:10.1039/c1cc15697g.

[3] R. Baby, B. Saifullah, M.Z. Hussein, Carbon Nanomaterials for the Treatment of Heavy Metal-Contaminated Water and Environmental Remediation, Nanoscale Res. Lett. 14 (2019). doi:10.1186/s11671-019-3167-8.

[4] Y. Wang, S. Hu, Applications of carbon nanotubes and graphene for electrochemical sensing of environmental pollutants, J. Nanosci. Nanotechnol. 16 (2016) 7852-7872. doi:10.1166/jnn.2016.12762.

[5] T. Kokab, A. Manzoor, A. Shah, H.M. Siddiqi, J. Nisar, M.N. Ashiq, A.H. Shah, Development of tribenzamide functionalized electrochemical sensor for femtomolar level sensing of multiple inorganic water pollutants, Electrochim. Acta. 353 (2020) 136569. doi:10.1016/j.electacta.2020.136569.

[6] Y. Patiño, E. Díaz, S. Ordóñez, E. Gallegos-Suarez, A. Guerrero-Ruiz, I. Rodríguez-Ramos, Adsorption of emerging pollutants on functionalized multiwall carbon nanotubes, Chemosphere. 136 (2015) 174-180. doi:10.1016/j.chemosphere.2015.04.089.

[7] M. Alvaro, P. Atienzar, P. De La Cruz, J.L. Delgado, H. Garcia, F. Langa, Sidewall functionalization of single-walled carbon nanotubes with nitrile imines. Electron transfer from the substituent to the carbon nanotube, J. Phys. Chem. B. 108 (2004) 12691-12697. doi:10.1021/jp0480044.

[8] Y.N. Zhang, Q. Niu, X. Gu, N. Yang, G. Zhao, Recent progress on carbon nanomaterials for the electrochemical detection and removal of environmental pollutants, Nanoscale. 11 (2019) 11992-12014. doi:10.1039/c9nr02935d.

[9] B.R. Adhikari, M. Govindhan, A. Chen, Carbon nanomaterials based electrochemical sensors/biosensors for the sensitive detection of pharmaceutical and biological compounds, Sensors (Switzerland). 15 (2015) 22490-22508. doi:10.3390/s150922490.

[10] T. Zhang, M.B. Nix, B.Y. Yoo, M.A. Deshusses, N. V. Myung, Electrochemically functionalized single-walled carbon nanotube gas sensor, Electroanalysis. 18 (2006) 1153-1158. doi:10.1002/elan.200603527.

[11] V. Sudha, S.M. Senthil Kumar, R. Thangamuthu, Simultaneous electrochemical sensing of sulphite and nitrite on acid-functionalized multi-walled carbon nanotubes modified electrodes, J. Alloys Compd. 749 (2018) 990-999. doi:10.1016/j.jallcom.2018.03.287.

[12] K. Elouarzaki, L.R.P. Mandoc, K. Gorgy, M. Holzinger, C.A. Amarandei, E.M. Ungureanu, S. Cosnier, Synthesis and electrochemical characterization of original "tEMPO" functionalized multiwall carbon nanotube materials: Application to iron(II) detection, Electrochem. Commun. 60 (2015) 131-134. doi:10.1016/j.elecom.2015.08.024.

[13] Y.F. Sun, L.J. Zhao, T.J. Jiang, S.S. Li, M. Yang, X.J. Huang, Sensitive and selective electrochemical detection of heavy metal ions using

amino-functionalized carbon microspheres, J. Electroanal. Chem. 760 (2016) 143-150. doi:10.1016/j.jelechem.2015.11.028.

[14] A.R. Thiruppathi, B. Sidhureddy, W. Keeler, A. Chen, Facile one-pot synthesis of fluorinated graphene oxide for electrochemical sensing of heavy metal ions, Electrochem. Commun. 76 (2017) 42-46. doi:10.1016/j.elecom.2017.01.015.

[15] H. Huang, T. Chen, X. Liu, H. Ma, Ultrasensitive and simultaneous detection of heavy metal ions based on three-dimensional graphene-carbon nanotubes hybrid electrode materials, Anal. Chim. Acta. 852 (2014) 45-54. doi:10.1016/j.aca.2014.09.010.

[16] F. von der Kammer, P.L. Ferguson, P.A. Holden, A. Masion, K.R. Rogers, S.J. Klaine, A.A. Koelmans, N. Horne, J.M. Unrine, Analysis of engineered nanomaterials in complex matrices (environment and biota): General considerations and conceptual case studies, Environ. Toxicol. Chem. 31 (2012) 32-49. doi:10.1002/etc.723.

[17] M.A. Deshmukh, M.D. Shirsat, A. Ramanaviciene, A. Ramanavicius, Composites Based on Conducting Polymers and Carbon Nanomaterials for Heavy Metal Ion Sensing (Review), Crit. Rev. Anal. Chem. 48 (2018) 293-304. doi:10.1080/10408347.2017.1422966.

[18] S. Su, S. Chen, C. Fan, Recent advances in two-dimensional nanomaterials-based electrochemical sensors for environmental analysis, Green Energy Environ. 3 (2018) 97-106. doi:10.1016/j.gee.2017.08.005.

[19] A. Saleh, G. Fadillah, O. Adi, Trends in Analytical Chemistry Nanoparticles as components of electrochemical sensing platforms for the detection of petroleum pollutants : A review, 118 (2019) 194-206.

[20] M. Pedrero, S. Campuzano, J.M. Pingarrón, Quantum dots as components of electrochemical sensing platforms for the detection of environmental and food pollutants: A review, J. AOAC Int. 100 (2017) 950-961. doi:10.5740/jaoacint.17-0169.

[21] Y. Shen, D. Rao, Q. Sheng, J. Zheng, Simultaneous voltammetric determination of hydroquinone and catechol by using a glassy carbon electrode modified with carboxy-functionalized carbon nanotubes in a chitosan matrix and decorated with gold nanoparticles, Microchim. Acta. 184 (2017) 3591-3601. doi:10.1007/s00604-017-2392-z.

[22] K. Lawrence, C.L. Baker, T.D. James, S.D. Bull, R. Lawrence, J.M. Mitchels, M. Opallo, O.A. Arotiba, K.I. Ozoemena, F. Marken, Functionalized carbon nanoparticles, blacks and soots as electron-transfer building blocks and conduits, Chem. - An Asian J. 9 (2014) 1226-1241. doi:10.1002/asia.201301657.

[23] S. Xiong, M. Wang, D. Cai, Y. Li, N. Gu, Z. Wu, Electrochemical Detection of Pb(II) by Glassy Carbon Electrode Modified with Amine-Functionalized Magnetite Nanoparticles, Anal. Lett. 46 (2013) 912-922. doi:10.1080/00032719.2012.747094.

[24] J. Du, L. Ma, D. Shan, Y. Fan, L. Zhang, L. Wang, X. Lu, An electrochemical sensor based on the three-dimensional functionalized graphene for simultaneous determination of hydroquinone and catechol, 723 (2014) 38-45.

[25] P. Kar, A. Choudhury, Carboxylic acid functionalized multi-walled carbon nanotube doped polyaniline for chloroform sensors, Sensors Actuators B Chem. (2013).

[26] K. Balasubramanian, M. Burghard, Electrochemically functionalized carbon nanotubes for device

applications, (2008) 3071-3083. doi:10.1039/b718262g.

[27] K. Sablok, V. Bhalla, P. Sharma, R. Kaushal, S. Chaudhary, C.R. Suri, Amine functionalized graphene oxide / CNT nanocomposite for ultrasensitive electrochemical detection of trinitrotoluene, J. Hazard. Mater. 249 (2013) 322-328.

[28] N.R. Devi, M. Sasidharan, A.K. Sundramoorthy, Gold Nanoparticles-Thiol-Functionalized Reduced Graphene Oxide Coated Electrochemical Sensor System for Selective Detection of Mercury Ion, J. Electrochem. Soc. 165 (2018) B3046–B3053. doi:10.1149/2.0081808jes.

[29] Z. Wang, E. Liu, D. Gu, Y. Wang, Glassy carbon electrode coated with polyaniline-functionalized carbon nanotubes for detection of trace lead in acetate solution, Thin Solid Films. 519 (2011) 5280-5284. doi:10.1016/j.tsf.2011.01.175.

[30] H. Dai, N. Wang, D. Wang, H. Ma, M. Lin, An electrochemical sensor based on phytic acid functionalized polypyrrole/graphene oxide nanocomposites for simultaneous determination of Cd(II) and Pb(II), Chem. Eng. J. 299 (2016) 150-155. doi:10.1016/j.cej.2016.04.083.

[31] S. Pang, X. Kan, Reliable detection of O-nitrophenol and p-nitrophenol based on carbon nanotubes covalently functionalized with ferrocene as an inner reference, New J. Chem. 43 (2019) 10517-10522. doi:10.1039/c9nj02276g.

[32] F. Hanif, A. Tahir, M. Akhtar, M. Waseem, S. Haider, M.F. Aly Aboud, I. Shakir, M. Imran, M.F. Warsi, Ultra-selective detection of Cd^{2+} and Pb^{2+} using glycine functionalized reduced graphene oxide/polyaniline nanocomposite electrode, Synth. Met. 257 (2019) 116185. doi:10.1016/j.synthmet.2019.116185.

[33] B. Zhang, J. Chen, H. Zhu, T. Yang, M. Zou, M. Zhang, M. Du, Facile and green fabrication of size-controlled AuNPs/CNFs hybrids for the highly sensitive simultaneous detection of heavy metal ions, Electrochim. Acta. 196 (2016) 422-430. doi:10.1016/j.electacta.2016.02.163.

[34] A. Safavi, E. Farjami, Construction of a carbon nanocomposite electrode based on amino acids functionalized gold nanoparticles for trace electrochemical detection of mercury, Anal. Chim. Acta. 688 (2011) 43-48. doi:10.1016/j.aca.2010.12.001.

[35] S. Chen, R. Huang, J. Yu, X. Jiang, Simultaneous voltammetric determination of hydroquinone and catechol by using a glassy carbon electrode modified with a ternary nanocomposite prepared from oxidized multiwalled carbon nanotubes, manganese dioxide and manganese ferrite, Microchim. Acta. 186 (2019) 3591-3601. doi:10.1007/s00604-019-3750-9.

[36] X. Li, S. Chen, X. Quan, Y. Zhang, Enhanced Adsorption of PFOA and PFOS on Multiwalled Carbon Nanotubes under Electrochemical Assistance, Environ. Sci. Technol. (2011) 8498-8505.

[37] Z. Meng, H. Zhang, J. Zheng, An electrochemical sensor based on titanium oxide-carbon nanotubes nanocomposite for simultaneous determination of hydroquinone and catechol, Res. Chem. Intermed. 41 (2015) 3135-3146. doi:10.1007/s11164-013-1420-9.

[38] A.A. Siong Luong Ting, Shu Jing Ee, P.C. Kam Chew Leong, Graphene quantum dots functionalized gold nanoparticles for sensitive electrochemical detection of heavy metal ions, Electrochim. Acta. (2015).

[39] C. Göde, M.L. Yola, A. Yılmaz, N. Atar, S. Wang, A novel electrochemical

sensor based on calixarene functionalized reduced graphene oxide: Application to simultaneous determination of Fe(III), Cd(II) and Pb(II) ions, J. Colloid Interface Sci. 508 (2017) 525-531. doi:10.1016/j.jcis.2017.08.086.

[40] M. Shaban, A.R. Galaly, Highly Sensitive and Selective In-Situ SERS Detection of Pb^{2+}, Hg^{2+} and Cd^{2+} Using Nanoporous Membrane Functionalized with CNTs, Sci. Rep. 6 (2016) 1-9. doi:10.1038/srep25307.

[41] J. Morton, N. Havens, A. Mugweru, A.K. Wanekaya, Detection of trace heavy metal ions using carbon nanotube modified electrodes, Electroanalysis. 21 (2009) 1597-1603. doi:10.1002/elan.200904588.

[42] S. Ramaraj, S. Mani, S.M. Chen, T. Kokulnathan, B.S. Lou, M.A. Ali, A.A. Hatamleh, F.M.A. Al-Hemaid, Synthesis and application of bismuth ferrite nanosheets supported functionalized carbon nanofiber for enhanced electrochemical detection of toxic organic compound in water samples, J. Colloid Interface Sci. 514 (2018) 59-69. doi:10.1016/j.jcis.2017.12.016.

[43] W. Deng, Y. Tan, Y. Li, Y. Wen, Z. Su, Z. Huang, S. Huang, Y. Meng, Q. Xie, Y. Luo, *S. Yao*, Square wave voltammetric determination of Hg(II) using thiol functionalized chitosan-multiwalled carbon nanotubes nanocomposite film electrode, Microchim. Acta. 169 (2010) 367-373. doi:10.1007/s00604-010-0366-5.

[44] B.C. Janegitz, L.H. Marcolino-Junior, S.P. Campana-Filho, R.C. Faria, O. Fatibello-Filho, Anodic stripping voltammetric determination of copper(II) using a functionalized carbon nanotubes paste electrode modified with crosslinked chitosan, Sensors Actuators, B Chem. 142 (2009) 260-266. doi:10.1016/j.snb.2009.08.033.

[45] N. Ruecha, N. Rodthongkum, D.M. Cate, J. Volckens, O. Chailapakul, C.S. Henry, Sensitive electrochemical sensor using a graphene-polyaniline nanocomposite for simultaneous detection of Zn(II), Cd(II), and Pb(II), Anal. Chim. Acta. 874 (2015) 40-48. doi:10.1016/j.aca.2015.02.064.

[46] L. Oularbi, M. Turmine, M. El Rhazi, Electrochemical determination of traces lead ions using a new nanocomposite of polypyrrole/carbon nanofibers, J. Solid State Electrochem. 21 (2017) 3289-3300. doi:10.1007/s10008-017-3676-2.

[47] T.Y. Libo Li, Dong Liu, Aiping Shi, Simultaneous Stripping Determination of Cadmium and Lead Ions Based on the N-doped Carbon Quantum Dots-Graphene Oxide Hybrid, Sensors Actuators B Chem. (2017).

[48] G. Zhu, H. Sun, B. Zou, Z. Liu, N. Sun, Y. Yi, K.Y. Wong, Electrochemical sensing of 4-nitrochlorobenzene based on carbon nanohorns/graphene oxide nanohybrids. Biosensors and Bioelectronics, 106(2018) 136-141. doi: 10.1016/j.bios.2018.01.058.

[49] T. Wu, C. Liu, B. Kong, J. Sun, Y. Gong, K. Liu, J. Xie, A. Pei, Y. Cui, Amidoxime-Functionalized Macroporous Carbon Self-Refreshed Electrode Materials for Rapid and High-Capacity Removal of Heavy Metal from Water, ACS Cent. Sci. 5 (2019) 719-726. doi:10.1021/acscentsci.9b00130.

[50] K.H. Wu, H.M. Lo, J.C. Wang, S.Y. Yu, B. De Yan, Electrochemical detection of heavy metal pollutant using crosslinked chitosan/carbon nanotubes thin film electrodes, Mater. Express. 7 (2017) 15-24. doi:10.1166/mex.2017.1351.

Chapter 5

Surface Plasmon Resonance Sensors for Concentration and Reaction Kinetic Detections

Xiaoying Wang, Mingqiang Ma, Xueliang Wang and Shoujuan Wang

Abstract

Surface plasmon resonance (SPR) is an optical phenomenon that occurs on the metal (normally gold or silver) film surface and the light that excited this phenomenon changes with the refractive index of materials on the metal surface. SPR sensors are constructed based on this phenomenon and are used in fields of biological and chemical analyses, drug screening, environmental monitoring, and so on. Here, we will make an introduction to applications of SPR sensors on reaction kinetic and concentration detections. To make this chapter readily comprehensible, we will divide it into three portions. The first part will be an abbreviated depiction of surface plasmon excitation and constructions of an SPR sensor. Then, we will aim at an introduction to the bimolecular interactions in SPR sensors. At last, we will make a summary on applications of SPR sensors.

Keywords: surface plasmon resonance, concentration, dissociation constant, kinetics, molecular reaction, sensors

1. Introduction

Surface plasmon phenomenon has been observed for over one century [1], and its applications in sensors have over 40 years [2, 3]. Surface plasmon resonance (SPR) sensors constructed based on this phenomenon are sensitive to the binding and unbinding of molecules on the sensor chip surface [4]. By online recording the reaction process, we obtained the SPR sensorgram and applied it to analyte concentration, especially molecular reaction kinetics analyses. These sensors are applied to the detections of the reactions between protein and protein, protein and DNA/RNA, small molecules and proteins/DNA/RNA, proteins/DNA/RNA/ small molecules and cells, and so on [5–8]. The application fields include chemical and biological analyses, food safety and environmental monitoring, and drug screening [9, 10].

Since the automatic sampling system and a series of commercial chips have been developed, the SPR instruments have been becoming operation-oriented design. And this is further expanding the application fields. But it still cannot maintain the molecular reactions occur at an ideal condition. As the reactions are occurred in solutions, the hydrodynamic conditions are a great variable factor [11]. To be better understanding of this technique and obtaining reliable data, we have to possess a good knowledge on the theory of molecular reactions in the SPR system.

2. SPR sensors

2.1 Excitations of surface plasmon

Surface plasmon (SP) phenomenon was first observed as anomalies in the metallic diffraction grating experiment by wood in 1902 [1]. In 1941, Fano concluded these anomalies on the metallic surface to the excitation of electromagnetic wave [12].

Figure 1 shows the relationship between the incident light angle and the reflected light intensity in the absence and presence of thin gold film [13]. The total reflection occurs at 39 degree in the prism and air system (**Figure 1A** and **B**) and a dip (the drop of reflection) is observed around 43 degree in the presence of gold film (**Figure 1C** and **D**).

This approximated semicircular prism (**Figure 1A** and **C**) makes it possible that the angle of incident light remains unchanged. This significantly simplifies the optical system. As a result, Kretschmann geometry [14], which couples a prism and attenuates the total reflection, is the most common used excitation approach. In another geometry, Otto also used a prism to implement the SP excitation [15], but placed the dielectric layer between the prism and metal layer.

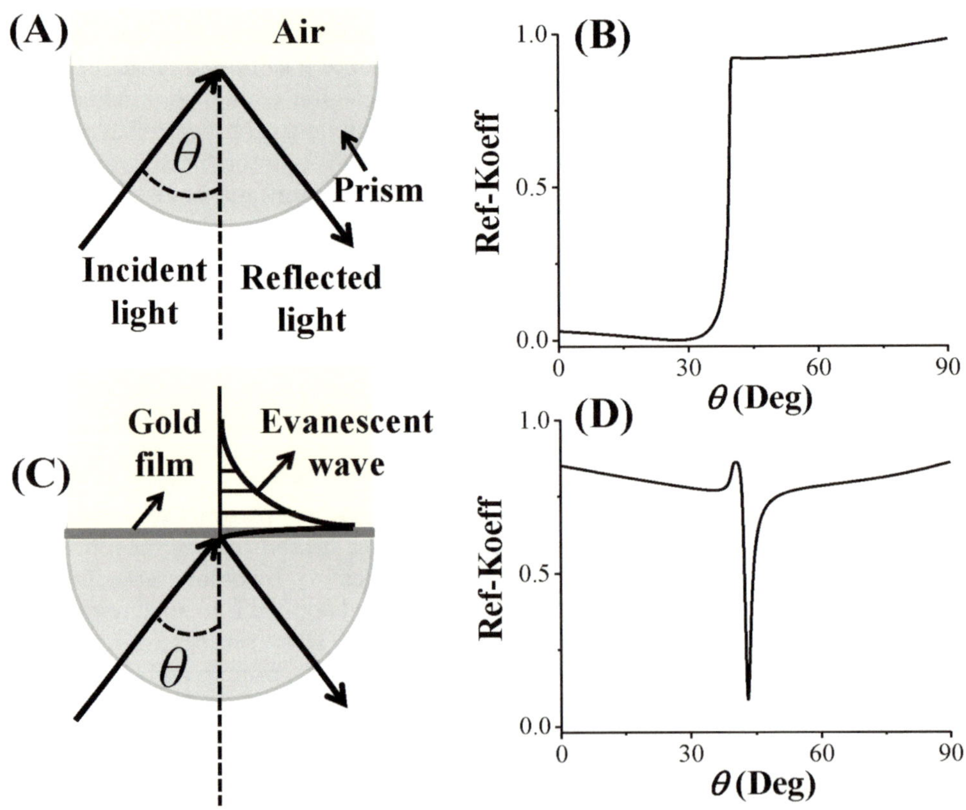

Figure 1.
(A) A polarized light illuminates the interface of air and a prism. (B) The intensity of reflected light corresponding to different incident light angle in panel (A). (C) a polarized light illuminate the interface of gold and prism. The dielectric material on the opposite side of the gold film is air. Panel (D) is the reflected light intensity in panel (C). The dielectric constants of air and the prism are 2.29 and 1, respectively. The thickness of the gold film is 50 nm, and its real and imaginary parts of dielectric constants are −12.3 and 1.29. The simulation was finished via the Winspall software.

As the SP wave intensity decays exponential along with and perpendicular to the propagation direction (**Figure 1C**), the SPR sensor can only be sensitive to the reflective index change next to the metallic surface and the sensitivity decays along with the distance from the metallic surface.

At last, the shape and position (resonance angle) of the dip (**Figure 1D**) are also affected by the reflective index of the materials next to the metallic film. For instance, the dip is much sharper in an air system than water, thus the sensor is more sensitive in an air solution than water.

Similar as Wood's observation, the metallic diffraction grating is also applied to accomplish the SP excitation (**Figure 2**). This method is based on the diffraction of light and is named as grating coupling [16].

The third method is the waveguide coupling method [4]. As is shown in **Figure 3**, a metal film is striped on the wave guiding layer to achieve the SP excitation, and a superstrate is placed on the other side of the metal film for analytic applications.

2.2 Constructions of an SPR sensor

The successful excitations of SP via convenient methods introduce a various applications of this optics. Surface plasmon was first used to characterize thin films in 1978 [2]. In 1982, Nylander and Liedberg applied it to gas detection and biosensors [17].

Nowadays, a series of operation-oriented SPR instruments have been developed. As a result, the main work for the construction of an SPR sensor is choosing an appropriate sensor chip and establishing a reproducible procedure.

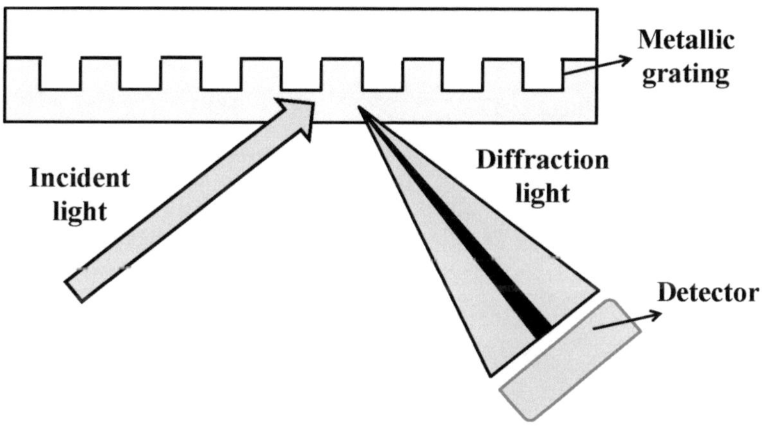

Figure 2.
An incident light with different wavelength illuminate the metallic grating and the narrow dark band is caused by the surface plasmon phenomenon.

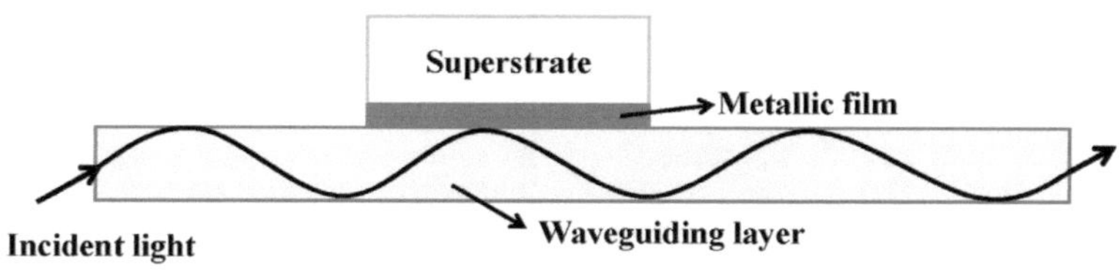

Figure 3.
Waveguide coupling method for surface plasmon excitation.

Nonetheless, a brief introduction on the structure of the SPR sensor is helpful on the construction of an adequate chip. **Figure 4** shows a typical SPR sensor, it consists of the portions for surface plasmon (SP) excitation, solutions delivery, signal detection and conversion, and data processing [18].

The solution delivery portion is applied for the transportation of solutions, which contain the analyte solutions and running buffers. It combines the SP excitation portion on the SPR sensor chip. In a typical Kretschmann geometry, the SP excitation accessories are placed next to the glass side of the sensor chip, and the hydrophobic fluidic channels, which have a width down to hundreds of micrometers, are placed next to the metallic film. The receptors pre-immobilized on the metallic side are reversed for the analytes in the solution transported by the fluidic channels. The reaction between the analytes and receptors causes the reflective index change next to the sensor chip surface (see Section 2.2.3). Furthermore, it causes the resonance angle shift. This shift is online collected by the detector and output as SPR signal. At last, we gather the SPR signal for data processing.

2.2.1 The dilution in the sample delivery

The dilution in the solution delivery process is a hindrance for obtaining reliable data. **Figure 5** shows the concentration distribution of an analyte and its corresponding SPR sensorgram in a system without any molecular reactions or physical adsorptions. This concentration change can be confused with the concentration change caused by the mass transport limitation, and miss leading the data processing procedure.

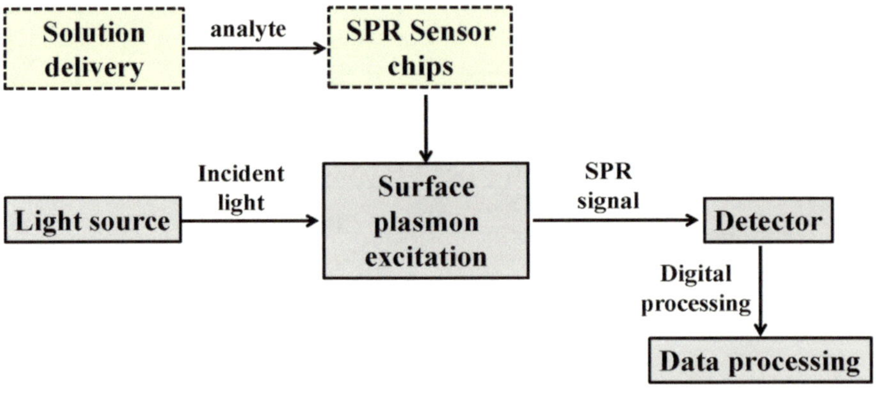

Figure 4.
The configuration of an SPR sensor.

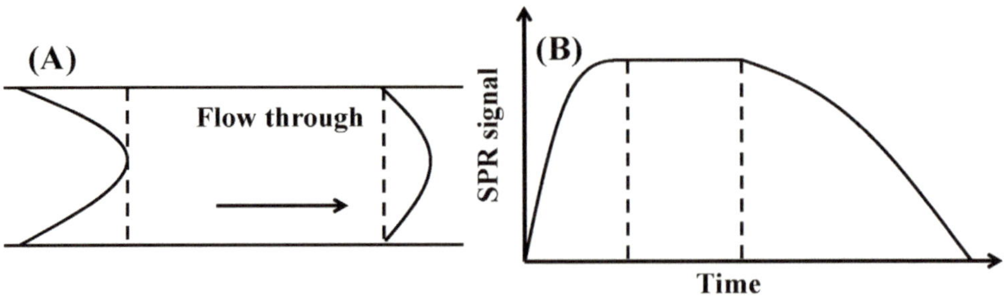

Figure 5.
(A) The dilution in a laminar flow tube during sample delivery. The arrow indicates the flow through direction. (B) The SPR sensorgram of the solution shown in panel (A).

To solve this problem, the microfluidic systems are applied. Also, the bi-direct flow system is efficient on cutting off the diluted samples [19, 20]. In reality, this problem can be alleviated by decreasing the sample delivery distance and increasing the sample delivery rate.

2.2.2 Constructions of SPR sensor chips

SPR sensor chip is the most critical accessory in the whole instrument for the construction of a sensor [21]. It contains a glass substrate and a thin metallic film. For simplify, this glass slide normally has the same reflective index as the prism in the instrument. To simplify the storage condition of the sensor chips and supply a longer shelf life, we usually choose gold film as a substrate. However, silver film supplies a more sensitive experimental condition.

Bare gold chip, containing glass slide and gold film, is the most common used sensor chip and is basic for the constructions of other chips. As this chip is susceptibly contaminated and is not so biocompatible, the surface modifications are essential. Besides, these modifications also satisfy the condition for the receptors immobilization (see Section 2.2.3).

Figure 6 shows the most common applied modification case, the self-assemble method based on the interactions of gold and thiol- or disulfide groups [22]. The uniform of the alkyl chains on the chip surface is maintained by the van der Waals interactive force. The active groups on the other side of the alkyl chains are essential for the immobilization of receptors.

The dextran chip, containing carboxyl groups on the alkyl chain, is applied to construct a series of chips for certain applications. These commercial sensor chips include nitrilotriacetic acid chips (histidine tag capture), streptavidin chips (biotin capture), and lipid capture chips. The chips rely on DNA hybridization or the interactions of streptavidin and biotin are also produced. Other commercial chips contain click chemistry chips, cell capturing chips, and hydrophobic chips [23].

2.2.3 Molecular reactions on the sensor chip

The molecular reactions on the SPR senor chip are different from the reactions in solutions. As is shown in **Figure 7**, receptors (*B*) are immobilized on the sensor chip surface and analytes (*A*) are transported to the fluidic channel on the top of the sensor chip to accomplish the reactions. The immobilization of *A* causes the change

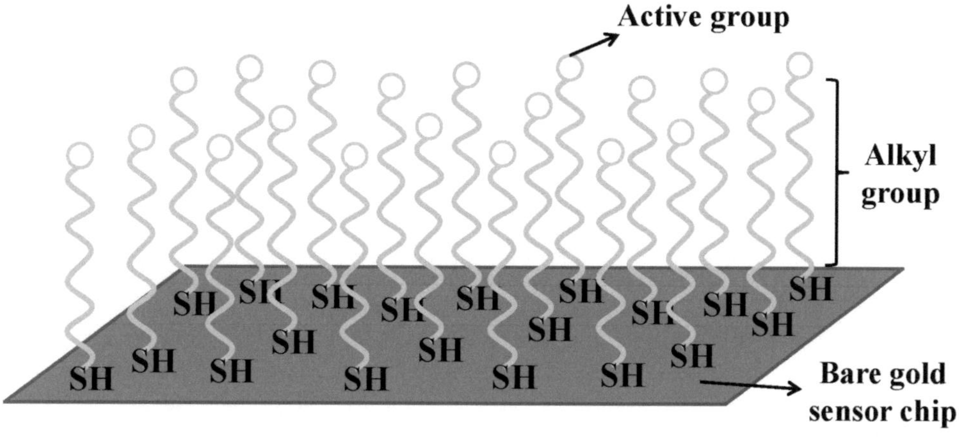

Figure 6.
The self-assembled monolayer on a bare gold sensor chip.

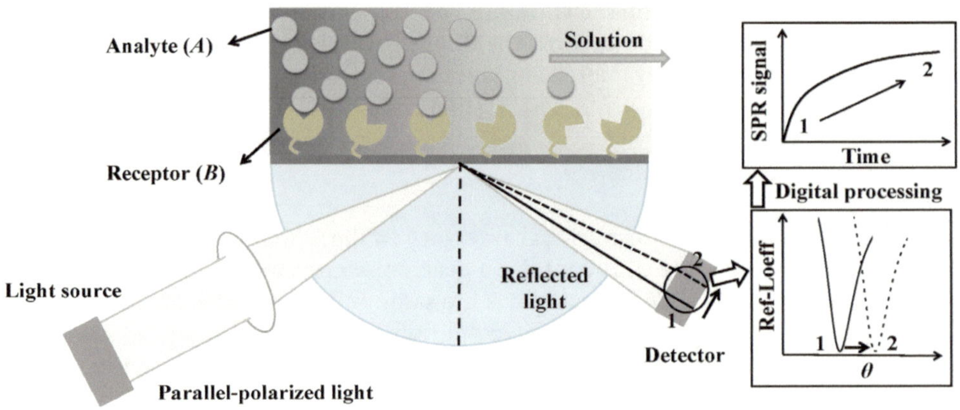

Figure 7.
A scheme shows the bimolecular reactions on the SPR sensor chip surface and the signal transformation in this process.

of reflective index next to the sensor chip surface and further leads to the dip shifting from position 1 to 2 (bottom right in **Figure 7**). This change is recorded by a camera (the detector) and transfers to the SPR signal [24]. By gathering the signal collected at a specific time interval, we obtained the SPR sensorgram and applied it for data processing.

3. Bimolecular interactions

The molecular reactions in reality solutions can be affected by a series of factors, like temperature, concentration, mass transport, and so on. To obtain a better signal, the temperature is normally remained at 25°C in the SPR system. But the temperature control usually allows a higher temperature interval to satisfy different reactions.

As the receptors (B) are pre-immobilized on the chip surface, whose total amount is inalterable, the molecular reaction is the receptors' consumption process. For the analytes (A), they are being transporting to react with B during the binding process. In an ideal condition, the analyte concentration can be maintained the same during the binding process. For the bimolecular reaction between B and A, it can be regard as a pseudo first-order reaction under this circumstance. It is the most common reaction in the SPR system and is a basic reaction for multiple molecules reactions (one B and more than one A).

3.1 Pseudo first-order reaction

Eq. (1) is the bimolecular reaction between B and A in the SPR system at an ideal condition.

$$B + A \underset{k_d}{\overset{k_a}{\rightleftarrows}} AB \tag{1}$$

where k_a and k_d are the association and dissociation rate constants.

If the binding rate of A is higher than its unbinding rate, the SPR signal increases. In reverse, the signal decreases. **Figure 8** are the SPR sensorgrams showing analytes (A) pass through the fluidic channel in the presence and absence of

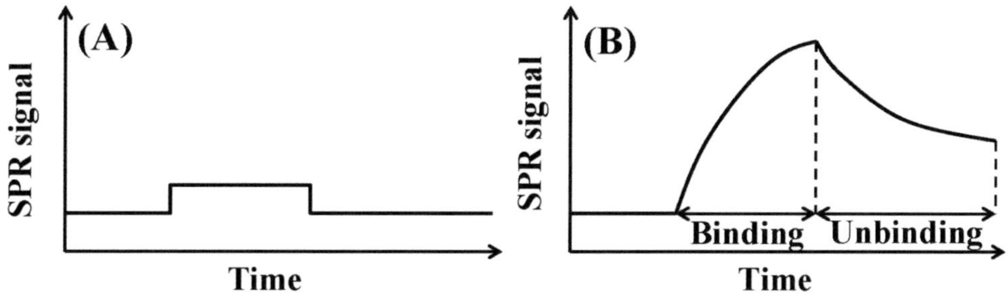

Figure 8.
SPR sensorgram showing analyte A *flow through an SPR sensor chip surface in the absence (A) and presence of* B *(B). On the sensor chip surface,* A *can only react with* B.

receptors (B) on the sensor chip surface. **Figure 8A** shows the concentration change of A without any reactions, and **Figure 8B** is the binding and unbinding of A on the sensor chip surface. By combining the sensorgram with a mathematical approximate model, we obtain the kinetic constants of the reaction.

In an ideal condition, the concentration of analytes, $[A]$ or C, maintains the same during the reaction. The pre-immobilized receptors can be divided into two portions, free receptors on the sensor chip surface (B) and receptors have already been combined with analytes (AB). The formation rate of AB can be expressed as [25, 26].

$$\frac{d[AB]}{dt} = k_a[A][B] - k_d[AB] \tag{2}$$

where $[B]$ is the concentration of free receptors and $[AB]$ is the concentration of AB. As $[AB]$ is proportional to the SPR signal R and the total amount of B is proportional to the highest SPR signal R_{max}, Eq. (2) can be further expressed as

$$\frac{dR}{dt} = k_a C(R_{max} - R) - k_d R \tag{3}$$

After the integral, Eq. (3) can be expressed as

$$R = \frac{k_a C R_{max}}{k_a C + k_d}\left[1 - e^{-(k_a C + k_d)t}\right] \tag{4}$$

At the end of the binding process, $[A]$ or C becomes to zero. Analytes begins to unbind from the sensor chip surface. Eq. (2) can be expressed as

$$\frac{d[AB]}{dt} = -k_d[AB] \tag{5}$$

or

$$\frac{dR}{dt} = -k_d R \tag{6}$$

After the integral, Eq. (6) can be expressed as

$$R = R_0 e^{-k_d t} \tag{7}$$

where R_0 is the SPR signal at the point $[A]$ becomes to zero.

By making a good match between the experimental and simulated sensorgrams, the reaction rate constant k_a and k_d can be obtained. Furthermore, the association and dissociation constants can be obtained via Eq. (8).

$$K_A = \frac{1}{K_D} = \frac{k_a}{k_d} \tag{8}$$

where K_A is the association constant and K_D is the dissociation constant.

At the condition, Eq. (4) is used for concentration detections, a higher amount of receptors have to be pre-immobilized to obtain a higher SPR signal. It also leads to a huge consumption of the analytes. At the point that the amount of analytes bound with receptors is higher than the analytes transported to the sensor chip surface, the concentration of the analyte next to the sensor chip surface changed.

By combining Eqs. (4) and (8), we obtain Eq. (9)

$$R = \frac{R_{max}}{1 + \frac{K}{C}}\left[1 - e^{-(k_a C + k_d)}\right] \tag{9}$$

where R_{max} and K_D are constant for a given reaction. This decreased analyte concentration (C) directly leads to the numerical increase of k_a. But the value of k_a should be the same for a given reaction at a certain temperature and pressure. Thus, the equations above cannot fit any more.

By introducing the mass transport coefficient into the reactions at a mass transfer limited condition [27], we obtain Eq. 10.

$$\mathrm{A} \underset{k_m}{\overset{k_m}{\rightleftarrows}} \mathrm{A} + \mathrm{B} \underset{k_d}{\overset{k_a}{\rightleftarrows}} AB \tag{10}$$

where k_m is the mass transport coefficient. For the laminar flow system in the SPR fluidic channel, k_m can be expressed as

$$k_m = 0.98(D/h)^{2/3}(f/bx)^{1/3} \tag{11}$$

where D is the diffusion coefficient, h and b are the height and width of fluidic channel, f is the volumetric flow rate, x is the distance from the receptor immobilization site.

Still, Eqs. (4) and (7) can be expressed as the following at the mass transport limited condition.

$$R = \frac{k'_a C R_{max}}{k'_a C + k'_d}\left[1 - e^{-(k'_a C + k'_d)t}\right] \tag{12}$$

and

$$R = R_0 e^{-k'_d t} \tag{13}$$

where k'_a and k'_d are the association and dissociation rate constants at the mass transport limited condition and they can be expressed as

$$k'_a = \frac{k_a k_m}{k_a[B] + k_m} \tag{14}$$

and

$$k'_d = \frac{k_d k_m}{k_a[B] + k_m} \tag{15}$$

Although the association and dissociation rate constants are affected by the mass transport, K_A and K_D can still be calculated via k'_a and k'_d.

$$K_A = \frac{1}{K_D} = \frac{k_d}{k_a} \tag{16}$$

Eq. (12) can be used for concentration detections at mass transport limited condition. Two methods are available, one is built on the equilibrium value R_{eq} and the other one is based on the curve at the beginning of the binding.

R_{eq} is a thermodynamic constant, and it cannot be affected by the mass transport limitation.

$$R_{eq} = \frac{k_a C R_{max}}{k_a C + k_d} = \frac{k'_a C R_{max}}{k'_a C + k'_d} = \frac{C R_{max}}{C + K_D} \tag{17}$$

or

$$\frac{1}{R_{eq}} = \frac{K_D}{R_{max}} \bullet \frac{1}{C} + \frac{1}{R_{max}} \tag{18}$$

Figure 9 are the SPR sensorgrams showing the bimolecular reactions at an ideal (solid line) and a mass transport limited (long dash line) condition. The mass transport limitation significantly prolongs the time to reach the equilibrium. As a result, the method based on R_{eq} is reliable, but not time consuming.

The method based on the curve at the beginning of the binding is attractive, as it is calibration free and high efficient. Although the calibration curve seems to be

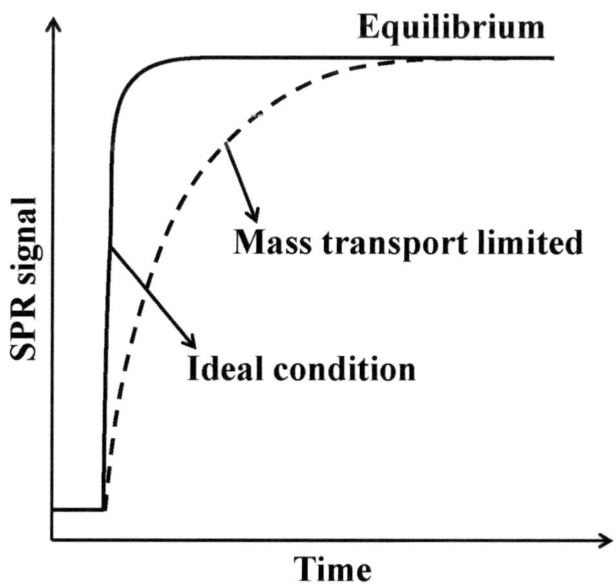

Figure 9.
SPR sensorgrams showing analytes binds to the receptors on the sensor chip surface at the condition with (solid line) and without (long dashed line) mass transport limitation.

ubiquitous and reliable for most of the concentration calculations, the standard samples normally cannot be obtained for a novel biomarker. This significantly hinders the development of the corresponding detection methods.

At a total mass transport limited condition, the binding rate of analytes is decided by the mass transport rate.

$$\frac{dR}{dt} = k_m(C_0 - C) \bullet M_W \bullet 10^9 \quad (19)$$

where C_0 is the analyte concentration in the bulk solution, and M_W is the molecular weight of the analyte. At the beginning of the binding process, C is zero. Thus, this equation can be expressed as

$$\frac{dR}{dt} = k_m C_0 \bullet M_W \bullet 10^9 \quad (20)$$

where the value of dR/dt approximately is equal to the slope of the curve at the beginning of the bimolecular interaction.

3.2 Other reactions

Bimolecular reaction is the basic reaction in some multistep reaction. For instance, the Eq. (21) can be expressed as Eq. (22), which contains two bimolecular reactions.

$$2A + B \underset{k_d}{\overset{k_a}{\rightleftarrows}} A_2B \quad (21)$$

$$A + B \underset{k_{d1}}{\overset{k_{a1}}{\rightleftarrows}} AB + A \underset{k_{d2}}{\overset{k_{a2}}{\rightleftarrows}} A_2B \quad (22)$$

4. Applications

The SPR sensors are available for both concentration and reaction kinetic detections [28]. Their application fields include biological and chemical analyses, drug screening, environmental and food safety monitoring [29].

The concentrations of the biomarkers are essential information for the early diagnose of disease [30]. The reaction kinetics is a crucial parameter for drug screening [31], biological molecules interactions, and chemical industry. The association/dissociation constant is helpful on estimating the strength of the bond [32, 33]. The SPR sensor equipped with different chips affords a friendly condition for the biological molecular reactions. And the chips modified with specific materials (fibronectin, polylysine, and so on) can satisfy the condition for cell culturing [34]. This enlarges the applications of the SPR sensors on biomarker concentration detections, reaction kinetics of molecules, and reaction kinetics between cell membrane and molecules [8].

Natural compounds are usually more biocompatible than synthetic ones, thus they are more suitable drug candidates. But natural compound normally has a low concentration and mixes with other chemicals. The low detection limit and the convenient kinetic detection method make SPR sensors to be widely used in small molecule drug screenings [35]. Furthermore, the high throughput [36, 37] and mapping technique allow a quick grasp of drugs from natural compounds [22, 23], and even reprocessing of approved drugs.

The contaminants in the environment and food are harmful to our healthy [38, 39]. The SPR sensors not only can detect the concentration of a certain contaminant, but also can make a rough screening of the unknown chemical and biological contaminants. For possible contaminants, the SPR sensor can grasp the molecules or cells for further analyses.

5. Summary

The binding and unbinding of analytes on the SPR sensor chip surface induces the refractive index change next to the thin metallic film. Surface plasmon resonance sensors are sensitive to this refractive index change and transfer this change to the shift of the resonance angle. By a real time recording of the angle change, we obtained the SPR signal for data processing.

The bimolecular reaction is a basic reaction on the sensor chip. By making a good match of the simulated and experimental SPR sensorgrams, we obtained the kinetic constants. Furthermore, we also use the SPR sensors on concentration detections. The concentration detection method based on the R_{eq} value is sensitive but time consuming. The novel calibration free method based on the curve at the beginning of the binding process is time saving. Besides, it efficiently avoids the demand of standard samples. Although its sensitivity is not high, it is still a promising method.

Right now, SPR sensors have been applied in the field of biological and chemical analyses, drug screening, environmental and food safety monitoring. As its integration with new technologies, the application fields of SPR sensors will be further enlarged.

Acknowledgements

Xiaoying Wang participated in the manuscript writing.

Conflict of interest

The authors declare no conflict of interest.

Author details

Xiaoying Wang[1*], Mingqiang Ma[2], Xueliang Wang[3] and Shoujuan Wang[1*]

1 State Key Laboratory of Biobased Material and Green Papermaking, Qilu University of Technology, Jinan, Shandong, China

2 Huadian Electric Power Research Institute Co., Ltd., Hangzhou, Zhejiang, China

3 College of Chemistry and Chemical Engineering, Heze University, Heze, Shandong, China

*Address all correspondence to: xiaoyingw@outlook.com and nancy5921@163.com

References

[1] Wood RWXLII. On a remarkable case of uneven distribution of light in a diffraction grating spectrum. Philosophical Magazine. 1902;**4**: 396-402. DOI: 10.1080/14786440209462857

[2] Pockrand I, Swalen JD, Gordon JG, Philpott MR. Surface plasmon spectroscopy of organic monolayer assemblies. Surface Science. 1978;**74**: 237-244. DOI: 10.1016/0039-6028(78)90283-2

[3] Gordon JG, Ernst S. Surface plasmons as a probe of the electrochemical interface. Surface Science. 1980;**101**:499-506. DOI: 10.1016/0039-6028(80)90644-5

[4] Homola J. Electromagnetic theory of surface plasmons. In: Homola J, editor. Surface Plasmon Resonance Based Sensors. Berlin, Heidelberg: Springer; 2006. pp. 3-44

[5] Stojanović I, Schasfoort RBM, Terstappen LWMM. Analysis of cell surface antigens by surface plasmon resonance imaging. Biosensors & Bioelectronics. 2014;**52**:36-43. DOI: 10.1016/j.bios.2013.08.027

[6] Kurinomaru T, Kojima N, Kurita R. Sequential assessment of multiple epigenetic modifications of cytosine in whole genomic DNA by surface plasmon resonance. Analytical Chemistry. 2019;**91**:13933-13939. DOI: 10.1021/acs.analchem.9b03423

[7] Peng T, Li X, Li K, Nie Z, Tan W. DNA-modulated plasmon resonance: Methods and optical applications. ACS Applied Materials & Interfaces. 2020;**12**:14741-14760. DOI: 10.1021/acsami.9b23608

[8] Bocková M, Slabý J, Špringer T, Homola J. Advances in surface plasmon resonance imaging and microscopy and their biological applications. Annual Review of Analytical Chemistry. 2019; **12**:151-176. DOI: 10.1146/annurev-anchem-061318-115106

[9] Yang HM, Teoh JY, Yim GH, Park Y, Kim YG, Kim J, et al. Label-free analysis of multivalent protein binding using bioresponsive nanogels and surface plasmon resonance (SPR). ACS Applied Materials & Interfaces. 2020;**12**: 5413-5419. DOI: 10.1021/acsami.9b17328

[10] Stojanović I, Ruivo CF, van der Velden TJG, Schasfoort RBM, Terstappen LWMM. Multiplex label free characterization of cancer cell lines using surface plasmon resonance imaging. Biosensors. 2019;**9**:70. DOI: 10.3390/bios9020070

[11] Štěpánek J, Vaisocherová H, Piliarik M. Molecular interactions in SPR sensors. In: Homola J, editor. Surface Plasmon Resonance Based Sensors. Berlin, Heidelberg: Springer; 2006. pp. 69-91

[12] Fano U. The theory of anomalous diffraction gratings and of quasi-stationary waves on metallic surfaces (Sommerfeld's waves). Journal of the Optical Society of America. 1941;**31**:213-222. DOI: 10.1364/JOSA.31.000213

[13] Homola J, Piliarik M. Surface plasmon resonance (SPR) sensors. In: Homola J, editor. Surface Plasmon Resonance Based Sensors. Berlin, Heidelberg: Springer; 2006. pp. 45-67

[14] Kretschmann E, Raether H. Notizen: Radiative decay of non radiative surface plasmons excited by light. Zeitschrift für Naturforschung A. 1968;**23**:2135-2136. DOI: 10.1515/zna-1968-1247

[15] Otto A. Excitation of nonradiative surface plasma waves in silver by the method of frustrated total reflection.

Zeitschrift für Physik. 1968;**216**: 398-410. DOI: 10.1007/BF01391532

[16] Ruemmele JA, Hall WP, Ruvuna LK, Van Duyne RP. A localized surface plasmon resonance imaging instrument for multiplexed biosensing. Analytical Chemistry. 2013;**85**:4560-4566. DOI: 10.1021/ac400192f

[17] Nylander C, Liedberg B, Lind T. Gas detection by means of surface plasmon resonance. Sensors and Actuators. 1982; **3**:79-88. DOI: 10.1016/0039-6028(80) 90644-5

[18] Nguyen HH, Park J, Kang S, Kim M. Surface plasmon resonance: A versatile technique for biosensor applications. Sensors (Basel). 2015;**15**:10481-10510. DOI: 10.3390/s150510481

[19] Wang S, Boussaad S, Tao N. Surface plasmon resonance spectroscopy: Applications in protein adsorption electrochemistry. In: Rusling JF, editor. Biomolecular Films: Design, Function, and Applications. New York: Marcel Dekker; 2003. pp. 213-251

[20] Wang X, Zhou F. Dual-valve and counter-flow surface plasmon resonance. Analytical Chemistry. 2018; **90**:4972-4977. DOI: 10.1021/acs. analchem.8b00277

[21] Masson J-F, Battaglia TM, Cramer J, Beaudoin S, Sierks M, Booksh KS. Reduction of nonspecific protein binding on surface plasmon resonance biosensors. Analytical and Bioanalytical Chemistry. 2006;**386**:1951-1959. DOI: 10.1007/s00216-006-0834-2

[22] Mrksich M, Sigal GB, Whitesides GM. Surface plasmon resonance permits in situ measurement of protein adsorption on self-assembled monolayers of alkanethiolates on gold. Langmuir. 1995;**11**:4383-4385. DOI: 10.1021/la00011a034

[23] Gedig ET. Surface chemistry in SPR technology. In: Richard BMS, editor. Handbook of Surface Plasmon Resonance (2). Cambridge, UK: The Royal Society of Chemistry; 2017. pp. 171-254

[24] Hinman SS, McKeating KS, Cheng Q. Surface plasmon resonance: Material and interface design for universal accessibility. Analytical Chemistry. 2018;**90**:19-39. DOI: 10.1021/acs.analchem.7b04251

[25] Karlsson R, Roos H, Fägerstam L, Persson B. Kinetic and concentration analysis using BIA technology. Methods. 1994;**6**:99-110. DOI: 10.1006/ meth.1994.1013

[26] Zhao H, Schuck P. Detailed analysis of kinetic binding traces with distributions of surface sites. In: Richard BMS, editor. Handbook of Surface Plasmon Resonance. Cambridge, UK: The Royal Society of Chemistry; 2017. pp. 149-170

[27] Pol E, Roos H, Markey F, Elwinger F, Shaw A, Karlsson R. Evaluation of calibration-free concentration analysis provided by biacore systems. Analytical Biochemistry. 2016;**510**:88-97. DOI: 10.1016/j.ab.2016.07.009

[28] Singh P. SPR biosensors: Historical perspectives and current challenges. Sensors and Actuators B: Chemical. 2016;**229**:110-130. DOI: 10.1016/j. snb.2016.01.118

[29] Rich RL, Myszka DG. Survey of the 2009 commercial optical biosensor literature. Journal of Molecular Recognition. 2011;**24**:892-914. DOI: 10.1002/jmr.1138

[30] Choi J-H, Lee J-H, Son J, Choi J-W. Noble metal-assisted surface plasmon resonance immunosensors. Sensors. 2020;**20**:1003. DOI: 10.3390/s20041003

[31] Nunes NM, de Paula HMC, Coelho YL, da Silva LHM, Pires ACS.

Surface plasmon resonance study of interaction between lactoferrin and naringin. Food Chemistry. 2019;**297**: 125022. DOI: 10.1016/j. foodchem.2019.125022

[32] Pelosi P, Zhu J, Knoll W. From radioactive ligands to biosensors: Binding methods with olfactory proteins. Applied Microbiology and Biotechnology. 2018;**102**:8213-8227. DOI: 10.1007/s00253-018-9253-5

[33] Juhász Á, Luty-Błocho M, Wojnicki M, Tóth GK, Csapó E. General method for kinetic and thermodynamic evaluation of a receptor model peptide-drug molecule interaction studied by surface plasmon resonance. Microchemical Journal. 2019;**147**: 311-318. DOI: 10.1016/j. microc.2019.03.048

[34] Kreysing E, Seyock S, Hassani H, Brauweiler-Reuters E, Neumann E, Offenhäusser A. Surface plasmon resonance microscopy: Correlating surface plasmon resonance microscopy of living and fixated cells with electron microscopy allows for investigation of potential preparation artifacts. Advanced Materials Interfaces. 2020;**7**: 2070026. DOI: 10.1002/ admi.202070026

[35] Honarmand S, Dabirmanesh B, Amanlou M, Khajeh K. The interaction of several herbal extracts with α-synuclein: Fibril formation and surface plasmon resonance analysis. PLoS One. 2019;**14**:e0217801. DOI: 10.1371/journal. pone.0217801

[36] Wegner GJ, Wark AW, Lee HJ, Codner E, Saeki T, Fang S, et al. Real-time surface plasmon resonance imaging measurements for the multiplexed determination of protein adsorption/desorption kinetics and surface enzymatic reactions on peptide microarrays. Analytical Chemistry. 2004;**76**:5677-5684. DOI: 10.1021/ ac0494275

[37] Wang Y, Zhang C, Zhang Y, Fang H, Min C, Zhu S, et al. Investigation of phase SPR biosensor for efficient targeted drug screening with high sensitivity and stability. Sensors and Actuators B: Chemical. 2015;**209**: 313-322. DOI: 10.1016/j.snb.2014.11.134

[38] Homola J. Surface plasmon resonance biosensors for food safety. In: Narayanaswamy R, Wolfbeis OS, editors. Optical Sensors: Industrial Environmental and Diagnostic Applications. Berlin, Heidelberg: Springer; 2004. pp. 145-172

[39] García-Aljaro C, Muñoz-Berbel X, Jenkins ATA, Blanch AR, Muñoz FX. Surface plasmon resonance assay for real-time monitoring of somatic coliphages in wastewaters. Applied and Environmental Microbiology. 2008;**74**: 4054-4058. DOI: 10.1128/aem.02806-07

Chapter 6

Crude Distillation Unit (CDU)

Serge-Bertrand Adiko and Rifat Radisovich Mingasov

Abstract

The chapter considers the technology of the crude distillation unit in general. The crude distillation unit is at the front-end of the oil refinery. The desalting process and distillation of crude oil are included in the crude distillation unit (CDU). The desalting process of crude oil is imperative to ensure the good quality of crude oil, that is, to remove impurities before its transfer to refining. This procedure minimizes or eliminates harmful substances such as sulfur, water, salts, and even mechanical impurities, which ensures a long operation of pipelines. However, the desalting process is only part of the distillation unit. The other phase of this unit is as already mentioned above is: distillation. Distillation is a more physical than a chemical process. The distillation process is characterised by mass-thermal transfer of materials, which leads to the obtaining of fractions. The distillation in the crude distillation unit is carried out consecutively in two ways: atmospheric and vacuum. In the Russian Federation, we classify oil refining plants in general as follows: fuel, fuel-oil, fuel-petrochemical, and fuel-oil-petrochemical. Also, regardless of the profile of the refinery, great importance is given to the crude distillation unit. The crude distillation unit, if well modeled and organized, makes it possible to obtain already more light products at this stage; therefore, there is a reduction of heavy residues, the refinement of which requires more expensive processes such as hydrocracking or catalytic cracking. The oil topping column K-1, within this framework, has been adopted in several oil refining plants in addition to the main column commonly referred to as K-2. The principal purpose of column K-1 is the separation of light gasoline and the major part of dissolved gases from crude oil. This first step of distillation at K-1 has the effect of normalizing the amount of gasoline hydrocarbons and stabilizing the operation of the main column K-2 despite possible fluctuations in the composition of crude oil. Besides, the CDU equipped with column K-1 demonstrates an increase in some valuable products such as methane, ethane, naphtha, etc.

Keywords: demulsifiers, crude oil, desalting process, desalter, electrostatic desalter, sump, separator, crude distillation units, atmospheric distillation units, vacuum distillation units, K-1 oil topping column, K-2 main atmospheric column

1. Introduction

Many crude oils usually contain besides the basic elements of its chemical composition hydrocarbons such as sulfur, oxygen, nitrogen, and mechanical impurities [1]. Oil also contains the gaseous phase as methane, ethane, and liquid phase. Gas typically dissolves in the liquid, whereas in the balance's case remains strong. It is noticed with the experience that natural gas fields are close to oil fields [2]. In general, there are four or three oil: aromatic, paraffin, naphthenic. But there is also a mixed type [1]. The classification of crude oil is carried out for various reasons. The criteria for classifying crude oil refer more or less to its quality [3]. For a better understanding of classificatory at the quality (quality/price), the Petroleum world has adopted a benchmark called stallion as Brent Blend produced in the North Sea, West Texas Intermediate (WTI) from the USA, Dubai Crude from Dubai... However, there are other possibilities to classify oil by density, sulfur content, and viscosity. Crude oil supplied to refineries must meet requirements [4]. The pre-treatment of oil from harmful impurities occupies an important place among the main processes associated with the production, collection, and transportation of oil to refineries or export [5]. The harmful impurities of oil can be divided into two groups: hydrophilic (lipophobic) and hydrophobic (lipophilic). The hydrophobic impurities are soluble in oil. These hydrophobic impurities are organometallic compounds (metalloporphyrin complexes), organic acids, among which the most undesirable organochlorine compounds. The Hydrophilic impurities include impurities that are inherently insoluble in oil, these impurities are water and inorganic salts dissolved in it, as well as solid salts, mechanical impurities (sand, clay), hydrogen sulphide, etc. These hydrophilic impurities are in another phase (water), which are dispersed in the oil in the form of drops of water [6, 7]. Therefore, the degree of pre-treatment oil in production areas is an imperative and significantly affects the efficiency and reliability of transportation by pipeline, tanker, rail, and even road transportation [2]. These requirements should facilitate the transportation and future processing of Petroleum. The requirements are summarized as follows: water content, the content of mechanical impurities, saturated steam pressure, the content of chloride salts, and organochlorine compounds [5]. Usually, the degree of pre-treatment of oil for transportation does not consider the content of chloride salts and the content of organochlorine compounds [8], literally because all salts are large in water and not in oil [9], in connection with the decision of the Russian Federation to join the world trade organization (WTO), the requirements for oil are unified following the requirements for prepared oil according to ASTM D1250-97 [4] and EN 224 [3]. However, countries such as Russia use the GOST R 51858-2002 standard for oil, established by oil and gas companies for transportation by oil pipeline [5] for delivery to oil refining plants in the Russian Federation and export. In Russia, the pipeline transport takes into account the above-mentioned points [3, 10]: salt content (100–900 mg/l), water content from 0.5 to 1.0% of mass, impurity content (not over 0.05% of mass).

1.1 Pre-treatment of crude oil

The crude distillation unit is the first phase of oil refining. That's what we usually hear. However, if we are very careful, we can notice that oil refining began at the level of oil wells with its pre-treatment; the first pre-treatment of crude oil. However, additional pre-treatment is essential before the first separation (fractions our cups) are obtained. This additional pre-treatment and the manufacturing of these first fractions are carried out at the crude distillation unit as shown in **Figure 1**. As well, in our humble opinion for a better understanding of this unit, before getting into the thick of the topic, it is important to make the list of the main devices that make up it: The desalination plant, furnaces, distillation columns (atmospheric and under vacuum). Furthermore, in this process, there are also auxiliary equipment that is also very important, such as pumps, valves, heat exchangers, coolers, automatic process control systems, etc. The Russian Federation, as a major participant in the world oil industry alongside the United States of America, Saudi Arabia, Canada, Argentina, and other countries, has developed its own standards incorporating international standards. In a similar approach, ordinarily, in Russia, they

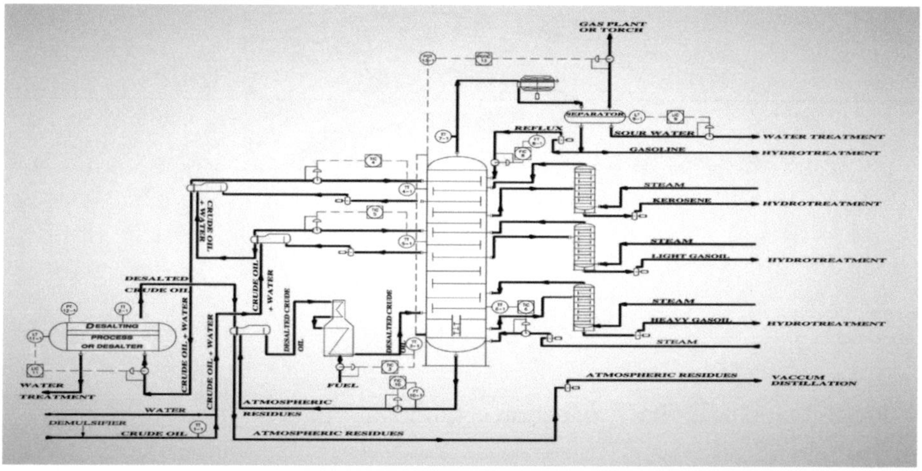

Figure 1.
P&ID diagram flow process of crude distillation unit without vacuum distillation. We would like to emphasize that reboilers were used in this scheme in this way, just for esthetics. Less flow to the reboilers is done differently.

appoint the distillation unit: Atmospheric-vacuum distillation unit (AVDU). AVDU is itself subdivided into two other units which are: Atmospheric distillations unit (ADU) and vacuum distillation unit (VDU). As a result, in this chapter, we tried to explain the operation of the atmospheric distillation crude, to model it, and to compare the usual model of the crude distillation unit with the model of the Russian Federation.

1.2 Description of the crude oil and demulsifier

1.2.1 Oil emulsion

An emulsion is a suspension of small droplets of a liquid (water) in a second liquid (oil crude) [11]. The liquids, in a stable emulsion, can only be separated by a treatment process which calls the desalting process [10]. The electrical conductivity of oil emulsions in the literature note that it has an ionic nature. A detailed theory of ionic conductivity of liquid dielectrics. Despite the low dielectric constant, the polar components of crude oil dissociate into ions and create a fairly significant amount of electrical conductivity. The reason of this manifestation is asphalt-resinous substances in oil [12]. Oil emulsion is illustrated in **Figure 2**.

1.2.2 The demulsifier (surfactant)

The demulsifier is a chemical additive to break the crude oil emulsion. The emulsifier destroys the emulsion by reducing the mechanical resistance of the protective shells formed on the surface of the water droplets. By maintaining a strict hydrophilic-lipophilic balance [6]. These emulsions are destroyed using synthetic surfactants (demulsifiers) added to the oil emulsion [13]. The criteria for selecting and evaluating the demulsifier in terms of technology, that is, yield, are as follows [8, 12, 14]:

- The velocity of separation of water from oil;
- Quality of destruction of emulsions (fullness of water separation);

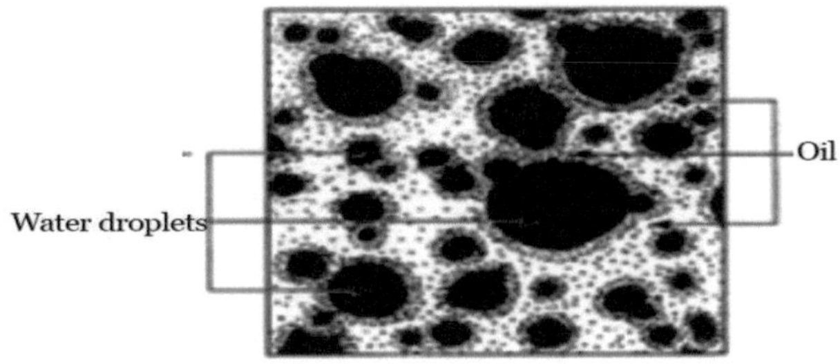

Figure 2.
Oil emulsion.

- Reliability in changing production conditions;
- The demulsifier must have a good quality/price or cost/efficiency ratio.

1.3 Mechanisms and devices of process of dehydration-desalination (pre-treatment) of oil crude

The process of dewatering of crude oil enters the oil production process. At the initial production period of a fresh field, wells often produce either anhydrous or low-water oil. However, over time, the water content of the oil produced increases at different rates and sometimes reaches 80–90% in old fields [1, 14]. The desalting (dewatering) processes depend on the type of oils and its content, any of the following typical processes of desalting of crude oil are combined a lot of methods like thermal, chemical, electrical and mechanical Often, a combination of Thermal and chemical methods with the mechanical or the electrical method is used to achieve effective Dehydration of crude oil [11]. The chemical treatment presents itself as a good option. It consists to use a demulsifier. The demulsifier must be introduced as early as possible (at the bottom of the well). This is an increased contact time and the interaction capacity of the demulsifier with emulsion for maximum destruction of the system [15, 16]. The injection of the demulsifying agent before the pump ensures proper contact with the crude oil and minimizes the formation of an emulsion [15]. The Industrial application of demulsifiers is not based on non-electrolytes, as their use is associated with excessive consumption or high price, as well as complex separation from oil after the deposition of the water layer. Preference is given to colloidal surface-active substances, among which are anionic, cationic, and nonionic types [13]. The number of steps (one, two or three) during oil desalination is determined by the characteristics of the initial oil emulsion and the salts contained therein For the desalination of oil emulsion independently of the stage: At the well or the refinery, the main types of equipment used for oil dewatering are electrostatics disasters, sumps, and separators [7], and they are detailed in **Figure 3** [7, 17]:

- thermo-chemical dehydrator;
- sump;
- ultrasonic;
- centrifugal;
- electrostatic desalter.

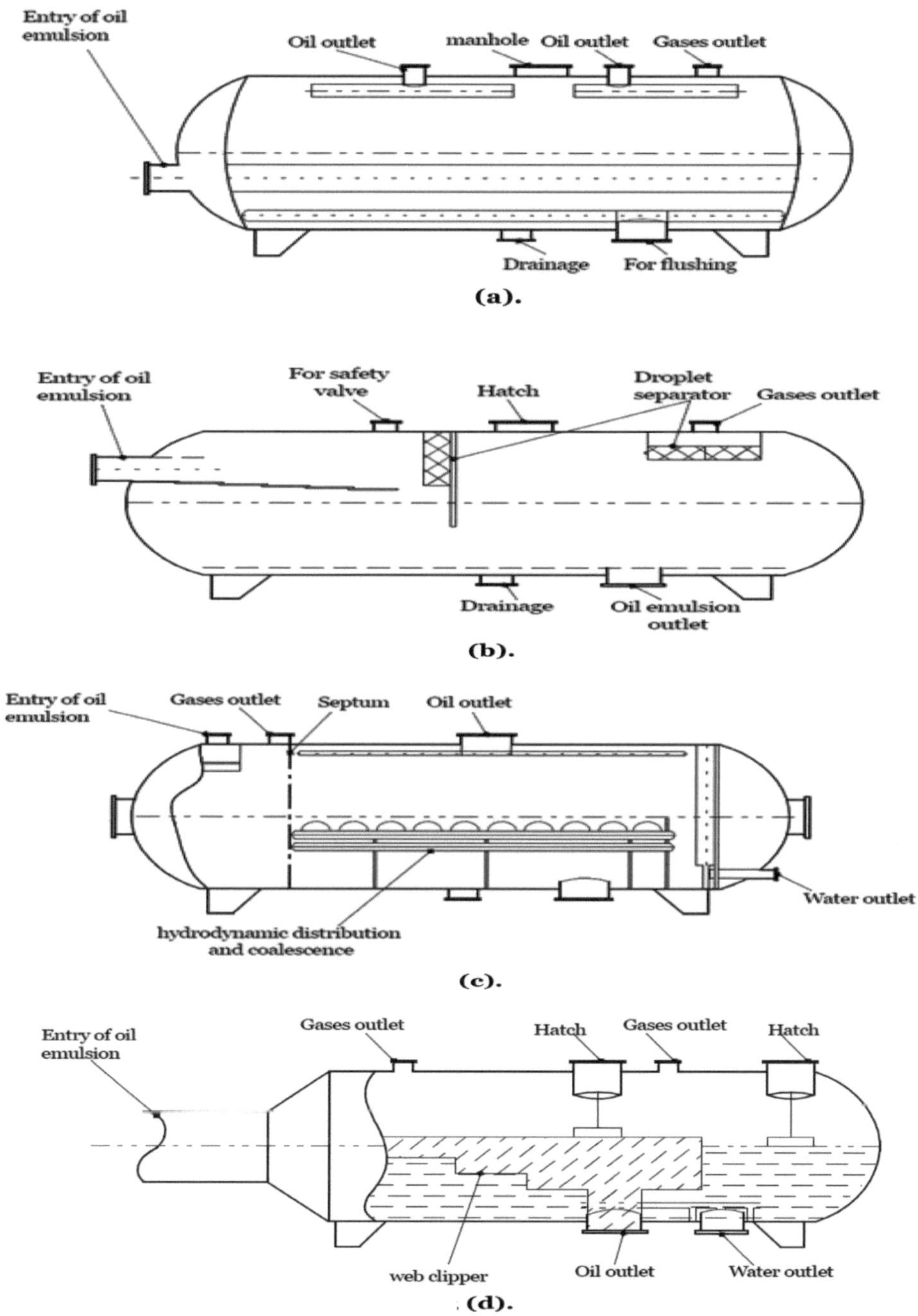

Figure 3.
Some examples of devices for crude oil dewatering, degassing, and desalting. (a) Sump. (b) Separator. (c) Three-phases separator. (d) Hydrodynamic phase separator.

Each of the above-mentioned apparatus has its own peculiarities and its own derivatives, according to the precise tasks to be accomplished:

- Advanced separators-vertical flow, horizontal unit with advanced electrostatic oil dehydration capabilities.

- Desalting device-vertical flow, horizontal device with improved capabilities of electrostatic dehydration and desalting of oil. The process can occur in one or two stages.

- Field separator-electrostatic coagulator, horizontal apparatus of electrostatic coagulator. It contains the initial section of the chimney of the heating section for preheating.

- Electromechanical separators-a horizontal device that is characterized by a combination of electrostatic grids, coalescing blocks, and the initial section of the chimney heating section for preheating.

The selection criterion of devices, it is recommended to consider the following parameters when choosing the equipment [8, 12, 18]:

- properties of formation fluid and its quantity;

- stream type;

- the resistance of emulsion;

- pressure into the device;

- temperature regime.

In the separators, they usually separate gas is usually from the oil in two or three steps under slight pressure or dilution. The separators of the first stage simultaneously play the role of buffer reservoirs and are usually located on the deposit. The separators of the second and third stages are mostly on the territory of central collection and distribution points (the sites of pre-treatment and pumping of oil) [7, 17]. The vertical separators are more productive in comparison with horizontal ones, but also the vertical separators have a higher price. They are suitable for enterprises with high production capacities, as well as if the emulsion contains many solid particles [18]. The horizontal separators are the best option for processing small volumes of material, as well as liquids with a high content of dissolved gas. They are in the greatest demand, as they are quite productive and affordable. To achieve maximum efficiency when using horizontal oil and gas separators, the oil is mixed in the separation process; the temperature is increased, and the pressure is reduced. To increase the depth and improve the quality of separation in Russia, introducing hot drainage water before separation with a three-phase separator is often used. Thus, they use heat, which increases the selection of gas, intensifies the process of demulsification of oil, improves the quality of spilled formation water, reduces investments, energy costs of processes, and significantly improves the conditions of Environmental protection. The three-phase separator can split oil, gas, and water at once. However, separators have limits. Even with a three-stage separation, complete separation of gas from oil is not achieved [8]. Besides, it should be noted that, despite the improvement in oil and gas separation techniques and technologies, field separators remain cumbersome and expensive devices [19]. Their operation is based on an inefficient gravitational principle and they are unproductive; the separators cease to function when the oil and gas mixture forms foam. The loss of energy contained in the flow of oil and gas, with a decrease in the pressure in the degassing in stages, leads to the need to use more pumping and compressor units for the collection and transportation of oil and gas through the pipeline [7, 8].

The final processes of dewatering and desalting are carried out at oil refineries with electrostatic desalter [20]. At the refinery, three types of electrostatic desalter are mainly operated as oil pre-treatment equipment: vertical, horizontal, and ball electrostatic desalter [1].

1.4 The salts in crude oil

The salts and minerals often present in the oil are mainly magnesium, calcium, and sodium chlorides with sodium chloride being the abundant type. These salts cause corrosion of equipment. For example, hydrogen chloride (HCl). Hydrogen chloride dissolves in the emulsion's water producing hydrochloric acid, an extremely corrosive acid. However, $NaCl = Na^+ + Cl^-$. Water, though slight, but dissociates into ions, and the equilibrium is established: $H_2O=H^++OH^-$, thus, in the salt solution, there is a mixture of Na^+ cations and OH^-, Cl^-anions. Ions in the solution move randomly and constantly collide with each other [21]. But these collisions of Na^+ and OH^- ions, H^+ and Cl^- ions do not lead to the formation of compounds, since NaOH is a strong base, and HCl is a strong acid. Since weak electrolytes are not formed when sodium chloride is dissolved in water, sodium chloride is not hydrolyzed. The concentration of H^+ ions is equal to the concentration of OH^- ions, so the color of the indicators does not change [13, 20]. All chemical compounds based on chlorine hydrolyze, except for NaCl, hydrolyze at high temperature to hydrogen chloride:

$$CaCl_2 + 2H_2O \rightarrow Ca(OH)_2 + 2HCl \quad (1)$$

$$MgCl_2 + 2H_2O \rightarrow Mg(OH)_2 + 2HCl \quad (2)$$

Any remaining salts are neutralized by the injection of sodium hydroxide which reacts with the calcium and magnesium chloride to produce sodium chloride because NaCl does not hydrolyze to the corrosive hydrogen chloride.

$$CaCl_2 + 2NaOH \rightarrow Ca(OH)_2 + 2NaCl \quad (3)$$

1.5 Description of dewatering of oil emulsions with electrostatic desalter

The electrical conductivity of oil emulsions is due to the ionic conductivity of oil, the conductivity of dispersed water droplets in oil. Under certain conditions, droplets of emulsified water in oil form conductive structures in the form of "chains", located along the field lines [12, 22]. The electrical conductivity of the system in such cases increases sharply (hundreds and thousands of times) in comparison with the electrical conductivity of anhydrous oil. Electrical Conductivity of hydrocarbons (oil), the hydrocarbons have three types of conductivity regimes in general [23]:

- The stable insulating regime, this regime is characterized by very high resistance. In this state, conduction is related to traces of dissolved water in the hydrocarbon.
- The semiconductor, this regime is independent of the presence of water in the hydrocarbon.
- Conductor, relatively stable. The regime, however, it disappears by sufficient removal or reduction of the applied electric field and reappears when it is restored to its initial value.

1.5.1 Description of electrostatic desalter and its process

The modern desalting units are designed with electric dehydrators only horizontal execution and are part of the equipment for preparing oil: both atmospheric and atmospheric-vacuum installations. The horizontal design of electrostatic desalter has several advantages, such as a large area of the electrodes and, accordingly, a greater amount of oil per unit of section of the device, a lower vertical speed of the moving flow of oil, which provides favorable conditions for settling water, as well as implementing processes with higher pressure and temperature [22, 24]. The dominant types of electrostatic desalter are shown in **Figure 4**. The increases of the electrical conductivity are related to traces of dissolved water in the oil; it has the effect of slowly eliminating this water, correlative elimination of an indefinite increase in resistivity. The work and efficiency of the electrostatic desalter are based on the Stokes formula [25] for the time of deposition of water droplets and the specific electrical conductivity (γ, Om m^{-1}) is the electrical conductivity of a substance measured between flat electrodes of the same area, located at a distance (L) in meter, and divided by the electrode area [12]. The principle of operation of the electrostatic desalter is quite simple. We have one inside two electrodes that have two opposite charges. The oil is supplied to the lower stage of the device, which provides its additional washing and passing through two (2) electric fields: weak and strong as shown in **Figure 5**. The charges of these electrodes change alternately, resulting in a separation of the different phases: oil, water, and gas [11, 26]. The water has a higher density than the oil and by the gravitational force, so the water is concentrated at the bottom of the electrostatic desalter. The oil and the gas without water, so without salts, come out from the upper part of the electrostatic desalter by manifold as shown in **Figure 6** [20]. In the literature, there are indications that the electrical conductivity of oil and petroleum products is largely determined by the content of polar surfactants (asphalt-resins) in them [25].

1.5.2 Desalting processes at a refinery

They can be one and two-stage (respectively, with one or two electrostatics desalters) [20]. To increase the efficiency of the operation, the demulsifiers are added to the crude oil. Two-stage installations are used to reduce the flow of freshwater when washing crude oil [13]. The desalter of this design achieves 90% salt removal. However, 99% salt removal is possible with two-stage desalters [7, 20] as shown in **Figure 7**. A second stage is also essential since desalter maintenance requires a lengthy amount of time to remove the dirt and sediment which settle at the bottom. Therefore, the crude unit can be operated with a one stage desalter while the other is cleaned.

1.6 Example of modeling of electrostatic desalter at refinery

The electrostatic desalter volume 96 m^3, which length-10 m, diameter –3 m. U (voltage of the electrodes) = 22000 v. Maximum emulsion pre-treatment capacity

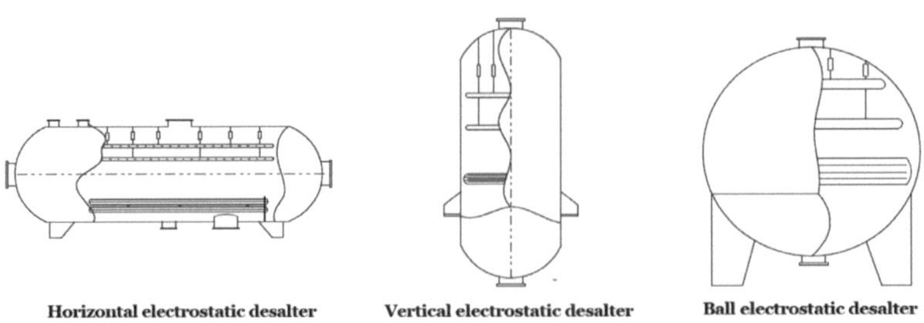

Figure 4.
The major types of electrostatic desalter.

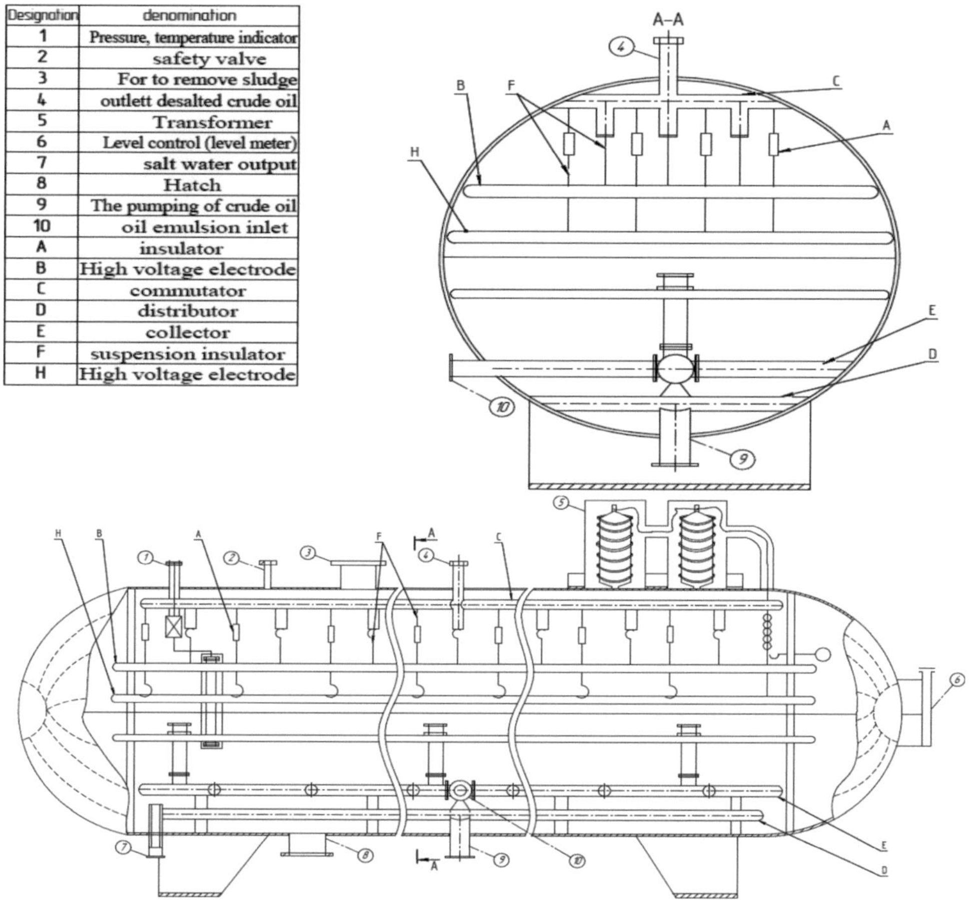

Designation	denomination
1	Pressure, temperature indicator
2	safety valve
3	For to remove sludge
4	outlett desalted crude oil
5	Transformer
6	Level control (level meter)
7	salt water output
8	Hatch
9	The pumping of crude oil
10	oil emulsion inlet
A	insulator
B	High voltage electrode
C	commutator
D	distributor
E	collector
F	suspension insulator
H	High voltage electrode

Figure 5.
Technical schema of electrostatic desalter design.

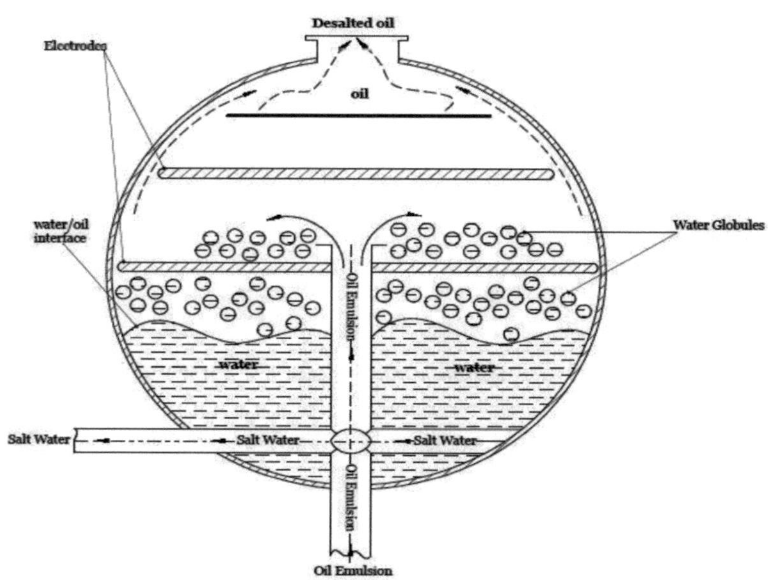

Figure 6.
The simplified work description of electrostatic desalter.

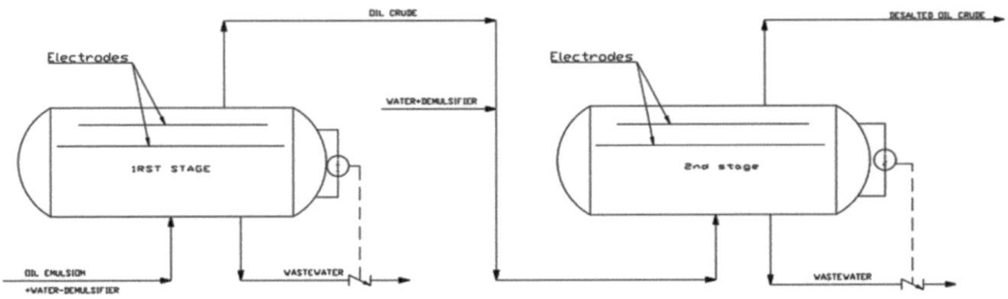

Figure 7.
Two-stage desalting flow processing of crude oil.

is 140 tons/h. The program used in our modeling is Mathcad, to calculate the material balance and mechanical calculations of electrostatic desalter. The characteristics of the feedstock are presented in **Table 1** [27, 28].

1.6.1 Materials balances of electrostatic desalter

The establishment of a material balance is necessary for modeling [7, 25, 29]. The material balance material allows us to have an idea of the results we need to get and correct some errors during the project in order to have a satisfactory performance of our facilities. To calculate the material balance of electrostatic desalter, it is necessary to know the data:

- the temperature at the entrance of feedstocks into the electrostatic desalter, $t_1 = 90°C$;
- the temperature at the outlet of feedstocks in the electrostatic desalter, $t_2 = 145°C$.

The quantity of oil emulsion chosen is 140 t/h. the water Content of the initial emulsion is 1% of the total mass. The consumption of the demulsifier (surfactant) is 30 g/t. Our work in this part is to make material balances of the pre-treatment of the oil at the refinery. Electrostatic desalter performs two procedures at the same time which are: dehydration and desalination. The descriptive diagram e of the desalting process is presented in **Figure 8**. The amount of freshwater or water processing supplied at the desalting stage is 10% of the mass. The reagent is fed only at the dehydration stage. We consider the reagent water-soluble. First, the status of "incoming" is drawn up. The

Indicators	Methods	Units of measure	Values
Density at 15°C	NF EN 12185	kg/m^3	884
Water content	NF EN 9029	% mass	1
Total acid number	ASTM D 664	mg KOH/g	0.619
Sulfur content	NF EN 8754	% m/m	0.254
API	ASTM D1298	—	29
Kinematic viscosity at 100°C	ASTM D445	cSt	1.9
Freezing point	NF EN ISO 3016	°C	−17

Table 1.
The physicochemical characteristics of Bonga crude oil.

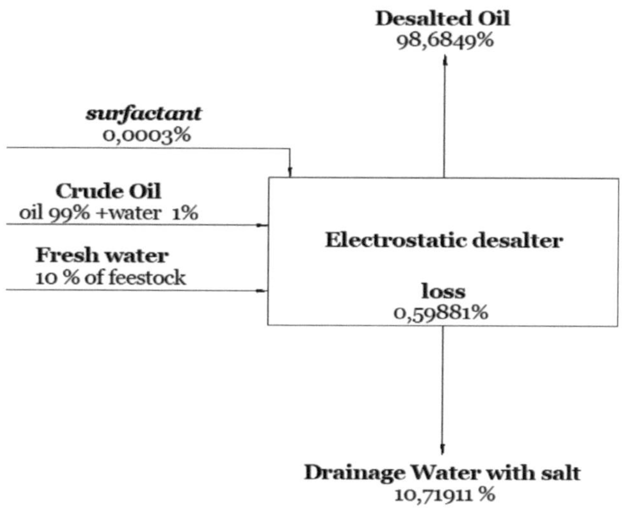

Figure 8.
The descriptive diagram of the desalting process.

status "Expense" is similar to the desalting- dewatering process. 140 t of emulsion – 1.4 t of water in emulsion = 138.6 t of oil. The status "Expense" is similar to the desalting-dewatering process. Loss is equal: $\frac{(138.6 \times 0.3)}{100}$ = 0.4191 t. The amount of dewatered oil or desalted oil will be 138.6 t – 0.4158 t = 138.0897 t. According to this task and Russian GOST R 51828-2002, the water content in the desalted oil shall not exceed 0.5% of the mass. Thus, the water contained in the desalted oil will be:

$$138.0897 \text{ t} - 99.95\% \text{ mass.} \tag{4}$$

$$\text{X} - 0.05\% \text{ mass.} \tag{5}$$

The amount of water X is 0.0691 t.

The water content-X is 0.0691 t in the desalted oil. The total water content phase in the "incoming" status is 14 t + 0.14 t = 14.14 t.

Based on the previous results and by the possibility that we can really have more losses. We arbitrarily decide to multiply our losses by 2; the total loss is: 0.4191 t + 0.4191 t = 0.8383 t. Consequently, the summary material balance of the installation is made. The total amount of drainage water is equal to the drainage water by oil treatment stage: 0.9332 t + 14.07 t = 15.0032 t. The material balance of the electrostatic desalter installation is presented in **Table 2**.

1.6.2 Modeling of the desalting process with electrostatic desalter

The exercise of modeling is a variant very often useful and used in the modeling of units and devices [7, 14, 19]. This task is interesting in case we do not have a simulator or simulation and modeling program like Hysys Aspen to solve this operation. In the framework of modeling the desalination unit, we use the ASTM D 341-2003 standard is based on Walter's equation and proposes the dependence of the kinematic viscosity of crude oil (hydrocarbon):

$$lg \times lg(\nu + 0,8) = a + b \times lgT, \tag{6}$$

$$a = lg \times lg(\nu_1 + 0,8) - b \times lgT_1 \tag{7}$$

Feedstocks	ton/h	% mass	Products	ton/h	% mass
1. Oil emulsion, including:	140.0000	100.0000	1. Stable oil, including:	138.1589	98.6849
• Oil	138.6000	99.0000	• Desalted oil	138.0897	98.6355
• Water	1.4000	1	• Water	0.06916	0.0495
2. Surfactants (Demulsifiers)	0.0004	0.0003	2. Drainage water	15.0032	10.7191
3. Water processing (freshwater)	14.0000	10	3. Loss	0.8383	0.5988
Total	154.0004	110.0003	Total:	154.0004	110.0003

Table 2.
The material balance of electrostatic desalter.

$$b = \frac{lg \times \left[\frac{lg(\nu_1+0,8)}{lg(\nu_2+0,8)}\right]}{lg\frac{T_1}{T_2}}, \tag{8}$$

where ν – viscosity, cSt, T is the absolute temperature, K; a and b are the constants determined by two known values of viscosity ν_1 and ν_2 at temperatures, respectively, T_1 and T_2. The empirical formula of formula Filonov – Reynolds is more convenient for analytical solutions:

$$\mu_t = \frac{1}{C} \times \left(C \cdot \mu_{t_o}\right)^{\chi} \tag{9}$$

where.

$$\chi = \frac{1}{1 + \alpha(t - t_o) \times lg\left(C \cdot \mu_{t_o}\right)}; \tag{10}$$

ν_t, ν_o – Dynamic viscosity of the oil at temperatures t and to, respectively, mPa·s; α and C are empirical coefficients.

$$\text{If } \nu_o \geq 1000 \text{ mPa} \cdot \text{s, then } C = 10, 1/\text{mPa} \cdot \text{s}; \alpha = 2.52 \cdot 10^{-3}\ 1/°\text{C}; \tag{11}$$

$$\text{If } 10 \leq \nu_o < 1000 \text{ mPa} \cdot \text{s, then } C = 100, 1/\text{mPa} \cdot \text{s}; \alpha = 1.44 \cdot 10^{-3}\ 1/°\text{C}; \tag{12}$$

$$\text{If } \nu_o < 10 \text{ mPa} \cdot \text{s, then } C = 1000, 1/\text{mPa} \cdot \text{s}; \alpha = 0.76 \cdot 10^{-3}\ 1/°\text{C}. \tag{13}$$

$$x = \frac{1}{[1 + 0.76 \times (90 - 100)] \times log\,(1000 \times 0.0014144)}$$

$\nu = \frac{1}{1000} \times (1000 \times 0.0014144)^{x} = 7.055 \times 10^{-4}$

In modeling of electrostatic desalter to determine his maximum performance and the required number of devices for desalting crude oil.

$$E_{Critical} = \theta \times \sqrt{\frac{2 \times \lambda}{\phi \times d}}, \text{v/cm}, \tag{14}$$

The interaction between water droplets can be increased by increasing the electric field strength, E.

$$EReal = \frac{U}{l}, \text{v/cm} \tag{15}$$

For to determine the *E-Critical* and *E-Real* for the desalter with the following data. Water content in oil – 1%; λ (interfacial surface tension) = 12 dyn/cm; d (the diameter of the droplets) = $1.5 \cdot 10^{-2.55}$ m or $4.228{\cdot}10^{-1}$ cm; φ (dielectric constant of the emulsion) = 2; l (distance between electrodes) = 20 cm; U (voltage of the electrodes) = 22000 v.

$$E_{Critical} = 382 \times \sqrt{\frac{2 \times 12}{2 \times 4.228 \cdot 10^{-1}}} = 2035.1\ \mathrm{V/cm}.$$

$$E_{Real} = \frac{22000}{20} = 1100\ \mathrm{V/cm}.$$

$E\text{-}_{Real} < E\text{-}_{Critical}$, consequently, the dispersion of droplets will not. The electrostatic desalter receives crude oil, the amount of Q = 162 m^3/h with temperature 90^0C (preheat). $\rho^{90}{}_{crude\ oil}$ = 838.4 kg/m^3, $\rho^{90}{}_{H20}$ = 965.3 kg/m^3, the kinematic viscosity of crude oil $\mu_{oil\ crude}$ = $7.055 \cdot 10^{-4}$ m^2/s, the diameter of the globules of water – $1.5{\cdot}10^{-2.55}$ meter. The residence time of the emulsion in the apparatus – 30 minutes is applied to the device volume 96 m^3, which length – 10.19 m, diameter – 3 m. The maximum surface deposition: $S = D \times L = 10.19 \times 3 = 30.57\ m^2$. The efficiency of the electric desalter depends on the values S/V, where S – the average area of the horizontal section of the device, m^2; V – the volume of the device, m^3. For effective sludge must be met the condition $\tau \geq \tau_W$, where τ - oil residence time in the apparatus, h; τ_W – time required for precipitation of water droplets, an hour.

$$\tau = \frac{h_{Em}}{U_{Em}}, \tag{16}$$

where h_{Em} – the height of the emulsion layer in the apparatus, m; U_{Em} – the velocity of the crude oil flow during its lower flow, m/h.

$$\tau_W = \frac{h_{Em}}{U_{Real}} = \frac{h_{Em}}{U_{water} - U_{wo}}, \tag{17}$$

where U_{water} – deposition rate of water droplets in a stationary medium, m/h; U_{wo} – real deposition rate of water droplets in the rising oil flow, m/h.

$$U_{water} - U_{wo} \geq U_{Em} \text{ or } U_{Water} \geq 2 \times U_{Em}. \tag{18}$$

The linear velocity of the oil in the electrostatic desalter must be at least 2 times less than the calculated rate of water droplet deposition. The deposition rate is calculated using the Stokes formula [7, 25]:

$$U_{Water} = \frac{d^2 \times g \times \left(\rho_{H_2O} - \rho_{crudeoil}\right)}{18 \times \mu_{crudeoil} \cdot \rho_{crudeoil}}\ \mathrm{m/s}, \tag{19}$$

where d – diameter of water drops, m; ρ_{H2O}, $\rho_{\ crude\ oil}$ – the density of water and oil, kg/m^3.

$\mu_{crude\ oil}$ – kinematic viscosity of oil at sludge temperature, m^2/s.

$$U_{Water} = \frac{\left(1.5.10^{-2,552}\right)^2 \times 9.81 \times (965.3 - 838.4)}{18 \times 7.055 \times 10^{-4} \times 838.4}\ \mathrm{m/s}$$

$$U_{water} = 0.0021\ m/s.$$

It is necessary to check the Reynolds number (Re) by the formula:

$$\Re U_{Water} \times \frac{d}{\nu_{crudeoil}}. \quad (20)$$

The condition must be met:

$$10^{-4} \leq Re \leq 0.4 \quad (21)$$

$$\Re 0.00208974 \times \frac{1.5.10^{-2,55}}{7.055\times10^{-4}} = 0.0125,\ Re < 0.4$$

h_{Em} is the height of the emulsion level in the devices, m = meter

$$h_{Em} = 0.75 \times D - h_1, \quad (22)$$

where h_1 the distance from the bottom of the electric Hydrator to the oil-water interface. We choose h_1 = 0.75 m × U_{Em} – the velocity of the oil flow during its lower flow, m/h, $h_{Em} = 0.75 \times 3 - 0.75 = 1.69$ m

$$U_{Em} = \frac{h_{Em}}{\tau} \quad (23)$$

$$U_{Em} = \frac{1,69}{0.75} = 2.25\text{m/h} = 6.259 \cdot 10^{-4}\ \text{m/s}.$$

Knowing the U_w, determine the U_{wo} and the required cross-section of the electric dehydrator (S). The actual deposition rate of water droplets in the rising oil stream will be:

$$UReal = U_{water} - U_{Em} \quad (24)$$

$$UReal = 2.08 \times 10^{-3} - 6.259 \times 10^{-4} = 1.464 \cdot 10^{-3}\text{m/s}.$$

The performance of the desalter:

$$J = UReal \times S, \text{m}^3/\text{s} \quad (25)$$

$$J = 1.464 \cdot 10^{-3} \times 30.57 = 0.045\ \text{m}^3/\text{s} = 161.3\ \text{m}^3/\text{h}.$$

The number of devices (N):

$$N = \frac{Q}{J}, \quad (26)$$

where Q – the quantity of emulsion supplied to the unit, m^3/h; J – The performance of the desalter, m^3/h.

The required number of parallel running for the electrostatic desalter:

$$N = \frac{162}{161.3} = 1.005$$

1.6.3 Mechanical modeling of electrostatic desalter (theory)

This task is as interesting as the one already solved. In case we do not have a simulator or simulation and modeling program like Ansys to solve this vital operation [1, 18, 29, 30]. The calculation of the strength of a cylindrical shell under internal pressure. The wall thickness is determined by the formula (25).

$$S_R = \frac{P \times D}{2 \times [\sigma] \times \phi - P}; \tag{27}$$

Permissible internal overpressure.

$$[p]_D = \frac{2 \times [\sigma] \times \phi \times (S - C)}{D + (S - C)} \tag{28}$$

P is the pressure in the device, mPa; S_R – the calculated value of wall thickness, mm; D- intern diameter of the shell, mm; $[\sigma]$ – permissible voltage, mPa (depends on the steel grade and operating temperature).

The steel grade is chosen depending on the properties of the processed medium. For butt and t-shaped double-sided seams performed by automatic welding, the coefficient of the strength of the weld $\phi = 1$, for the same manual stitches, $\varphi = 0.9$. Increase taking into account corrosion C is determined by the formula:

$$C = \mathrm{V} \times \mathrm{T}, \tag{29}$$

V is the rate of corrosion (usually take 0.1–0.2 mm/year); T – the service life of the device (usually take 10–12 years). For materials resistant to the processed medium, in the absence of data on permeability, it is recommended to take C = 2 mm. Condition for reliable operation $P < [P]_D$ must be observed.

The choice of corrosion resistance of materials is made concerning this environment is resistant to corrosion.

$$S \geq S_R + C \tag{30}$$

The wall thickness calculated using this formula is rounded up to the nearest standard sheet thickness (4, 6, 8, 10, 12, 14, 18, 20 mm). The calculation of shells loaded with external over-pressure consists in determining the permissible external pressure since the wall thickness of the shell was determined earlier. Permissible external pressure:

$$[p]_{pp} = \frac{[P]_p}{\sqrt{1 + \left(\frac{[P]_p}{[P]_E}\right)^2}}, \tag{31}$$

$[P]_p$ durability allowable pressure corresponding to the strength condition:

$$[P]_D = \frac{2 \times [\sigma] \times \phi \times (S - C)}{D + (S - C)}. \tag{32}$$

i – coefficient of stability (for the operating conditions of i = 2.4), l_R – the estimated length of the shell.

$$l_R = l - 2 \times h_0 + \frac{H_D}{3}, \tag{33}$$

E is the modulus of elasticity. l = length of the cylindrical part of the corpus; h_O – the height of the bottom flanging, H_D – the height of bottom edge. Allowable pressure of conditions of stability within the limits of elastic deformation:

$$[P]_E = \frac{18 \cdot 10^{-6} \times E}{i} \times \frac{D}{l_R} \left[\frac{100 \times (S - C)}{D}\right]^2 \times \sqrt{\frac{100 \times (S - C)}{D}} \quad (34)$$

Condition for reliable operation $[P]_E \leq [p]_{pp}$ must be observed.

1.6.4 Resolution example based on our electrostatic desalter

P – pressure in the device, mPa = 1.1; D-inner diameter of the shell, mm = 3000. The increase in corrosion C is determined by the formula (27).

$$C = 0.2 \times 20 = 4 \text{ mm}$$

V is the rate of corrosion (usually take 0.1–0.2 mm/year); T – the service life of the device (usually take 10–12 years), but we take 20 years. For materials resistant to the processed medium, in the absence of data on permeability, it is recommended to take C = 2 mm. According to the table of corrosion resistance of materials, we choose the steel grade 316Ti, S31635-USA (ASTM/AISI), or 1.4571, X6CrNiMoTi17-12-2-Germany (DIN/WNr). $[\sigma]$ = 168.6 mPa. The coefficient of the strength of the weld is assumed $\phi = 1$. The service life is 20 years. The corrosion rate is 0.2 mm/year. Increase taking into account corrosion C = 4 mm.

$$S_R = \frac{1.1 \times 3000}{2 \times 168.6 \times 1 - 1.1} = 9.819 \text{ mm.}$$

$$S = 9.819 + 4 = 13.819 \text{ mm.}$$

The wall thickness calculated using this formula is rounded up to the nearest standard sheet thickness (4, 6, 8, 10, 12, 14, 18, 20 mm). We assume S = 14 mm.

$$[p]_D = \frac{2 \times 345 \times 1 \times (14 - 4)}{3000 + (14 - 4)} = 1.12 \text{ mPa.}$$

Condition for reliable operation $P < [P]$ (1.1 mPa < 1.12 mPa) observed.

The calculation of shells loaded with external over-pressure consists in determining the permissible external pressure since the wall thickness of the shell was determined earlier.

$[P]_p$ the allowable pressure corresponding to the strong condition.

$$[P]_p = \frac{2 \times 168.6 \times 1 \times (14 - 4)}{3000 + (14 - 4)} = 1.12$$

E is the modulus of elasticity, i – coefficient of stability (for the operating conditions of i = 2.4), l_R – the estimated length of the Desalter. $E = 1.86 \cdot 10^5$ mPa; l = 10,190 mm; H_D = 1500 mm; h_O = 25 mm.

$$l_R = 10190 - 2 \times 25 + \frac{1500}{3} = 10640 mm$$

Allowable pressure of conditions of stability within the limits of elastic deformation:

$$[P]_E = \frac{18{\cdot}10^{-6}{\cdot}\times 1.86{\cdot}10^{-5}}{2.4} \times \frac{3000}{10640} \times \left[\frac{100\times(14-4)}{3000}\right]^2 \times \sqrt{\frac{100\times(14-4)}{3000}} = 0.0252\ \text{mPa}$$

Permissible external pressure:

$$[p]_{pp} = \frac{1.12}{\sqrt{1+\left(\frac{1.12}{0.0252319}\right)^2}} = 0.0252\ \text{mPa.}$$

Condition for reliable operation $[P]_E \le [p]_{pp}$ (0.0252 mPa ≤ 0.0252 mPa) observed.

1.7 Summary

The desalting and dewatering of crude oil take begins on the oil fields. This operation is part of the processing of crude oil, and a good pre-treatment of crude oil implies:

- degassing – removal of gases from crude oil;
- dewatering – removal of water from crude oil.

If the desalting of crude oil is carried out qualitatively, the oil has almost no harmful effect on the equipment. Most of the impurities that cause corrosion of the metal are in the remains of formation water. Therefore, the fundamental task of desalting is to remove water drops from the oil. This is a fairly complex process, because the water in crude oil is in the form of droplets with a size quantity often. The improvement and appreciation of the oil pretreatment process level are characterized by:

- providing effective sludge at high viscosity and density of oil through the use of a sump with intermediate partitions of variable height;
- increasing the degree of dehydration and desalination of oil due to effective washing with water from salts;
- improving the efficiency of the sump and improving the quality of the prepared oil due to more complete and qualitative removal of the intermediate layer from it, as well as due to a more uniform distribution of fluid flow rates when entering the sump and its cross-section;
- ensuring uniform receipt of production of wells for the installation of oil treatment and prevention of failures of its work through the use of the reservoir;
- improving the performance of the electrostatic desalter, that is, expanding the range of workloads, improving the efficiency of desalination and dehydration, as well as reducing the cost of oil pre-treatment.

The poor pre-treatment (desalting process) of crude oil can lead to considerable extra costs. We are talking about the high cost of transportation if the product is not cleared of unnecessary substances that give it extra volume and weight as well as financial investments in equipment. After all, oil, which is not derived from salt, can very quickly damage the pipeline. The requirements for the oil content of the water and especially for chlorides more stringent on the refineries. The oil content of water before processing must be no more than 0.1% of the mass and for salts-no more than 5 mg/l [3, 31]. These related requirements using much more expensive equipment (columns, heat exchangers, reboilers, etc.). Also, these the requirements induce to reduce the energy consumption, to reduce the corrosion of the equipment, to increase the life of the catalysts, improves the quality of the petroleum products. If these requirements not performed, the oil is necessarily subjected to desalting and dehydration at the electric desalting plant (electrostatic desalter). The surfactants are added if the crude oil contains a lot of suspended solids. The crude oil is heated usually to a temperature of 50 to 90°C to reduce viscosity and surface tension for easier mixing and separation of water.

2. Atmospheric Distillation Unit (ADU)

2.1 Introduction

The distillation is heating a liquid solution or mixture of liquid stream to produce steam, and then collecting and condensing that steam. In the simplest case, the products of the distillation process are limited to the upper distillate and the bottom, the composition of which differs from the feed. Thus, the major purpose of distillation is the separation or splitting into two or more products from a feed [32]. The distillation or rectification is one of the oldest and most common methods of chemical separation [33]. Historically, one of the most famous uses is the production of spirits from wine. More simply, in other words, the target of distillation is usually to remove the light component from the heavy component mixture, or vice-versa, to separate the heavy product from the light component mixture [32]. One of the components of crude oil refining is distillation, therefore a stage of refining. Oil refining is a rather complex technological process that begins with the transportation of oil from the field to refineries. The fractionation or distillation column exists in practically all oil refining technologies through reforming, thermal cracking, hydrocracking, etc. [20]. Before receiving the petroleum products common to users, the oil goes through several stages before becoming a finished product. Some of these stages are [1, 9, 20, 21]:

- pre-treatment of crude oil;
- distillation atmospheric and vacuum;
- catalytic reforming, isomerization;
- alkylation, visbreaking, coking;
- catalytic cracking, hydrocracking;
- treatment of petroleum products (hydrotreatment, desulphurization, etc.)

2.2 Description and characteristics crude distillation unit (CDU)

The oil emulsion is a complex mixture of naphthenic, paraffin, aromatic carbohydrates, which have different molecular weight and boiling point, and sulfur, oxygen, and nitrogen organic compounds. At distillation receive a big range of oil products and semi-products. The principle of the process is based on the difference between the initial boiling point of the components. As a result, the oil is divided into fractions up to fuel oil and tar and even base oil. The distillation of oil can be carried out [34]:

- Single evaporation or flash vaporization
- Multiple evaporations
- Gradual evaporation

The single evaporation of oil is a one-step separation technique. The single evaporation process involves heating the oil to increase the temperature and enthalpy to the true boiling point (TBP) of the vapor-liquid mixture. The process of multiple evaporations is a sequence of single evaporation with a gradual increase in the heating temperature. Distillation by gradual evaporation is a slight change in the oil's state with every single evaporation. The main devices in which the distillation of oil are distillation columns, reboilers, furnaces, etc. [1].

The difficulty of separating crude oils is mainly determined by the volatility of the key components or by the difference in the boiling point of the key components [34]. The closer the relative volatility is to one per mole, the more plates or plates are needed and more irrigation is needed to achieve the same purity of distillate and residue. For key components, two adjacent components are taken for key components, one belonging to the distillate, the other to the residual product. The relative volatility of these adjacent components and their proportion in crude oil are the main criterion determining the difficulty of sufficiently clear separation from crude oil [34]. More the tangent of the angle of inclination of the TBP increase, the easier the separation conditions are. The wider the crude oil TBP intervals, more the distillate and the residue obtained during the process are clean. As we all know, crude oil from the bowels of the Earth contains salt as dissolved salt in a tiny drop of water that forms water into an emulsion [1, 9]. This water cannot be completely separated by gravity or mechanical means. The deep separation of water from the emulsion occurs with electrostatic at the plant, before the distillation process [20].

After desalting process, the crude oil is heated in heat exchangers or reboilers (preheating of crude oil). Of course, preheating is not sufficient, since the oil must be partially evaporated to the extent that all products except atmospheric residues must be in the vapor phase when the oil enters the atmospheric column. Thus, the furnace is required to raise the temperature between 330 and 385°C depending on the components of oil [35]. The partially evaporated crude oil is transferred to the flash zone column located at a point below the distillation column and above what is called the stripping section. The main distillation column is generally up to 50 m with a 30–50 valve [34]. The size of the column is determined by the number of plates and the amount of steam. Besides, the amount of steam is determined by the content of crude oil in volatile elements or compounds. As a result, the rising steam in huge amounts and at top flow rates, requires a large diameter column above the flash zone [1, 34]. At the bottom of the section, water vapor is injected into the column to remove the atmospheric residue of any light hydrocarbon and reduce the partial pressure of hydrocarbon vapors in the flash zone [20]. This causes the true boiling point of hydrocarbons to decrease and causing more hydrocarbons to boil and raise the column to eventually thicken and removed as lateral flows. As hot vapors grow from the flash zone, they ascend into the column through the plates to the upper zone of the column [36]. A portion part of the light fraction of naphtha or gasoline returns to the column in the form of reflux. This reflux allows controlling the quality of the distillate and the pressure in the column. The main products of atmospheric petroleum distillation are [33]:

- Gas
- Gasoline (35–200°C)
- Kerosene (220–275°C)
- Diesel or gas oil (200–400°C)
- Atmosphere residue (above 350°C).

The distillation process is carried out in columns with plates inside and it is as follows: hot vapors, rising on the column, in contact with the cooler liquid (reflux) flowing down. There is a cooling of the vapors, hence condensation; a transition from vapors to liquid. This is the process of mass transfer and heat exchange; the contact efficiency is provided by the plates. Most of the side currents (light fractions) pass through the striping section, where they are again dispersed to control their starting and ending boiling point, while the liquid is heated, the low boiling components evaporate [36]. The temperature and pressure have a great influence on the processes (movement of fluids) due to the plates inside the column [32]. The distillation plates are designed to create close contact between Steam and liquid during the grinding process. Mainly apply grid, groove, cap, s-shape, valve, and other types of plates. The design of the plates, besides the close contact between steam, and liquid must ensure sufficient performance of the column, have a low hydraulic resistance to the flow of steam. For example, the movement of vapors in distillation columns depends on the types of plates shown in **Figures 9** and **10** [32]. The materials used in the design are also of great importance because they influence the metal capacity of structures, ease of assembly, and cleaning.

2.3 Description of the distillation column and its operation

The distillation is designed to separate gas and liquid mixtures comprising two or more components or fractions. We can classify distillation columns according to various characteristics [32, 34]:

- by technological application;
- by an internal device (plates) ensuring contact between the steam and the liquid;
- columns of atmospheric vacuum.

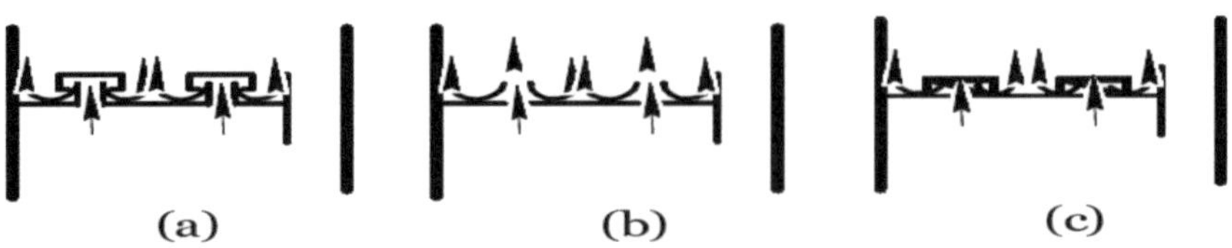

Figure 9.
The types of plate in the column: (a) cap, (b) grid, (C) valve.

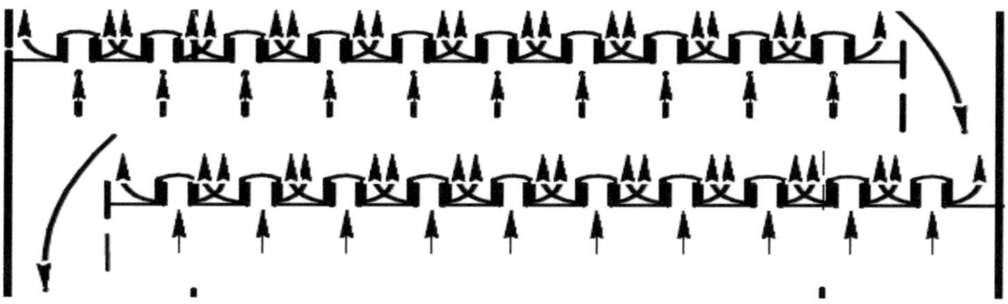

Figure 10.
The movement of steam and liquid flows on the cap plates.

The quality of distillation depends on the number of plates in the column and the amount of irrigation. The speed of movement of vapors in the column and the distance between the plates is great significance. The design of distillation plates is important in the column's performance. Thus, under practical conditions, the heat exchange between steam and liquid on the plate does not reach the equilibrium state, so the concept of plate efficiency (energy conversion efficiency) was introduced. It depends on the design and operating conditions and usually varies between 0.4 and 0.8 [32]. The distillation column comprises three (3) zones: the separation zone, the crude supply (feeding zone), and the heating zone. The feeding zone is generally below the separation zone. However, in some columns like the stripping columns, the feeding zone is located in the upper area of the column. Also, we can say that the separation zone in this type of column is non-existent. In short, this type of column is a column stabilization.

In oil refining, the first oil separation process is very often operated with complex columns or main columns [20]. The oil is sent to the atmospheric distillation column and after the desalination process. The oil is supplied to the feeding area in the form of steam, liquid, or vapor-liquid mixture. This area is called evaporative. Above the entrance (feeding zone) to the oil is the separation zone of the column, and below-the heating zone. In the evaporation zone of the column, there is single evaporation of the oil heated long before in the furnace or the heat exchanger. The purpose of atmospheric distillation is the primary separation of various "Fractions" of hydrocarbons: fuel gases, liquefied petroleum gas, naphtha, kerosene, diesel fuel, and from the bottom of the distillation; the atmospheric residue (heavy hydrocarbon residue). The residues from the atmospheric distillation column or atmospheric residues are sent to the vacuum distillation column. The atmospheric residues in the vacuum column (under the vacuum condition) and temperatures above 400°C give new fractions (heavy gasoil) [1, 32].

2.4 The plates for distillation columns of oil plants

In oil refining, a fairly large number of plate types are widespread; new types are being developed. This situation is explained by the desire to develop more productive and economic plates than those used previously. The choice of the types of plate depends on a load of steam and liquid, the physical properties of steam and liquid [34]. In order to ensure optimal conditions for the operation of columns in various technological processes, it is necessary to use plates of different designs [32]. In atmospheric columns, steam and liquid charges are moderate. While in vacuum columns, large vapor loads and small liquid loads are observed, that is low volumetric expenditures of liquid and high linear velocities of vapors. Thus, for some varieties of columns, several types of plates are recommended, depending on the particular conditions of operation and use of these columns. These operating conditions of the column provide a comparative assessment of the different types of plates under these operating conditions as shown in **Table 3**. The comparison is carried out on the following characteristics and indicators [34]:

- efficiency;
- operating stability range;
- hydraulic resistance;
- ability to work on contaminated raw materials;
- cost (or weight) of the column.

As already noted above, there is a multitude of types of plates. Then we will talk about the most frequent plates in oil plants.

2.4.1 The lattice or grid plates

The plates with holes, usually have holes on 3/16 inch to 1 inch. Pairs come out of the holes to get a multi-hole effect. The Steam or vapor velocity prevents liquid from passing through the holes. The number and size of the hole are based on the fall of steam in the column. The fluid flow is transported down the tower using triggers, a dam, and a bypass device on the side of the plate. The Minimum capacity sieve tray capacity is about 70% [30].

2.4.2 The cap plates

The cap plate is a "classic" plate, although it is still very widespread nowadays, it is gradually replaced by plates of other, more effective types. The dominant part of the cover plate is a steel disc (or cover plate) with holes for steam pipes. Usually provided with slots for passing steam, the steam will be in contact with the liquid bubble formed on the next tray. It can operate at low steam and liquid velocities [36] (less than 2 gallons per minute per foot of average flow width). The fittings are welded to the disc. Above the nozzles, caps with a diameter of 60 or 80 mm are installed. Caps have slots with a height of 15; 20 or 30 mm.

Types columns or application	Plates types	Restrictions on column diameter
AVDU units or CDU	C or S-shaped plates	No
Atmospheric distillation of CDU or ADU, with including stripping	C or S-shaped plates	No
	Valve plates	No less 3 m
	Jet plates	Not over 3.2 m
Vacuum distillation of CDU or VDU	Jet plates with bumpers	No
Any installation, especially if there is no reliable design data	Cap plates	No

Table 3.
Types of plates common in refining oil plants.

2.4.3 The valve plates

The valve plates are mainly used in CDU or AVDU and gas plants. The dominant element of the plate is a valve of round or rectangular shape that closes the holes at the base of the plate of appropriate shape. Structurally, the valve is made so that it cannot be raised to a certain height. The arrangement of the valves of round shape on the plane of the plate is the same as the plugs on the cap plate. The valve plate is used that rises with increasing steam velocity and then decreases as the stream velocity decreases, which stops the liquid from flowing. The valve can be round or rectangular, with or without a frame. The disk rises with an increase in the velocity of steam [30, 36].

2.4.4 The jet plates

The jet plate consists of a flat sheet, on which there are offset slots made in the form of a tongue with the cut-out part bent upwards. All slots are bent at a side angle. The plate has a drainage device of the usual design, but does not have a drain bar at the outlet of the liquid from the plate [32].

2.5 Examples of material balances of crude distillation unit and with its major blocks or units like the furnace

2.5.1 Determination of the yield (material balance) of the furnace

The role of the furnace is to heat the oil to a temperature in the range of 330–385°C in order to achieve a vapor-liquid balance before the distillation column [1, 7, 25, 29]. The most popular furnace in the Russian Federation is the tube furnace, as shown in **Figure 11**. The tube furnace is a unit designed to heat and force chemical processes inside the chamber, which are achieved by using the heat generated when fuel is burned. To calculate the heat balance of the atmospheric column, it is necessary to determine the heat of the furnace by the vapor phase and the liquid phase, therefore determine the material balance of the furnace. For example, the furnace material balance we used Bonga crude and the following data [25]:

- the temperature at the entrance of raw materials into the convection chamber t_1 = 250°C;
- the temperature at the outlet of raw materials in the convection chamber t_2 = 380°C;
- the pressure at the exit of the furnace, p = 2 bars;
- the data of the molar and mass composition of desalted crude are presented in **Table 4**.

We use the method of gradual approximation to calculate the percentage of the top product's distillate, with the Eq. (35).

$$\text{Fn (emol)} = \sum_i \frac{Cmol_i}{1 + emol \times (K_i - 1)} \quad (35)$$

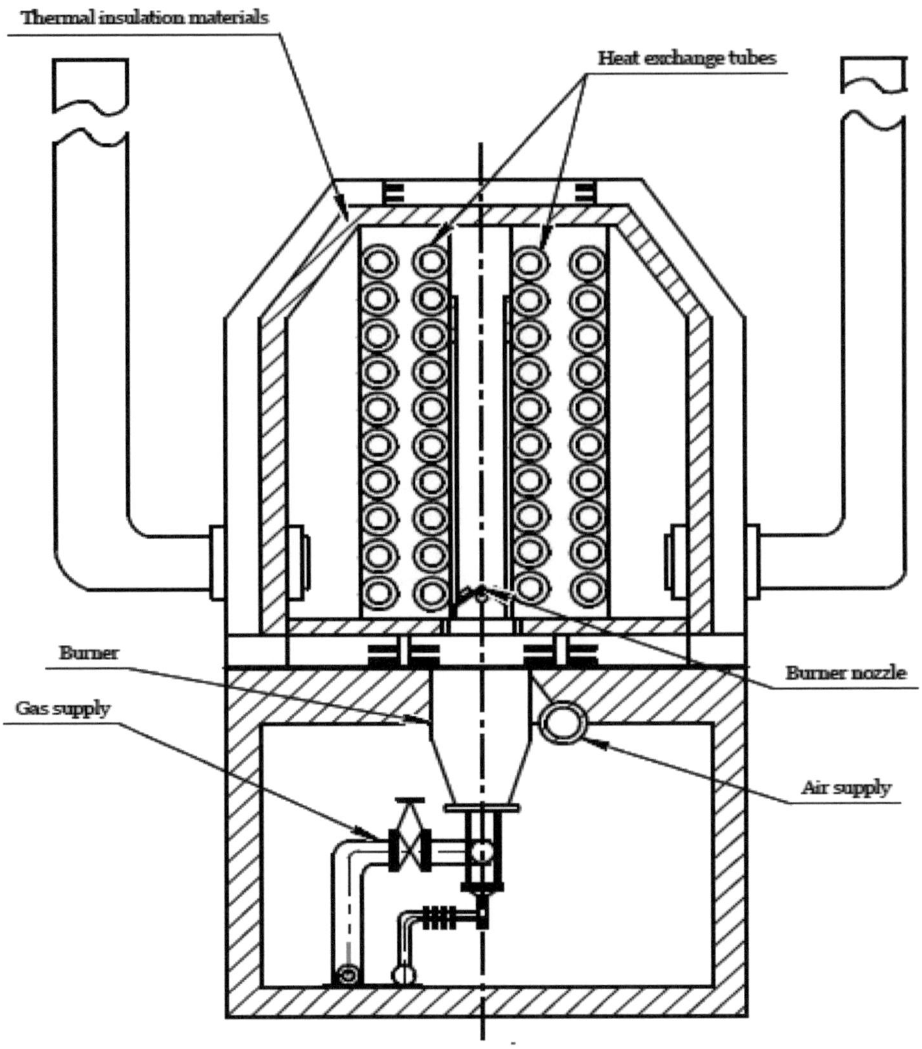

Figure 11.
Cross-sectional with heat exchange chamber of a tube furnace.

Where, Cmol-molar composition of the upper product (distillate); emol- the mole fraction of the distillate; Ki-phase equilibrium constant of components i under conditions. The graph or **Figure 12** shows the proportion of distillation through the formula (33) of Bonga crude – 138158.9 kg/h.

From the graph (**Figure 12**) we determine that the molar fraction of distillation is emol = 0.855. The molar composition of the liquid is also found by the formula (36).

$$Y_i = K_i \times Y_i, \tag{36}$$

The mass fraction of distillation is calculated by the formula (37):

$$\text{emas} = emol \times \frac{MsrP}{MsrC}, \tag{37}$$

where MsrP is average molecular weight of the vapor phase; MsgC is average molecular weight of feedstocks.

Components	kg/kmol	kg/h	% mass	Mole fraction	Steam
C_2H_6	30	22.11	0.02	0.00	1028.85
C_3H_8	33	388.23	0.28	0.01	404.90
Iso-C_4H_{10}	58	563.69	0.41	0.01	135.04
N-C_4H_{10}	58	886.98	0.64	0.02	126.43
Iso-C_5H_{12}	72	124.34	0.09	0.00	85.12
N-C_5H_{12}	72	31.78	0.02	0.00	84.19
75–105	95.1	4863.19	3.52	0.08	103.15
105–125	107.72	3550.68	2.57	0.05	25.25
125–145	118.72	4738.85	3.43	0.06	19.53
145–165	130.52	9035.59	6.54	0.10	15.72
165–185	143.12	2528.31	1.83	0.03	12.51
185–205	156.52	5526.36	4.00	0.05	9.84
205–225	170.72	5567.80	4.03	0.05	7.66
225–245	185.72	7046.10	5.10	0.06	5.88
245–265	201.52	4324.37	3.13	0.03	4.47
265–285	218.12	6769.79	4.90	0.05	3.35
285–305	235.52	7046.10	5.10	0.04	2.47
305–325	253.72	8469.14	6.13	0.05	1.80
325–345	272.72	15017.87	10.87	0.08	1.30
345–375	297.6	21456.08	15.53	0.11	0.92
>375	443.91	30132.46	21.81	0.10	0.58
H_2O	18	69.08	0.05	0.01	0.05

Table 4.
Data of the molar and mass composition of desalted crude.

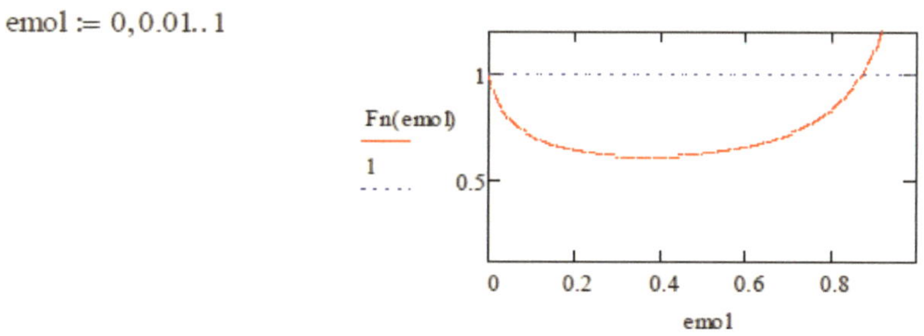

Figure 12.
Graph of the vapor-liquid state isotherm.

Molecular masses are calculated according to the rule of additivity (38):

$$\text{MsrP} = \sum M_i \times Y_i \tag{38}$$

Accordingly, MsrP = 180.54; MsrC = 206.77. emas = 0.747.

The amount of steam and liquid mG phases is determined by the formulas (39), (40):

$$mP = F \times emas, \quad (39)$$

$$mG = F \times (1 - emas), \quad (40)$$

where F (feedstock) is the exiting flow from the electrostatic desalter, and entering the tube furnace.

The mass of the vapor phase = 103,104 kg/h, the mass of liquid phase = 35054.88 kg/h. The mass composition of crude oil, vapor, and liquid phase after furnace are presented in **Table 5**.

2.5.2 Examples of material balance of atmospheric distillation unit (ADU)

The production capacity of the atmospheric distillation unit for processing crude oils is 1,159,200 tons/year, with an initial load of 140 kg/h. This material balance as shown in **Table 6** was made based on the physicochemical characteristics of Bonga crude oil and the typical CDU model without the K-1 oil topping column.

Components	Feedstock		Products			
	Crude oil composition		Vapor phase composition		Liquid phase composition	
	kg/h	% mass	kg/h	% mass	kg/h	% mass
C_2H_6	22.11	0.02	22.10	0.02	0.05	0.00
C_3H_8	388.23	0.28	388.05	0.38	1.47	0.00
Iso-C_4H_{10}	563.69	0.41	562.91	0.55	4.85	0.00
N-C_4H_{10}	886.98	0.64	885.67	0.86	8.16	0.00
Iso-C_5H_{12}	124.34	0.09	124.07	0.12	1.37	0.00
N-C_5H_{12}	31.78	0.02	31.71	0.03	0.35	0.00
75–105	4863.19	3.52	4827.44	4.68	135.74	0.00
105–125	3550.68	2.57	3517.01	3.41	112.85	0.00
125–145	4738.85	3.43	4683.13	4.54	169.45	0.00
145–165	9035.59	6.54	8902.49	8.63	368.15	0.01
165–185	2528.31	1.83	2481.17	2.41	118.91	0.00
185–205	5526.36	4.00	5394.59	5.23	303.92	0.01
205–225	5567.80	4.03	5396.30	5.23	362.68	0.01
225–245	7046.10	5.10	6762.91	6.56	550.50	0.02
245–265	4324.37	3.13	4095.48	3.97	410.07	0.01
265–285	6769.79	4.90	6293.92	6.10	787.63	0.02
285–305	7046.10	5.10	6384.01	6.19	1014.91	0.03
305–325	8469.14	6.13	7401.03	7.18	1519.84	0.04
325–345	15017.87	10.87	12475.46	12.10	3365.64	0.10
345–375	21456.08	15.53	16251.21	15.76	6314.27	0.18
>375	30132.46	21.81	6154.41	5.97	19501.56	0.56
H_2O	69.08	0.05	68.95	0.07	2.51	0.00
Total	138158.9	100	103,104	100	35054.88	100

Table 5.
The material balance of tube furnace.

Feedstocks	ton/year	% mass	Products	ton/year	% mass
1. Oil emulsion	1,159,200	99.56	1. Gases (LPG)	4938.72	0.42
2. Chemical agent (demulsifiers)	3.48	0.00	2. Fractions 35–145	69109.88	5.93
			3. Fractions 145–230	125052.05	10.74
			4. Fractions 230–320	201243.36	17.28
3. Water processing (freshwater)	115.92	0.01	5. Fractions 320–350	231212.84	19.86
			The target products. (sum. fractions)	631556.87	54.25
			6. Associated product (Atmospheric residues)	413475.76	35.51
4. Water vapor	4968	0.43			
			7. Drainage water	119254.770	10.24
Total	1164287.39	100	Total	1164287.39	100

Table 6.
The material balance of atmospheric distillation unit with operating time 340 days by year (365 days −20 days) • 24 h = 8280 h.

2.6 Crude distillation unit (CDU) used in Russia compared to the typical model

Based on the evaluation of the concept of the technological scheme for the construction of a technological scheme of crude distillation unit at refineries in Russia. The topping column K-1 in most cases is a simple design as shown in **Figure 13** [28], although it fractionates the crude oil into the many components like ethane, light gasoline, etc.. There are schemes in which light gasoline is displayed under the distillate in the top of the column, and heavy gasoline on the side flanks of the column. The topping column K-1 collects 50–60% of the potential of light gasoline. The residues of column K-1 are therefore the raw material of the main atmospheric column. The composition of this raw material is weighted in such a way that an excessively high feed temperature is required, permissible temperature (380°C). The singularities of the topping column K-1 operation are as follows [1, 28]:

- The low yield of rectified gasoline (5–15% of the mass of the column load) makes it difficult to collect the gasoline fraction from the oil;
- Extremely high liquid loading in the heating zone of the pre-evaporation column due to the low steam load worsens the conditions for steaming light fractions from the residue under the action of hot jet reflux;

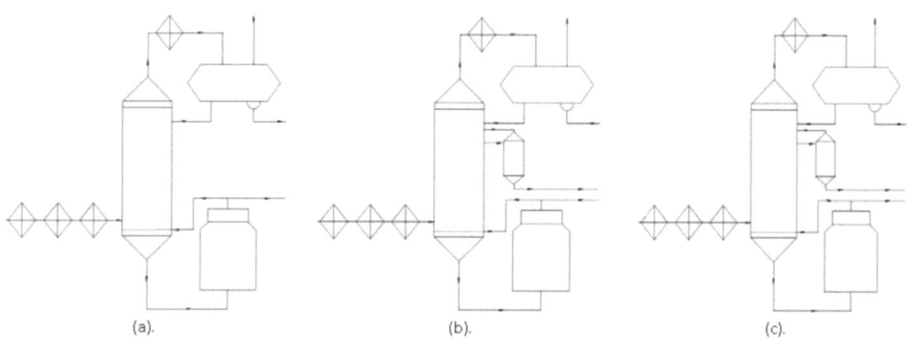

Figure 13.
Possible schemes of operation with oil topping column K-1. (a) Column K-1 with back topped crude. (b) Column K-1 with heavy gasoline and back topped crude. (c) Column K-1 without back topped crude.

- The relatively small enthalpy of crude oil (220°C) introduced into the column does not create sufficient heat flow for distillation, therefore, it is necessary to introduce additional heat in the form of hot jet reflux;

- Maintaining high pressure in the column to ensure air and water condensation of light gasoline fractions in condensers-refrigerators (separators).

The purpose of this column is to extract from the oil the remains of dissolved gas (if the oil has not been stabilized) and a light gasoline fraction with a boiling point of 85°C. After atmospheric residue of K-2 shipped to the vacuum column K-5 for to obtain vacuum distillates (heavy gasoil) and tar. In order to illustrate our argument, a typical P&ID flow diagram process of the atmospheric distillation unit in Russian refineries is presented in **Figure 14**.

2.7 Summary

A general technological approach to the atmospheric distillation unit (ADU) can be described as below [28].

The oil topping column K-1:

- Top temperature, °C – no higher than 245;

- Low temperature, °C – no higher than 150;

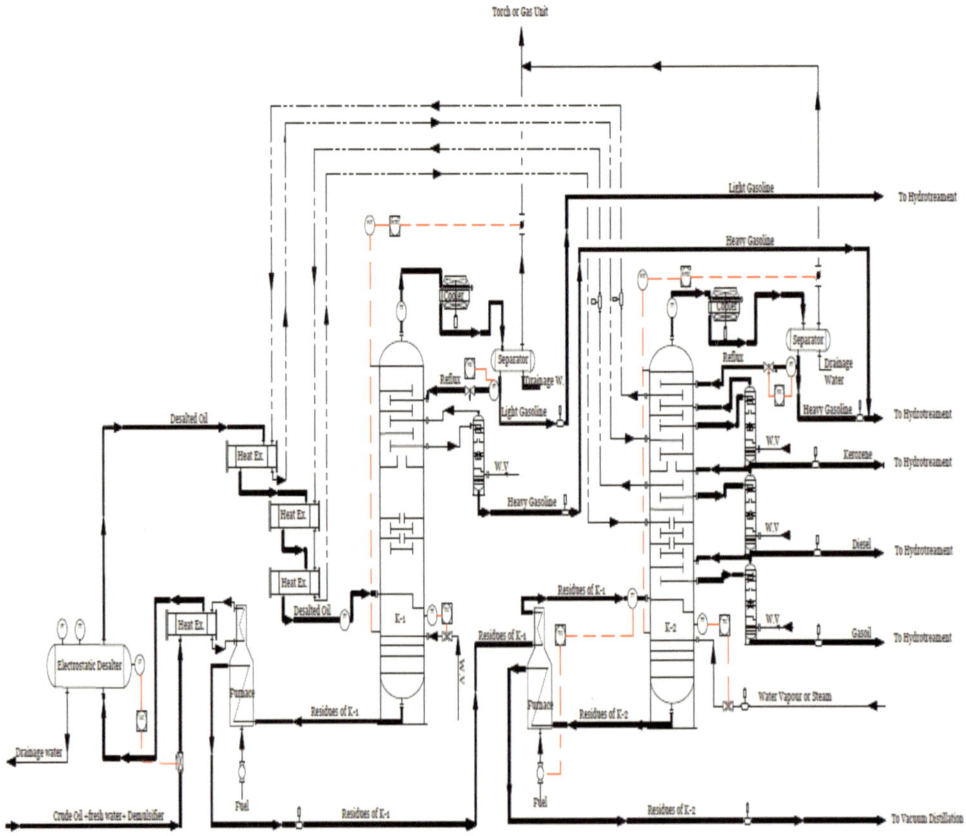

Figure 14.
The typical P&ID flow diagram process of atmospheric distillation unit at refineries in Russian Federation.

- Pressure, bar – no more than 3.0.

The typical products of K-1 are;

- Gases
- Light straight run naphtha (light gasoline or light naphtha)
- Heavy gasoline (or military jet fuel)
- Topped crude

The main atmospheric column K-2:

- Top temperature, °C – no higher than 350;
- Low temperature, °C – no higher than 150;
- Pressure, bar – no more than 1.4.

The typical products of K-2 are:

- Heavy gasoline (or military jet fuel)
- Kerosene (light distillate or jet fuel)
- Middle distillates or light gas oil (LGO) – Diesel
- Heavy distillates – atmospheric gas oil (AGO) or heavy gas oil (HGO)
- Crude column bottoms -atmospheric residue or topped crude

The distillation processes are the processes of oil separation into more or less homogeneous fractions without chemical conversion of its constituent substances. The process of separation of liquid substances by their boiling temperatures is atmospheric and vacuum. The atmospheric distillation can be carried out in the following ways:

- With a single separation – furnaces and separation of distillate in one (main) distillation column k-2;
- Double separation in two distillation columns- in the pre-evaporation column k-1 with separation of light gasoline fractions and the main column-2;
- Gradual separation.

The oil topping column K-1, in atmospheric distillation, collects 50–60% of the potential of light gasoline from oil [28]. Otherwise, the composition of oil for the atmospheric column K2 is so weighted that an excessively high supply temperature is required, permissible temperature (380°C). The oil topping column K-1 is especially important in the general crude distillation unit technology in Russia. We know that the choice of flow process technology is dictated by the physicochemical properties of the oils and their compounds. Indeed, the choice of structure of

Types of CDU	Crude distillation unit with K-1	Crude distillation unit without K-1
Advantages	• Flexibility of technology for adaptation on any crude oil • Increased reliability with more products (range of products more) • Stabilization of oil for K-2 (main column) • Increased durability and viability of the equipment by avoiding the overload of k-2 (main column)	• Simplicity of construction compares to those with k-1 • Known technology is widely used • Reasonable project cost • Less scrap metal
Disadvantages	• The increase in the complexity of the design • The increase in the cost of the project • More scrap metal	• Modeling more stringent for to determine oil • No flexibility of technology for adaptation on any crude oil • Not enough products

Table 7.
Summary of advantages and disadvantages of different types of technology.

atmospheric vacuum distillation units or atmospheric distillation units is determined by the characteristics of crude oil. Finally, according to our studies, the distillation unit with the oil topping column K-1, is more flexible and versatile, which on the whole allows processing any feedstock. **Table 7** shows the conclusions on the different types of atmospheric distillation [28].

3. Conclusion

The crude oil is a complex mixture of paraffin, naphthenic, aromatic, and other hydrocarbons with different molecular masses and boiling temperatures. Also, the oil contains sulfur, oxygen, and nitrogen-containing organic compounds. Therefore, to obtain from oil commodity products for various purposes, apply methods of separation of oil into fractions or groups of hydrocarbons. The modern dewatering-desalting processes are designed with electrostatic desalter only by horizontal geometry design: Atmospheric and atmospheric-vacuum. The horizontal design of electrostatic desalter has a lot of advantages, such as a large area of electrodes and, accordingly, a larger amount of oil per unit section of the device, a lower vertical speed of the moving oil flow, which provides favorable conditions for water sedimentation, as well as implementing processes with higher pressure and temperature. The electrostatics desalters are characterized by:

- More favorable conditions for the deposition of water drops;
- Specific performance, 3 times higher than vertical and ball structures in the presence of low specific gravity and lower cost of the device;
- Simplicity of design, a small amount of electrical equipment, ease of installation, availability of maintenance and maintainability;
- Ability to work with high pressure and temperature parameters.

However, the pretreatment of crude oil is not only concerned with its desalting and dewatering, but also with its degasification and packaging into a semi-product under the standard like ASTM D 1250-97, API Petroleum Measurement and GOST R 51858.

The distillation processes are the processes of oil separation into more or less homogeneous fractions (cups) without chemical conversion of its constituent substances. The process of separation of liquid substances by their boiling temperatures is atmospheric and vacuum. The oil distillation via single and multiple evaporations. In industrial conditions, the processes are carried out on the equipment of continuous action. During the single-evaporation distillation, the oil is heated to a certain temperature and all fractions that have passed into the vapor phase are collected. The distillation of the oil by multiple evaporations consists in that the oil is first heated to a temperature allowing to separate the light gasoline fraction. Then the crude without the fraction of light gasoline is heated to a higher temperature, and fractions that boil at about 375°C (that is, fractions of heavy gasoline, jet fuel, and diesel fuel) are also collected. In the residue from distillation, atmospheric residues are obtained. The atmospheric residues are distilled under vacuum, from which the lubricating oil fractions are obtained. Finally, the tar is obtained as residues from vacuum distillation. In other words, oil is consistently heated three times, each time separating the vapor phase from the liquids. The resulting steam and liquid phases are subjected to rectification in columns. Thus, industrial processes of oil distillation are based on a combination of distillation with single and multiple evaporations and subsequent rectification of the steam and liquid phases. We want to emphasize that this is the case, we use the K-1 oil topping column. However, if the process is executed without column k1, the same process is shorter and with the results already mentioned above. In the distillation column, there are distillation plates on which the vapors rising along the column are in contact with the flowing liquid (reflux). Reflux is created due to the fact that part of the upper product returns to the liquid state to the upper plate and flows down, enriching the rising vapors with low-boiling components. One of the ways to increase the concentration of high-boiling components in the residue from oil distillation is to introduce an evaporator into the lower part of the distillation column. As such, you can use water vapor, inert gas (nitrogen, carbon dioxide, petroleum gas), gasoline vapor or kerosene. The water vapor is most widely used as an evaporator for oil refining. Its presence in the distillation column reduces the partial pressure of hydrocarbons, and therefore their boiling point. As a result, the lowest-boiling hydrocarbons in the liquid phase, after single evaporation, pass into a vaporous state and, together with water vapor, rise up the column. In many literatures, it is recommended to use super-heated water vapor and enter it into the column with a temperature equal to the temperature of the feedstock or slightly higher. Usually, the water vapor used after steam pumps and turbines at a pressure of 2–3 bar is superheated in a tube furnace and introduced into a column with a temperature of 350–450°C. The use of wet vapor steam is not practiced, since its temperature and pressure are interrelated, for example, when introducing a column of 350°C, its pressure is equal to 170 bar. Besides, with increasing pressure, the cost of saturated water vapor increases sharply, so it is not economical to use it. If the wet vapor steam enters a column of low-pressure saturated water vapor, for example, 10 bar and, accordingly, with a temperature of 180°C, then part of the heat will go to heat it.

Abbreviation

GOST	Russian Interstate-standard
ASTM	American Society for Testing and Materials
NF EN ISO	French Norms European Norms International Organization for Standardization
ρ^{90}	density at 90°C

$\mu_{oil\ crude}$	kinematic viscosity of crude oil
AVDU	atmospheric-vacuum distillation unit
ADU	atmospheric distillations unit
CDU	crude distillation unit
Refinery	petroleum refinery plant
VDU	vacuum distillation unit
K1	oil topping atmospheric column
K2	main atmospheric column
TBP	true boiling point
t/h	ton/hour
kg/h	kilogram/hour
WV	water Vapor
P&ID	piping and instrumentation diagram

Author details

Serge-Bertrand Adiko[1*] and Rifat Radisovich Mingasov[2]

1 National University of Oil and Gas, Gubkin University, Moscow, Russian Federation

2 Kazan National Research Technological University, Kazan, Russian Federation

*Address all correspondence to: sergefriends2002@yahoo.fr

References

[1] Akhmetov SA. Technology of deep processing of oil and gas. In: Textbook for High Schools. Ufa: Guillem; 2002. p. 672

[2] Eden B, Laycock PJ, Fielder M. Oilfield Reservoir Souring. HSE; 1993. p. 95

[3] Sharifullin AV, Terenteva NA. Analysis of the Quality of Oil and Petroleum Products and Metrological Evaluation of Measuring Instruments. Laboratory Workshop. Kazan: KSTU; 2010. p. 141

[4] A. D 1250. Standard guide for use of the petroleum measurement tables. ASTM International West Conshohocken, PA; 2017

[5] R. GOST, 51858-2002. Crude petroleum. Gen. Specif. (Izd. Stand. Moscow, 2006); 2002

[6] Kang W, Xu B, Wang Y, Li Y, Shan X, An F, et al. Stability mechanism of W/O crude oil emulsion stabilized by polymer and surfactant. Colloids and Surfaces A: Physicochemical and Engineering Aspects. 2011;**384**:555–560

[7] Grechuhina AA, Leginski AA. Oil Treatment Plants. Kazan: Kazan State Technical University; 2011. p. 83

[8] Kasparyants KS. Field preparation of oil and gas. Moscow, Nedra; 1973. p. 376

[9] Treese JSA, Pujadó PR, Jones DSJ, Handbook of Petroleum Processing. Springer; 2015. p. 1910

[10] Lipsky VK, Demidova ME. Pipeline Transport of Oil and Petroleum Products. Textbook for High Schools UMK. PSU; 2007. p. 312

[11] Chen TY, Mohammed RA, Bailey AI, Luckham PF, Taylor SE. Dewatering of crude oil emulsions 4. Emulsion resolution by the application of an electric field. Colloids and Surfaces A: Physicochemical and Engineering Aspects. 1994;**83**:273–284

[12] Baykov NM, Pozdnyshev GH, Mansurov RI. Collection and Field Preparation of Oil, Gas and Water. Moscow: Nedra; 1981. p. 261

[13] Abdurahman HN, Yunus RM, Jemaat Z. Chemical demulsification of water-in-crude oil emulsions. Journal of Applied Sciences. 2007;**7**:196–201

[14] Ahmadi S, Khutoryansky FM, Soltani B. Modeling of an electric desalting plant for the preparation of heavy oils. In: Oil and Gas Technology (journal). 2017. pp. 3–9

[15] Naeger DP, Perugini JJ. Crude Oil Desalting Process; 1991. US Patent N° 4,992,210

[16] Naeger DP, Perugini JJ. Crude Oil Desalting Process; 1992. US Patent N° 5114566; A

[17] Ishmurzin AA, Khramov RA. Processes and Equipment of the Drainage System and Preparation of Oil, Gas and Water System. Ufa, UGNTU; 2003. p. 145

[18] Kim SF, Usheva NV, Samborskaya MA, Moises OE, Kuzmenko EA. Modeling of water-oil emulsions destruction processes for large-capacity oil preparation technologies. In: Fundamental Research. Academy of natural Science; 2013. pp. 626–629

[19] Usheva NV, Moises OE, Kim SF, Gizatullina SN. Influence of technological parameters on the processes of oil dewatering and desalination, Izvestiya Vuzov. Chemistry and Chemical Technology. 2014;**57**(11):101–103

[20] Fahim MA, Al-Sahhaf TA, Elkilani A. Fundamentals of Petroleum Refining. Elsevier; 2009. p. 513

[21] Speight JG. The Chemistry and Technology of Petroleum. 5th Edition. CRC press; 2014. p. 953

[22] Galvin P. Design principles for electrical coalescers. Proc. Extr. 84, EFCE. 2013;**43**:101–113

[23] Bruninghaus L. Électrisation et conduction électrique des hydrocarbures liquides. Journal de Physique et le Radium. 1930;**1**:11–36

[24] Noik C, Chen J, Dalmazzone CSH. Electrostatic demulsification on crude oil: A state-of-the-art review. In: Int. Oil Gas Conf. Exhib. China: Society of Petroleum Engineers; 2006

[25] Tanatarov MA, Akhmetshina MN, Faskhutdinov RA, Voloshin ND, Zolotarev PA. Technological calculations of oil refineries. Moscow: Chemistry publishing house; 1987. p. 352

[26] Mousavi SH, Ghadiri M, Buckley M. Electro-coalescence of water drops in oils under pulsatile electric fields. Chemical Engineering Science. 2014; **120**:130–142

[27] E. Intelligence, The International Crude Oil Market Handbook, Research Report; 2005

[28] Adiko S-B, Akhmadiyarov AA, Varfolomeev MA. Crude distillation unit and the technology adopted in Russia. International Multidisciplinary Scientific GeoConferences. 2019;**19**: 751–758

[29] Sardanashvili AG, Lvova AI. Examples and Tasks of Oil and Gas Processing Technology. Moscow: Chemistry; 1980. p. 256

[30] Laptev AG, Mineev NG, Malkovsky PA. Modelling and modernization of separation devices in oil and gas processing. Kazan: Printing house; 2002. p. 250

[31] Ivanova YV, Kuzmina RI, Kozhemyakin IV. I18 Chemistry of oil: Textbook-method. Manual for students of chemical science, training in spec. 250400. Chemical technology of natural energy carriers and carbon materials. CHI-Saratov: Sarat Publishing house. UN-TA; 2010. p. 56. ISBN: 5-292-03605-06

[32] Glagoleva OF, Kapustina VM. The Technology of Oil Refining. Part 1. Primary Oil Refining. Moscow: Chemistry; 2006. p. 399

[33] Gary JH, Handwerk GE, Kaiser MJ. Petroleum Refining: Technology and Economics. 5th ed. Boca Raton: CRC Press; 2007. p. 488

[34] Bagirov IT. Modern Installations of Primary Oil Refining. Moscow: Chemistry; 1974. p. 240

[35] Hsu CS, Robinson PR. Springer Handbook of Petroleum Technology. Springer; 2017. p. 1182

[36] Laptev AG, Farakhov MI, Mineev NG. Fundamentals of Calculation and Modernization of Heat and Mass Exchange Installations in Petrochemicals. Kazan: Kazan State Energy University; 2010. p. 574

Chapter 7

CO_2-Philic Surfactants Structural Morphology Prerequests for CO_2 Philicity for Foam Durability for EOR Applications

Muhammad Sagir, Muhammad Bilal Tahir, Muhammad Pervaiz, Muhammad Hassan Qasim, Sami Ullah and Reema Ansar

Abstract

In oil fields CO_2-EOR have extravagant interest because of its increasing microscopic sweep efficiency. As a balance and dense solution over a huge range of temperature and pressure, carbon dioxide can improve viscosity reduction and oil swelling because of all partial miscibility with heavy oils and proportion miscibility with light oils, and also carbon dioxide has mutual solubility with water and hydrocarbons that these properties result in increasing microscopic sweep efficiency in mechanism of CO_2 injection. However, due to CO_2 lower viscosity than water and oil, CO_2-EOR efficiency is limited that causes near well-bore conformance issue and mobility contrast problem such as gravity override and fingering. The carbon dioxide philic surfactant blends traditionally includes foam booster, foam stabiliser and foaming surfactant. An integrated property-performance analysis of blends of anionic surfactants includes carbon dioxide philic groups such as twin-tailed carbonyl group, propylene-oxide and methyl group of CO_2-philic groups under the circumstances of reservoir showed that carbon dioxide philic surfactants can be the mixture of CO_2 problems. Here we will underline that modifying the surfactant tail can be a proper path to surpass foaming performance.

Keywords: EOR, foam, CO_2-philic surfactants

1. Introduction of foam

Foam is the gas dispersion in liquid. The liquid phase is called exterior stage and the stage of gas is spread in interior stage. We can see the foam formation in **Figure 1**. We called thin fluid film lamella. In Foam we can separate a lamella into internal gas bubbles. The Point where the lamella contact with each other is called 'Plateau border. It is shown in **Figure 1**. There is a density difference between the medium and gas bubbles due to which the system rapidly split in to two layers and the gas bubbles moved to the top [1]. Pure liquid could not foam. They only foam when they have a surface-active material [1, 2]. We find that whenever a bubble of gas is injected under the surface of liquid, its outburst instantly. If we use diluted surfactant solution then a restoring force is created that effort to launch the equilibrium,

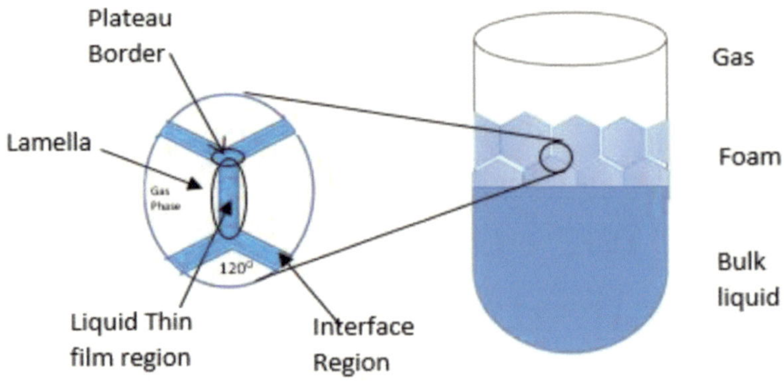

Figure 1.
A schematic of three bubbles meeting at a plateau border.

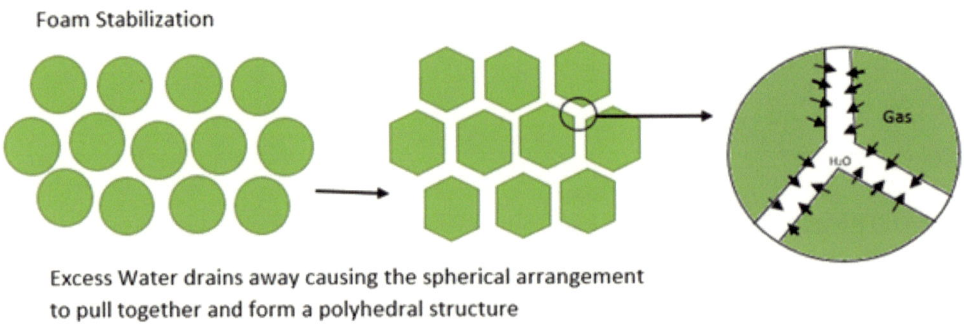

Figure 2.
Foam stabilisation and formation of polyhedral structure foam.

as the air/liquid interface enlarge the surface equilibrium changes, this leads to form polyhedral structure as shown in **Figure 2**.

2. Types of foam

According to the structure, a foam may be distinguished in two main types [2, 3]:

2.1 Spherical foam

We can say it as a provisional dilute dispersion of bubbles in the liquid. It involves of gas bubbles separated by thick films of viscous liquid formed in recently prepared systems.

2.2 Polyhedral gas cells

In polyhedral gas cells thin flat "walls" are produced on ageing with connection points at plateau borders [2].

Mainly, the spume can be categorised as:

2.2.1 Bulk foam

The bulk foam is an "agglomeration of gas bubbles parted from One another by thin liquid films. In bulk foam the total volume of gas (discontinues phase) is relatively much larger than the thin liquid films (continues phase). It is regarded as a single stage of homogenous. The gas content in bulk foams is high which 60–97% is usually. In bulk form foams are produced when gas links a liquid in the presence of powered agitation.

2.2.2 Foam in porous media

In porous media foam occurs as a distinct micro gas bubbles which is contact along the wetting fluid of aperture walls. Thin liquid films (lamella) make the bridge between these micro gas bubbles in order to separate them. In permeable media, the behaviour of spume and bubble size reliant on pore throat and pore size distribution. Mostly foam spread as bubble train in matrix of reservoir rock. In many cases, specific foam bubbles in reservoir matrix rock may be numerous pore bodies in length. Foam bubbles are mostly larger than pore bodies in porous media. As Foam present in the form of bubble trains in reservoir-rock porous media where the Plateau border of the foam lamellae is made at the pore wall and has, for stationary non flowing foam in the pore body, an angle of about 90° in the middle of the liquid lamellae and the pore wall [4–6].

Foam may also have classified as:

- Transient foam: All foams are thermodynamically unstable reason is the high interfacial free energy. Affording to breakdown kinetics of foam films, the foam films may be of two types: The foams having very short life time only in seconds are called unstable foams. They are mostly produced using "mild" surfactants, e.g., pine oil, phenol. Short-chain alcohols, aniline [7, 8].

- Stable foam: Second type of foam is permanent type with lifetime of hours, called metastable foams. These foams are generated by using good foaming agents (surfactants) proteins solid particles or long-chain fatty acids.

3. Condition to form foam

To create a foam three condition are required.1st Condition is the mechanical work so that the surface area is increased. This can be done by, dispersing a high volume of gas in the liquid, agitation, or inserting a gas in the liquid [9, 10].

The second condition is surfactants that must be available so that surface tension decreases. And the last is foam must be produced more rapidly than it break.

4. Stability of foam

Lamella's stability is the stability of foam. It depends on different factors which are rate of capillary drainage, mechanical deformations, surfactant concentration, like gas diffusion, aqueous phase salinity, anti-foaming effect of oil, disjoining pressure [1, 4, 11–13].

4.1 Drainage of film

In film drainage process the lamella undergoes a thin method. Which goes toward the super thin or rupture of liquid films.it is also called structure of black spot. Gravity drainage and capillary suction can be done by two means. The drainage of gravity is usually occur in lamella that is thick .in this process due to the gravity the film is moved downward .we can slow this process by decreasing the liquid in the foam and increasing the bulk viscosity. The two indicators are the lamella's time to reach the crucial broadness and the limited broadness beneath which the lamella coalesces. The signals reliant on different factors like viscosity, surfactant solubility, elasticity of surface, adsorption on the plane and ratio of gas to liquid. Usually as the attentiveness of the surfactant extended, the critical thickness decreases [14, 15].

4.2 Gas diffusion

With the side pressure of lamella could not be equal so the gas is dissolved in lamella the by diffusion it escape .it is common when the porous media trapped the bubbles. When the moving lamella is reshaped continuously then the inter bubble diffusion is complicated. In high flow rates the gas diffusion may be negligible. The main foam film drainage tool is capillary drainage. The reason of the occurrence is the capillary tube (at plateau border suction). The curvature at centre of film is comparatively higher than the radius of the curvature at the plateau border. The reason is the fact at the middle of foam lamella the film border is nearly equal. While, at plateau boundary, the curvature is further curved [16].

4.3 Oil effect

The oil-foam contacts are significant as the oil occurrence has incompatible belongings on stability of foam. However, the damaging consequence on spume strength by the oil is prevalent. Oil interaction with surfactant creates some problems like it causes in lamellae liquid depletion which brings alteration in wet ability and this scattering of oil on lamella become reasons for destabilisation of the interface. Surfactant and oil solution makes the emulsion and they break the structure of foam. Three coefficients typically used to describe the mechanisms of oil destabilising foam. The spreading coefficient, S: the entering coefficient E, and the bridging coefficient B. Coefficients are described as: [17–21].

Spreading (S), entering (E), and bridging (B) coefficients are described to assess the possibility of oil droplet to enter the gas-water surface. Eqs. 1–3 explaining the foam destabilisation.

$$E = \sigma_{gw} + \sigma_{ow} + \sigma_{og} \tag{1}$$

$$S = \sigma_{gw} - \sigma_{ow} - \sigma_{og} \tag{2}$$

$$B = \sigma^2{}_{gw} + \sigma^2{}_{ow} + \sigma^2{}_{og} \tag{3}$$

where, σ_{gw}, σ_{ow}, and σ_{og} stand for gas/water, oil/water, and gas/oil surface.

If the E, entering coefficient is in positive, a drop of oil is predicted to be drawn up in the lamellar area among two bubbles. This will breaching the air/water interface causes the film to drop the foam become stable ability and skinny to rupture point. When bridging coefficient, B, is positive, the oil droplet bridges the lamellar area among the two neighbouring bubbles. When the spreading coefficient, S, is positive, the droplet of oil in the lamella area is predicted to extent like a lens above a foam. The extent of a droplet of oil above a foam lamella reasons the lamella's foam to break (**Figure 3**).

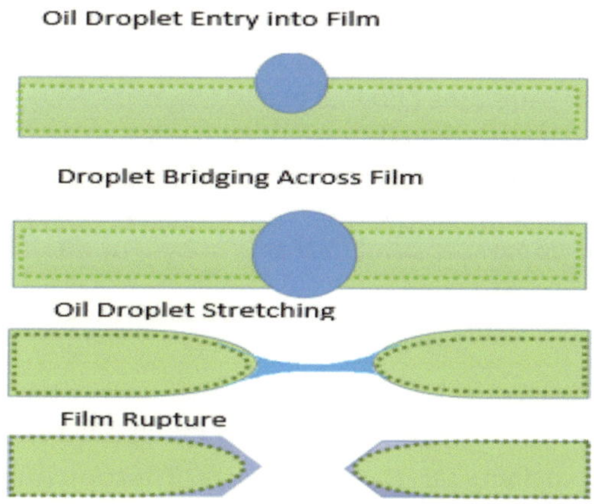

Figure 3.
Oil destabilising foam mechanism.

4.4 Effect of temperature

The stability of foam depends on temperature. A high temperature results in decreased foam drainage time and therefore causing the foam stability decrease with the temperature increase.

5. Role of surfactant in foam generation and stability

The generation is not the serious challenge alone; the important ones are foam quality, form and its stableness mainly when it is in proximity with oil. Surfactants is used for foam spread have low endurance with salinity and results in extreme adsorption on carbonate rocks. The surfactants are capable of playing central roles in enhancement of oil recovery, not only in foam generation but also in IFT reduction. The modern type of surfactants CO_2-phillic surfactants are used for CO_2 control application movability and for the stability of foam in the creation. As the traditional surfactants, these have two surfactants that have well defined areas, tail and head; nonetheless, surfactants tail has a capability for stabilising the CO_2 gas. The reference of foam stability, surfactants that are non-iconic are minor but their stability at high temperatures is a problem. A foam usually absorbs on the rock matrix, deteriorates over time, and has a higher deterioration at high temperatures in the existence of oil. When the Carbon dioxide gas is used the problems become more severe. This phenomena was particularly created to produce fresh surfactants with an affinity for CO_2 gas under controlled conditions and to defeat the problems that arise from traditional surfactants. The surfactants novel can produce much balance spume at a higher temperature and in the existence of oil with less adaption problems.

6. Problems with conventional surfactants in foam stability

The creation of spume, alone, is not a serious problem; the highlighted ones are quality of spume, formation of spume and its balance is particularly when it is in contact with oil. The surfactants used for spume creation have low endurance with salinity and result in extreme adaption on rocks of carbonate. With reference to stability of foam, the surfactants of non-ionic are lesser but the problem is their stability at high temperatures. Usually, a foam that absorbs on the rock matrix becomes worse over time, and has a higher decay at increased temperatures if present with oil. These problems become worse when the CO_2 gas is present. This phenomena was particularly created to produce fresh surfactants with an affinity for CO_2 gas under controlled conditions and to defeat the problems linked with surfactants that is conventional. The surfactants of novel can produce much balance foam at an increased temperature and in the existence of oil with lesser adaption problems.

7. CO_2-philic surfactants

Like conventional surfactants, the surfactants of carbon dioxide phallic are pure amphophilic compounds but rather than lipophilic and hydrophilic parts, they are made up of carbon dioxide-phobic fragments and carbon dioxide philic. Usually the ends of these surfactants are attracted toward CO_2, and are known as the surfactants of CO_2-philic segments, and CO_2-phobic parts are head group of the surfactant [22]. Phobic segments of carbon dioxide are usually selected as a traditional group of hydrophilic, formerly we have identified the segment of tail of the carbon dioxide-philic surfactant. The distinctive form of a carbon dioxide philic surfactant is illustrated in **Figure 4**.

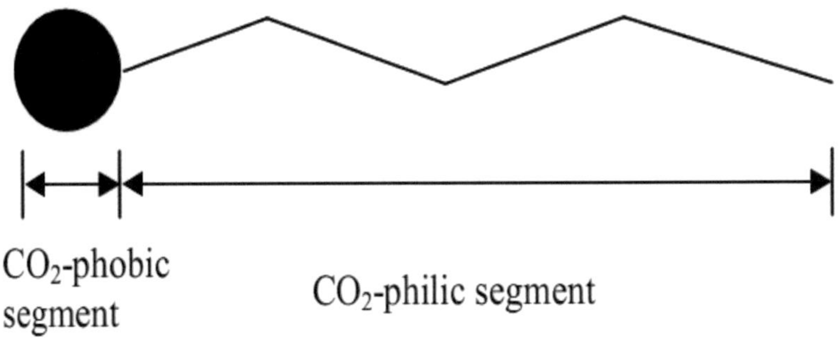

Figure 4.
Structure of a CO_2-philic surfactant.

8. CO_2-philic surfactants for foam

The leading group of researchers have informed that non-fluorinated, hydrocarbon-based systems can be created in a way that they are CO_2-philic in nature. These surfactants resolve the above mentioned foam problems [23–26]. Since CO_2 is a weak solvent, the polar and high molecular weight substances are only partially soluble, but CO_2 can dissolve in other few volatile and low molecular weight solvents. The carbon dioxide is a Lewis acid because it has an accepting electron nature, in spite of the fact that it has low polarisable properties [27]. The CO_2 can take part in Lewis acid–base interactions because of CO_2 having an electron accepting property. Many researchers have proved this kind of carbon dioxide bonding with other stuff like polymers and surfactants, etc. Fin and Lei Hong have expressed this kind of collaboration in their phenomena.

Fin also stated in his work that ab initio molecular simulation research have shown that the O_2 (ester or ether) in the side chain play an major role in promoting philicity of carbon dioxide because of the carbonyl oxygen. They recognised the three different ways the CO_2 linked with a hydrocarbon end of the molecule.

They need the Isopropyl acetate molecule as a sample. In the picture below, the red colour imitate oxygen, black represents carbon and white is used for atoms of hydrogen [17, 28–32].

9. Requirements for CO_2-philic surfactants for CO_2-philicity

Maximum attraction with the CO_2 and minimum intermolecular attractions are the basic requirements involved in designing a CO_2-philic molecule. Few of the Carbon dioxide philic appearance are splitting, less molecular weight hydrophobes, tip tail, and presence of groups of carbonyl, methyl, propylene oxide (PO), tert-butyl tip and a minimum number of methylene groups [22, 27, 33, 34]. A detailed discussion of the important factors that favour the carbon dioxide philicity of a surfactant will be followed in the next few sections.

9.1 Branches

The surfactants branching *is* a key factor for the carbon dioxide philicity of the hydrophobic part. It is because of the effect that when chain length decreases, the CMC (aqueous) increases; while, an increase in branching increases the solubility in CO_2. According to Ben Tan branching in the diacid as well as diol moiety has the increasing effect on solubility, and acyl chains branching increases the solubility up to 20 times.

9.2 Number of tails

It is observed that in case of CO_2-philic compounds the solubility is greatly affected by the tail number. With the increase in the number of tails the dissolved surfactants in the carbon dioxide increases. A huge and emergent part of literature is fixated on the phenomena of interfacial of the carbon dioxide/H_2O interface and proposes that there should be more contact with the interface for a double tail surfactant and as a result offers more stability for the micro emulsion. When a third chain is added in the surfactant structure the surfactant's solubility is increased in carbon dioxide even more [35–38].

9.3 Tail length and tip

In the past the relationship of CO_2-philicity with the tail length has been widely studied and the interdependency of CO_2-philic properties and surfactant tail structure was observed. At various temperature and pressure values, different double tail fluorinated surfactants have been studied for the phase behaviour [6, 7, 39]. This study leads to specifications for the optimization of tail length which is suitable for the maximum water/carbon dioxide emulsion formation at micro level. The phase behaviour for the oligomers is altered by the end-group modification of the oligomer PVAc-OH. Audrey DuPont examined P and T phase stability, chain structure effect and the aggregation structure. Surfactant free volume and surfactant packing are the parameters to view the effects of the chain lengths. Carbon dioxide solubility in esters is significantly affected by small structural changes in them. Depending upon the number of carbon atoms we can observe even/odd effects on solubility of carbon dioxide. According to observations made by Bray Christopher the acyl chain length is important with the carbon dioxide solubility for the molecule. With the increase in length of the chain by 10 carbon atoms the solubility increases in a systematic way. The influence of minor structural changes to the solubility of CO_2

molecule proposes that a qualitative as well as quantitative study of property–structure relationship is possible, that leads to the ability to predict properties associated with carbon dioxide solubility of molecules [8, 16].

9.4 Methyl groups, PO groups and methylene groups

CO_2-philicity is also favoured by the increase in methyl group number. Other factors that are considered in surfactant development are a smaller chain length, low molecular weight, lower no. of methylene and the propylene oxide groups.

9.4.1 Carbonyl groups

Beckman and Styranec have formulated CO_2-philic compounds by the use of only oxygen, hydrogen and carbon comprising precursors. They observed the polyether solubility was significantly influenced by the side chain or by adding the carbonyl group in its backbone. Addition of acetate group in the side chain gives an increase in solubility to a certain limit after that limit the solubility tends to decrease. According to the studies of Fink et al., the solubility of CO_2-philic compounds in carbon dioxide was significantly influenced by the addition of a good number of ester-functional side chains. The effect of numerous fluorine and vinyl-acetate groups in the side chain was studied by Bilal Baradie. He observed the polyether solubility was strongly changed with the side chain or with the addition of the carbonyl group [9, 40].

9.4.2 Molecular weight

A lot of research studies have claimed that carbon dioxide solubility greatly depends on the MW (weight of molecular) of the compound, as PVAc. At low pressures oligomer PVAc-OH (Mw $<$ 3000 g/mol) is found to be dissolved in carbon dioxide but as the molecular weight increases it decreases in solubility. It was determined by Tan et al. Polymer's solubility such as PEC and PEE is significantly dependent upon their molecular weight as well as chemical structure of those polymers. A series of trials was conducted by Matthew B. Miller* by the mixing of different solvents with various extents of carbon dioxide to measure the compatibility for both mixture components having bubble point as basis. It is observed that species with low MM (molar mass) having minimum one atom of O_2 in ether, acetate groups/or carbonyl would have most favourable interaction with carbon dioxide through Lewis base/Lewis acid interactions [17].

10. CO_2-philic surfactants as potential CO_2 gas mobility control agent

Although carbon dioxide has many advantages as being not-toxic, inexpensive, and gas that is not flammable overall process wipe-out efficiency is limited by the minimum viscosity, low density and the increase movement of the carbon dioxide. Most significantly, the poor process sweep efficiency that is caused due to the low density resulting in high velocity of carbon dioxide gas results in sticky fingering, an soon breakthrough of the inserted carbon dioxide gas and override of gravity. This undesired ratio of movement brings the process to a week wipe-out efficiency and subsequently oil recovery is less anticipated.

To reach the high recovery of oil, concerns regarding carbon dioxide movement have to be solved. By the use of foam the carbon dioxide movement can be efficiently controlled [7–10]. The velocity of carbon dioxide can be reduced by foam and it brings down the breakthrough of the inserted gas along with the reduction in the production of gas cap gas. With many advantages of foam it also has a few disadvantages, as instability beneath reservoir situation, For example, maximum salinity, maximum temperature, and particularly in the existence of oil. Incorporation of CO_2-philic surfactants promotes the foam stability. Surfactants are very distinctive molecules having a H_2O soluble part (head) as well as an oil soluble part (tail). From the generation of foam to IFT reduction, the surfactants tends to play vital roles in increasing oil recovery. For stable foam generation in carbon dioxide mobility control applications the CO_2-philic surfactants are employed. Like the other surfactants, these surfactants also have two different parts, tail and head; therefore, the tail for these surfactants has an empathy for carbon dioxide gas to balance the spume [41–45].

Author details

Muhammad Sagir[1]*, Muhammad Bilal Tahir[2], Muhammad Pervaiz[3], Muhammad Hassan Qasim[1], Sami Ullah[4] and Reema Ansar[1]

1 Department of Chemical Engineering, University of Gujrat, Pakistan

2 Department of Physics, University of Gujrat, Pakistan

3 Department of Chemistry, Goverment College University, Lahore, Pakistan

4 Department of Chemistry, King Khalid University, Abha, KSA

*Address all correspondence to: sagir.utp@gmail.com

References

[1] Schramm LL. Emulsions, Foams, and Suspensions: Fundamentals and Applications. Germany: Wiley-VCH, Weinheim; 2005

[2] Prud'homme RK. Foams: Theory: Measurements: Applications. Vol. 57. CRC Press; 1995

[3] Stevenson P. Foam Engineering: Fundamentals and Applications. Wiley; 2012. pp. 544

[4] Kovscek CJRAR. Foams: Fundamentals and Applications in the Petroleum Industry. Vol. 242. American Chemical Society; 1994

[5] Muhammad M, Tan IM, Lukman I, Muhammad N, Muhammad S, Rizwan A, et al. Influence of PZC (point of zero charge) on the static adsorption of anionic surfactants on a Malaysian sandstone. Journal of Dispersion Science and Technology. 2014;**35**(3):343-349

[6] Muhammad S, Tan IM, Muhammad M, Lukman I, Muhammad N, Rizwan AM. Synthesis of a new CO_2 philic surfactant for enhanced oil recovery applications. Journal of Dispersion Science and Technology. 2014;**35**(5):647-654

[7] Muhammad S, Tan IM, Muhammad M, Lukman I, Muhammad N, Rizwan AM, et al. Novel surfactant for the reduction of CO_2/brine interfacial tension. Journal of Dispersion Science and Technology. 2014;**35**(3):463-470

[8] Muhammad S, Tan IM, Muhammad M, Muhammad P, Suleman TM, Khurram S. CO_2 mobility control using CO_2-philic surfactant for enhanced oil recovery. Journal of Petroleum Exploration and Production Technology. 2016;**6**(3):401-407

[9] Hosna TS, Tan IM, Muhammad S, Mushtaq M. Static and dynamic foam/oil interactions: Potential of CO_2-philic surfactants as mobility control agents. Journal of Petroleum Science and Engineering. 2015;**135**:118-126

[10] Khurram S, Lidija Č, Muhammad S, Nadeem A, Imtiaz RM, Ruqia N, et al. An ecological feasibility study for developing sustainable street lighting system. Journal of Cleaner Production. 2018;**175**:683-695

[11] Belhaij A, Al-Mahdy A, Osama. Foamability and Foam Stability of Several Surfactants Solutions: The Role of Screening and Flooding. 2014

[12] Schramm LL. Surfactants and their applications. Annual Reportson the Progress of Chemistry, Section C. 2003;**99**:3-48

[13] Muhammad M, Tan IM, Ismail L, Lee SYC, Nadeem M, Sagir M. Oleate ester-derived nonionic surfactants: Synthesis and cloud point behavior studies. Journal of Dispersion Science and Technology. 2014;**35**(3):322-328

[14] Rizwan AM, Tan IM, Lukman I, Muhammad M, Muhammad N, Muhammad S. Kinetics and equilibria of synthesized anionic surfactant onto berea sandstone. Journal of Dispersion Science and Technology. 2014;**35**(2):223-230

[15] Mushtaq M, Tan IM, Nadeem M, Devi C, Lee SYC, Sagir M. A convenient route for the alkoxylation of biodiesel and its influence on cold flow properties. International Journal of Green Energy. 2014;**11**(3):267-279

[16] Muhammad S, Tan IM, Mushta M, Muhammad N. CO_2 mobility and CO_2/brine interfacial tension reduction by using a new surfactant for

EOR applications. Journal of Dispersion Science and Technology. 2014;**35**(11):1512-1519

[17] Talebian SH, Sagir M, Mumtaz M. An integrated property–performance analysis for CO_2-philic foam-assisted CO_2-enhanced oil recovery. Energy & Fuels. 2018;**32**(7):7773-7785

[18] Talebian H, Tan IM, Sagir M. Static and dynamic foam/oil interactions: Potential of CO_2-philic surfactants as mobility control agents. Journal of Petrol Science and Engineering. 2015

[19] Gold S, Eastoe J, Grilli R, Steytler DC. Branched trichain sulfosuccinates as novel water in CO_2 dispersants. Colloid & Polymer Science. 2006;**284**:1333-1337

[20] Chen K, Grant N, Liang L, Zhang H, Tan B. Synthesis of CO_2-philic xanthate-oligo(vinyl acetate)-based hydrocarbon surfactants by RAFT polymerization and their applications on preparation of emulsion-templated materials. Macromolecules. 2010;**43**:9355-9364

[21] Nnang-Avomo TI, Leon Carrera MF, Escobar-Alvarez E, Rodriguez Morillas N, Mancera-Gonzalez A, Guitian-Lopez J. Application of an integrated methodology for pre-filtering of EOR technologies. SPE-169944-MS. 2014

[22] Fink R, Hancu D, Valentine R, Beckman EJ. Toward the development of "CO_2-philic" hydrocarbons. 1. Use of side-chain functionalization to lower the miscibility pressure of polydimethylsiloxanes in CO_2. The Journal of Physical Chemistry B. 1999;**103**:6441-6444

[23] Sagir M, Tan IM, Mushtaq M, Pervaiz M, Tahir MS, Shahzad K. CO_2 mobility control using CO_2 philic surfactant for enhanced oil recovery. Journal of Petroleum Exploration and Production Technology. DOI: 10.1007/s13202-015-0192-8

[24] Sagir M, Tan IM, Mushtaq M, Nadeem M. CO_2 mobility and CO_2/brine interfacial tension reduction by using a new surfactant for EOR applications. Journal of Dispersion Science and Technology. 2013;**35**(11):1512-1519

[25] Sagir M, Tan IM, Mushtaq M, Ismail L, Nadeem M, Azam MR. Synthesis of a new CO_2 philic surfactant for enhanced oil recovery applications. Journal of Dispersion Science and Technology. 2013;**35**(5):647-654

[26] Sagir M, Tan IM, Mushtaq M, Ismail L, Nadeem M, Azam MR, et al. Novel surfactant for the reduction of CO_2/brine interfacial tension. Journal of Dispersion Science and Technology. 2013;**35**:463-470

[27] Fan X, Potluri VK, McLeod MC, Wang Y, Liu J, Enick RM, et al. Oxygenated hydrocarbon ionic surfactants exhibit CO_2 solubility. Journal of the American Chemical Society. 2005;**127**:11754-11762

[28] Muhammad S, Tan IM, Muhammad M, Talebian SH. FAWAG using CO_2 philic surfactants for CO_2 mobility control for enhanced oil recovery applications. In: SPE Saudi Arabia Section Technical Symposium and Exhibition. Society of Petroleum Engineers. 2014

[29] Ullah S, Bustam MA, Ahmad F, Nadeem M, Naz MY, Sagir M, et al. Synthesis and characterization of melamine formaldehyde resins for decorative paper applications. Journal of the Chinese Chemical Society. 2015;**62**(2):182-190

[30] Muhammad M, Tan IM, Umer R, Muhammad S, Mudassar M. Effect of pH on the static adsorption of foaming surfactants on Malaysian sandstone. Arabian Journal of Geosciences. 2015;**8**(10):8539-8548

[31] Sagir M, Mushtaq M, Tahir MB, Tahir MS, Ullah S, Shahzad K, et al. CO_2

foam for enhanced oil recovery (EOR) applications using low adsorption surfactant structure. Arabian Journal of Geosciences. 2018;**11**(24):789

[32] Sami U, Humbul S, Suleman TM, Muhammad S, Shabbir M, Al-Sehemi AG, et al. Reactive kinetics of carbon dioxide loaded aqueous blend of 2-amino-2-ethyl-1,3-propanediol and piperazine using a pressure drop method. International Journal of Chemical Kinetics. 2019;**51**(4):291-298

[33] Farzaneh SA, Sohrabi M. A review of the status of foam application in enhanced oil recovery. Presented at the SPE-164917-MS. 2013

[34] Xing D, Wei B, McLendon WJ, Enick RM, McNulty S, Trickett K, et al. CO_2-soluble, nonionic, water-soluble surfactants that stabilize CO_2-in-brine foams. 2012

[35] Sami U, Azmi BM, Ali AM, Al-Sehemi AG, Muhammad S, Kareem Firas AA, et al. Synthesis, and characterization of metal-organic frameworks-177 for static and dynamic adsorption behavior of CO_2 and CH_4. Microporous and Mesoporous Materials. 2019;**288**:109569

[36] Bilal TM, Muhammad S, Naeem A. Enhanced photocatalytic performance of CdO-WO_3 composite for hydrogen production. International Journal of Hydrogen Energy. 2019;**44**(45):24690-24697

[37] Muhammad S, Hosna TS. Screening of CO_2-philic surfactants morphology for high temperature-pressure sandstone reservoir conditions. Journal of Petroleum Science and Engineering. 2020;**106789**:186

[38] Muhammad S, Muhammad M, Suleman TM, Bilal TM, Ravoof SA. CO_2-philic surfactants, switchable amine-based surfactants and wettability alteration for EOR applications. Surfactants for Enhanced Oil Recovery Applications. 2020:89-102

[39] Azam MR, Tan IM, Ismail L, Mushtaq M, Nadeem M, Sagir M. Static adsorption of anionic surfactant onto crushed Berea sandstone. Journal of Petroleum Exploration and Production Technology. 2013;**3**(3):195-201

[40] Sagir M, Tan IM, Mushtaq M, Talebian SH. FAWAG using CO_2 philic surfactants for CO_2 mobility control for enhanced oil recovery applications. In: SPE Saudi Arabia Section Technical Symposium and Exhibition. Society of Petroleum Engineers. 2014

[41] Sagir M, Tan IM, Mushtaq M. CO_2-philic surfactant as possible mobility control agent in EOR applications. AIP Conference Proceedings. American Institute of Physics. 2014;**1621**(1):699-704

[42] Sagir M, Tan IM, Mushtaq M, Talebian SH. Static adsorption of new CO_2 philic surfactant onto Berea sandstone. In: ICIPEG 2014; Springer, Singapore. 2015. pp. 129-135

[43] Sagir DM, Tahir MS. CO_2 foam for CO_2 mobility control using a unique surfactant structure for EOR applications. In: 2nd International Conference on Engineering Sciences. 2015

[44] Sagir M, Mushtaq M, Tahir MS, Tahir MB, Shaik AR. CO_2 philic surfactants, switchable amine-based surfactants and wettability alteration for EOR applications. In: Surfactants for Enhanced Oil Recovery Applications. Cham: Springer; 2020. pp. 89-102

[45] Sagir M, Talebian SH. Screening of CO_2-philic surfactants morphology for high temperature-pressure sandstone reservoir conditions. Journal of Petroleum Science and Engineering. 2020;**186**:106789

Chapter 8

Analytical, Bioanalytical, Stability-Indicating Methods: Key Part of Regulatory Submissions

Mahesh Mukund Deshpande

Abstract

According to the International Conference for Harmonization (ICH), the validation and verification data must be included in the Electronic Common Technical Document. The validated analytical procedure gets automatically Food and drug approved (FDA) if it is part of New drug application (NDA), Abbreviated new drug application (ANDA) or Biologic license application (BLA). The analytical, bioanalytical and stability-indicating methods are essential part of all above said regulatory submissions. There are certain ways to generate these analytical methods like U.S. pharmacopeia/National Formulary which are Food and drug approved. The validated analytical method can also be submitted by any researcher or agency which can gets the food and drug approval. It is necessary that the methods which are Food and drug approved can only be applied to the various drugs and drugs products. In the current chapter, the meaning and requirements of analytical methods, procedures, acceptance criteria and evaluation of stability indicating methods, need, recommendations for bioanalytical methods are discussed in detail. The analytical techniques like HPTLC, HPLC, Spectrophotometry and Hyphenated techniques are also discussed as these are playing important role in validation of these methods.

Keywords: noncompendial and compendial analytical methods, forced degradation analytical techniques, bioanalytical methods, analytical techniques

1. Introduction

The United States of America (USA), Europe and Japan were developed the Common Technical Document (CTD) which should be implement while applying for registration of pharmaceutical product for human use. For the development of these guidelines International conference for Harmonization (ICH) plays important role and currently these are becoming the part of ICH guidelines [1]. The identity, strength, quality, purity and potency are important points of Investigational new drug application (IND), New drug application (NDA), Abbreviated new drug application (ANDA). The related analytical methods of drug substance and product should be included in NDA and ANDA. A Complete description, total manufacturing process including analytical procedures should ensure compliance with standards and potency should be part of Biologics license application (BLA). It is must to meet all standards of guidelines provided for analytical procedures. All these parameters should be suitable for their purpose wherever applicable. Detail Analytical procedures including detail validation parameters are the important part of Electronic Common Technical Document Specification as per International conference of Harmonization (ICH). The analytical procedure is Food and Drug Approved (FDA) if it a part of Approved NDA, ANDA or BLA. These methods can be generated from FDA recognized sources like U. S. Pharmacopeia/National Formulary (USP/NF) or if anyone submits validated procedure that will be accepted by FDA. The only validation or verification data of FDA approved methods of new products are considered for applications to various drug products (**Figure 1**).

Every manufacturer must generate large amount of corrected data for safety and efficacy of drug for commercial viewpoint. As it is mandatory to follow Current Good Manufacturing practices (cGMPs) for manufacturing purpose, likewise each analytical activity must follow Good Analytical Practices.

Method Validation, calibrated instrument, and training are three important tasks of Good analytical practices (GAPs). Commercially available dosage form is an outcome of several steps which are systematically carried out during product development. It is very important that all steps should be carried in systematic manner to ensure complete drug development stage. In recent years there is special focus on efficiency and efficacy of drug product and for this clinical study is most important task but apart from this there are various behind the scene activities are associated with drug development process without which pharmaceutical drug development is not possible. Among these behind the scene activities Method Development and Validation has its own uniqueness to ensure the drug development.

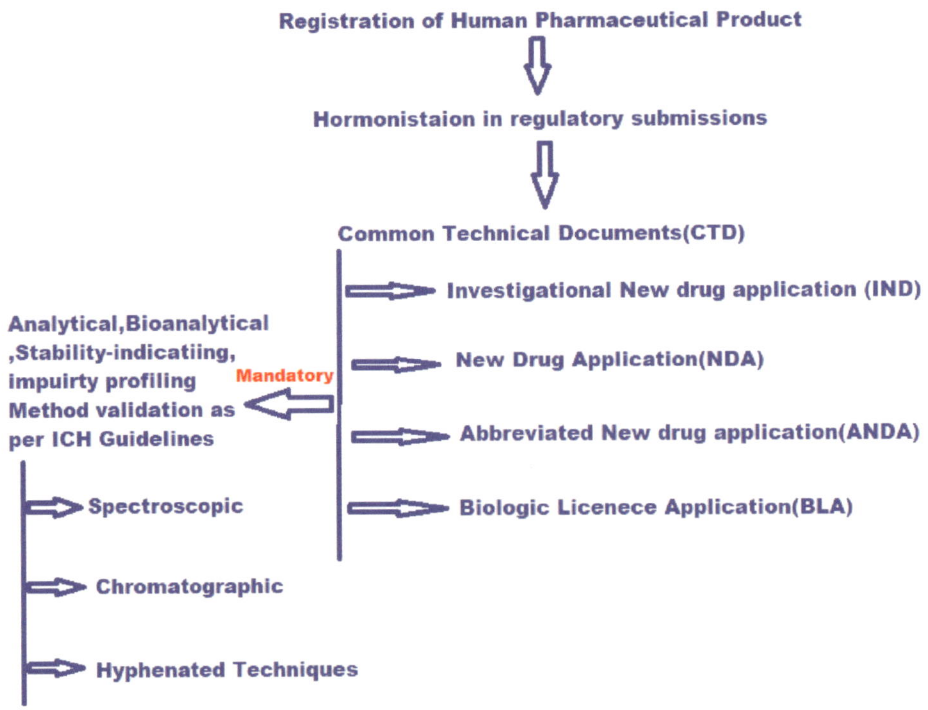

Figure 1.
Flow chart showing essential requirements of registration of pharmaceutical product.

2. Noncompendial analytical method validation

The objective and plan of work should be clearly defined prior to start the work. This data is based on scientific findings from the method development and Optimization. The validation of results should be obtained by approved protocol. Then sponsor must follow cGMP's which includes detail procedure, validation characteristics and acceptance criteria by the use of qualified instrumentation. All the protocols of drug substance and product analytes in respective matrices should be prepared and followed. All the results of validation studies, application should be included.

3. Compendial analytical procedures

The analytical procedure official in pharmacopeias, are cross checked for implementation stage and its suitability should be checked. The verification protocol should include details of data which explains suitable analytical procedure official in USP/NF for drug product or drug substance.

The following points are to be included in the verification protocol

1. The compendia method which should be verified with acceptance criteria's.

2. All parameters related to each aspect of method that is reagent, equipment, validation characteristics that is specificity, Limit of quantitation (LOQ), Precision, accuracy, should be included in validation are covered by procedure and extent of verification. There is no need to include robustness study for compendia assays if there is no deviation. For a BLA, if the methods are already specified in FDA regulations, there is must to take pre-approval from FDA to change in analytical method.

3.1 Statistical analysis

The statistical analysis is important work after finishing the method development and validation. The statistical values of validation are compared with the predetermined acceptance criteria.

The statistical parameters used are based on proper principle and required for evaluation of parameter. The methods like analysis of variance (ANOVA) for analysis of regression analysis, (R^2-Correlation coefficient) to measure the linearity are applied for studying validation characteristics. In case of observed data is not distributed then it is transformed normal distribution or distribution free approach. By using validates software or independent verification for correctness the data can be analyzed.

3.2 System suitability requirement for potency assay

Before starting actual analysis of standard sample, it is necessary to check whether system is working properly or not. This important task can be completed by analyzing system suitability. In this all integral system that is equipment, electronics, analytical operations and samples are evaluated. These system suitability parameters are depending on method or procedure under validation.

System suitability can be evaluated according to following points:

1. The system suitability measures the performance of given system of samples on a given day. 2. The variable parameters like chromatographic columns, column aging, mobile-phase variations, changes in instrumentation are checked whether they are working properly or not.

2. It is nothing but the part of method validation. The experience and information obtained at the time of method development, which is helpful to determine system suitability of final method.

3. At every time when system is used for performing the assay there is necessary to use system suitability test. For longer period if it is continuously in use then there is need to reevaluate the system suitability at proper intervals.

4. The system suitability means criteria and parameters obtained collectively which can explain the system are working properly [2].

The important aspects of pharmaceutical development program are analytical method development, validation method transfer but it is fact that they are less considered in sense of total contribution in development process, time and economy. At the time of drug development phases, all the analytical method related activities are interrelated. In early development stages they are related and occur one after another in coming phases of development. During drug development process the changes may require to be performed in current methods and these changes in methods again requires validation or method transfer treatment.

If one's objective of method development and validation is achieved, then it can prove that the laboratory facilities are accurate and fit for further development process that is one can say optimized. Method validation is the "process of demonstrating that analytical procedures are suitable for their intended use." Both method validation and methods transfer have important share in drug development and further changes in methods. To generate supportive data during manufacturing and quality control, these methods provides a valuable data by comparing with specifications including all types stability study, Safety, characterization and drug performance can obtain with these supportive methods.

Method development is the simultaneous process as the gradual development of drug product continues. The system suitability parameters are set of tests to checks the proper working of the system. After performing robustness with proper statistical data collection one can set the criteria for final system suitability of the method.

These methods focus on active pharmaceutical ingredient (API) behavior. As the knowledge about API and drug product goes on progress the analytical methods become more refined. The important aspects of analytical method are that should be robust, simple and meeting the regulatory guidelines. Various trial an error experiments are to be carried out to develop the method. The performance criterion's to be finalized before the final validation of method. Forced degradation study which is integral part of stability-indicating method and system suitability tests are one of the key points of method development and final validation [3].

3.3 Impurity profiling

The International Conference for Harmonization (ICH) guidances are available that are related to the qualification of impurities in new drug substances that are produced by chemical synthesis. These impurities can be addressed in two perspectives that are chemistry aspects and safety aspects. The chemistry aspects explain the identification and classification of impurities, the various analytical procedures and setting of specifications. In the safety aspects explains the qualification of impurities which are not addressed in clinical trials. In this aspect the threshold limits are defined. This ICH guidance classifies impurities in three classes as Organic, Inorganic and Solvent. Each class of impurities should be properly reported, with all aspects, developed during synthesis, storage of the new drug product. The analytical procedure including validation reports related to impurities should be properly reported [4]. In concern with the above discussed guidelines related to qualification of impurities in new drug substances produced by chemical synthesis, impurities which are classified as degradation product developed with the reaction with excipients or container closure system may termed as reaction products. All the observed degradants during manufacturing and stability study should be reported. All the data related to their identification, specification, analytical procedures for quantification, their limits of detection and quantification should be reported [5].

4. Stability-indicating method

To ensure safety, efficacy and quality of drug product there are need of stability indicating methods. The Food and Drug Administration (FDA) defines the stability indicating method as a validated analytical procedure that accurate and precisely measure active ingredients (drug substance or drug product) free from process impurities, excipients and degradation products [6].

To obtain forced degraded samples for assessing selectivity of method, method development and method validation are three important steps of stability indicating methods.

4.1 Importance of forced degradation

As per guidance document available for stability indicating method does not contain any explanation about extent up to which the degradation should be carried out. There are no certain guidelines regarding stress degradation. Therefore, it is always necessary to keep all experimental conditions of degradation with more reality and deliberate degradation.

The main purpose of forced degradation is to obtain stability of drug. It should provide information about route of degradation and utility towards the stability indicating [6].

Forced degradation can be able to judge excipients or non-drug substances. It also provides information useful for structure elucidation of degradation products. Importantly it gives data related to thermal, hydrolysis, and oxidation and photo degradation behavior of drug substance and drug product. It is very important to know about chemical behavior of drug product and drug substance in formulation development, manufacturing and packaging. The data helps in quality improvement [7].

4.2 Acceptance criteria for forced degradation

There are vast discussions among the various scientists that what should be the limits of stress testing? Generally, values in between 5–20% are proper and acceptable for chromatographic assays. The acceptable stability limit for small molecules are 90% of label claim and generally employed by pharmaceutical scientist that is 10% degradation is optimum for use in analytical validation. There are some experiments in which very little or no degradants are obtained due to exceptional stability of molecule under study in such case accelerated storage 40^0 c for 6 months should be carried out. If positive result is obtained, then the stability of drug is noted. But overstressing the drug substance may produce false results [7].

According to the recent recommendations of Food and Drug administration (FDA) and ICH guidelines, the stability indicating property of analytical method can be obtained by carrying out forced degradation study. The pathway of decomposition from API, solution and formulation also be determined. The structural information, their characterization and isolation of major degradants are the important part of the new drug approval (NDA). The use of forced degradation study is primary to understand molecular chemistry of drugs, its stability indicating properties and its degradation products and their pathways. In most of the cases the Hydrolysis, Oxidation, Photolysis, Racemization, and Decarboxylation are the type of reactions that are responsible for decomposition of most the drugs. However, the regulatory guidelines do not define the procedures to carry out degradation study. Therefore, there are various approaches to carry out forced degradation study [8].

4.3 Stress conditions

In pharmaceutical industry thermal, hydrolysis, oxidation and photo degradation are generally employed. If one must serve the purpose of degradation, the expected degradation should be achieved. The optimum percentage of degradation should be obtained in all types of conditions or in minimum of one according to FDA guidelines. If no degradation is achieved, then in that case, all reports related degradation experiment carried out should be produced. It is important fact to obtain the degradation as per expected level. The degradation in between 5 and 20% is recommended [9].

4.4 Hydrolysis

By using acid and base the hydrolysis studies are performed. Generally, a chemical reaction is carried out with water to obtain decomposed analyte. The wide pH range that is from 2 to 12 is used for acid and base, which is to be used for hydrolysis purpose. For acid hydrolysis generally Hydrochloric acid (HCl) or Sulfuric acid (H_2SO_4) is used and for base hydrolysis Sodium or Potassium hydroxide is used. According to stability of molecule, the concentration of acid or base is decided. One can used more than one stress conditions to obtain desired degradation. If the desired degradation is not achieved at room temperature, then higher temperature is used. After the degradation process completed the degraded samples are neutralized by same acid or base so as it can easily injected in HPLC column without any harm to silica stationary phase. For water insoluble samples alcoholic acid or base are used for obtaining degradation [9].

4.5 Oxidation

To carry out oxidation degradation of drug substance or drug product, generally hydrogen peroxide is used. Apart from hydrogen peroxide metal ions, oxygen and radical initiators can also be employed. The oxidizing agent, its quantity requirement, properties are depending on the drug substance under study. If hydrogen peroxide is used as a degradant, in that case combination of stress should not be employed. If elevated temperature is used in case of Hydrogen Peroxide, it leads to hydrolysis instead of oxidation because in Hydrogen Peroxide O-O are not stable and they may decompose at ambient temperature also. Due to heating, these bonds break faster and oxidation occurs. Sodium metabisulfite solution is used for neutralization of oxidation degraded samples [9].

4.6 Heat

Active pharmaceutical ingredients, Dosage form with or without humidity can be undergoing thermal degradation. The sample is exposed to heat. (weather there is humidity or absence of humidity as mentioned earlier) In case of liquid, humidity is completely avoided. While applying stress to liquid samples especially for injections, oral solutions, and syrups as further diluting the samples, the precaution should be taken because these types of samples may loss water and concentration of actual sample. By obtaining multiple time results, the detail information about primary and secondary degradation can collected. If any molecule is so stable that it cannot generate degradation, in such situation the energy analogous to the accelerated stress condition is to be applied to express efforts taken for obtaining degradation [9].

4.7 Photo stability

The exposure to light is one of the important degradation steps to obtain degradation caused by light. This degradation is evaluated by obtaining any unacceptable change due to light. The recommendations related to photo stability are described in ICH guidelines Q1B.The UV–VIS light exposure with not less than 1.2 million hours to achieve degradation of sample. The samples are preferably exposed to cool white fluorescent light and near ultraviolet lamp. The natural light can be used, if specific instrument is not available, but there will be intensity problem as it is varying with time, weather conditions, pollution etc. due to which natural light becomes not suitable for degradation [9].

4.8 Evaluation of results

After generating forced degradation samples obtained by accelerated stress conditions, their evaluation is the important task. For evaluation purpose, each sample should be studied individually. The Chromatographic techniques including High performance liquid chromatography (HPLC), Ultra-performance liquid chromatography (UPLC), UHPLC and Capillary Electrophoresis are commonly used techniques for this important task. The most important work is development and validation of stability-indicating method which can be able to separate every degradation product from each other and from drug. Therefore, peak purity is important in sense of selectivity determinations of the method. One more important parameter is sensitivity, which can be helpful to asses' impurities at lower level. There are chances that the impurity peak may get depressed at the time of method development which is co-eluted. Many times, it may happen that two unknown impurity get merged due to which false results are appear for stability, therefore it is important to implement such analytical method that can have capacity to resolve each unknown impurity and that is helpful to control out of specification results [9] (**Figure 2**).

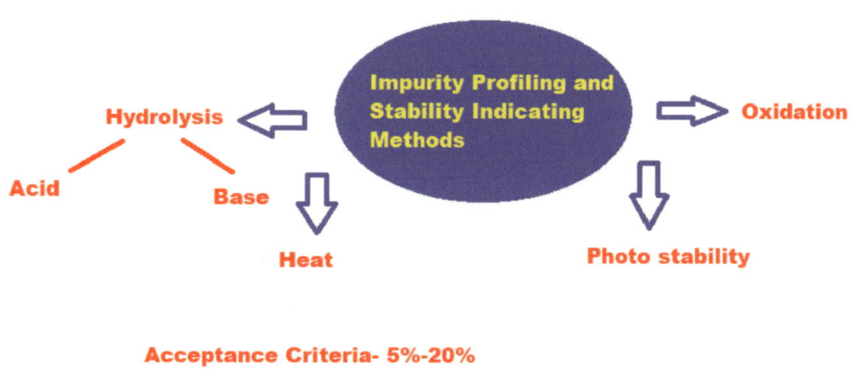

Figure 2.
Important parameters and acceptance criteria for impurity profiling and stability indicating methods.

5. Bioanalytical method development and validation

There were various regulatory agencies had done serious efforts to regulate bioanalytical method development and validation. Almost from last three decades there were large progresses in this area. The various regulatory agencies that were worked can be listed as US FDA, American association of pharmaceutical scientists (AAPS), Health protection Branch HPB, Association of analytical chemists (AOAC), Center for Veterinary medicine (CVM), U. S. Department of Health and Human Services Food and drug Administration, Center for Drug Evaluation and Research (CDER), European Medicine Agency (EMA), China Food and Drug Administration (CFDA), European Bioanalytical forum (EBF), Global CRO Council (GCC), The Brazilian health regulatory agency (ANVISA, Brazil). To regulate and harmonize bioanalytical method development and validation first workshop was held in 3–5 December 1990, report of which was published in pharmaceutical research and in other journals. On basis of the reports of this workshop, the FDA was issued draft guidance on bioanalytical method development and validation in January 1999.The second FDA guidance was published in May 2001 on the basis of workshop which was held in January 2000.The recommendations for bioanalytical method development and validation for macromolecules was published in 2006. The recommendations for regulation and harmonization of bioanalytical methods were again refreshed in 2006. In 2010, a draft guidance was published by EMA for development and validation of bioanalytical methods. As per above discussion this can be concluded that there were serious efforts carried out to regulate bioanalytical method development and validation by the various abovementioned regulatory agencies.

6. Need of bioanalytical method development and validation

The various manufacturers are applying for Investigational new drug application (IND), New drug application (NDA), Abbreviated new drug application (ANDA) to FDA. There was harmonization in this process related to human clinical pharmacology, bioavailability (BA) and Bioequivalence (BE), pharmacokinetic evaluation (PK), non-human pharmacology and toxicology studies and preclinical studies, which should be included in abovementioned applications. To obtain the data related to abovementioned requirements, there is need of development and validation of bioanalytical methods in biological matrices such as blood, serum, plasma or urine [10].

The most recent FDA guidance document on bioanalytical method development and validation was released in May 2018. Before this there was a guidance documents in 2001 and its revision in 2013 were released. The overall previous recommendations remain same, only the following points are revised.

- The validation criteria for dilution and carryover
- There was clarification on the number of Quality control (QC) samples and replicates
- There will be no acceptance criteria for QCs for accuracy and precision
- The QCs should have to cover the sample concentration range
- The LLOQ should be evaluated for interference for each run

- There will be the further acceptance criteria within the different batches
- The internal standard (IS) and the drift should be monitored

In this document following clarity regarding Ligand Binding Assay (LBA) was added

- The accuracy and precision runs
- The control of each sample should be included with clear definition
- The consistency in standard calibrator preparations

The significant change in final document, the incurred sample reanalysis section was added which includes endogenous compounds, biomarkers, diagnostic kits, bridging data and dried blood spots [11]. The guidance documents (M10) on Bioanalytical Method Development and Validation was released by International Council for Harmonization of technical requirements for pharmaceuticals for Human Use (ICH), in February 2019. Simultaneously, American association of Pharmaceutical scientists (AAPS), European bioanalysis forum (EBF), Japan Bioanalysis forum (JBF), China Bioanalysis forum (CBF) were organized a workshop of industry, academia, and health authorities to discuss this draft guidance. The objective of these discussions was the M10 guidelines which are for Bioanalytical Method Development and validation which are part of regulatory submissions. This guidance document explains the validation of Bioanalytical Methods form, which the concentration of analyte is determined from biological fluids. The concentration was obtained from pivotal nonclinical pharmacokinetic studies which are useful for taking the decisions over the regulatory submission including all phases of clinical trials [12].

6.1 Key principles of bioanalytical method validation and establishment

- Accuracy, precision, selectivity, sensitivity, reproducibility and stability are the fundamental parameters that ensure the acceptability of bioanalytical method.
- There should be specific protocol, study plan, report or SOP for bioanalytical method development and validation.
- How the analyte is being get affected by environmental, matrix, or procedural variables? Every step, including time of collection of matrix and overall investigation time, should be clarified.
- The physiological nature of samples gives variable matrix. When there are Liquid Chromatography-Mass spectrometry-Mass Spectrometry (LC-MS-MS) based procedures, then protocol should be designed to avoid matrix effect, matrix may change during method validation.
- It is necessary to validate bioanalytical method for the intended use or application.
- There should be written method validation report to claim the results.

- The same biological matrix as the matrix in the intended samples should be used for validation purposes. It is necessary in case of limited availability of matrix like bone marrow.

- The stability at the time of matrix during collection and storage should be assessed before analysis.

- The stability of analyte in matrix from dosed subjects should be finalized in case of potentially labile metabolites.

- The parameters like accuracy, precision, reproducibility, response function, and selectivity of method for endogenous substances, metabolites, and known degradation products should be set for biological matrix.

- In case of selectivity the evidence should be produced that substance being quantified is the intended analyte.

- The concentration range of analyte should be defined on standard samples including their statistical parameter which clears the standard curve.

- To define concentration and response relationship an enough sample should be analyzed. This relationship should be continuous and reproducible. For this purpose, the standard used should be from dynamic range and nature of the concentration-response relationship. Generally, six to eight concentrations excluding blank can be used to define standard curve. In case of nonlinear concentrations more standard may be recommended.

- There should be proper demonstration to show the ability to dilute samples originally above the upper limit of the standard curve by accuracy and precision parameters in the validation.

- In case of high throughput analyses like multiplexing, multicolumn and parallel systems, enough Quality control (QC) samples should be assessed to prove control of the assay. Based on the run size, the number of QC samples should be determined.

There should be proper placement of the QC samples in the run.

- There is a need to set a specific acceptance criterion for bioanalytical method to be considered as a valid method. That should be achieved for accuracy and precision for validation of QC samples over the range of standards.

6.2 Specific recommendations for bioanalytical method validation

- There should be minimum six standard points for matrix based standard curve excluding blank, which may be single or replicates and should cover the entire range of expected concentrations.

- Standard curve should explain the concentration-response relationship with appropriate weighting and statistical tests for goodness of fit.

- The Lower limit of quantitation, (LLOQ) should be measured with acceptable accuracy and precision which is the lowest concentration of the standard curve. By using a least five samples independent of standards and its coefficient of variation, the LLOQ can be established. The LLOQ should not be confused with the limit of detection and/or the low-Quality Control (QC) samples. The upper limit of quantification will be defined by highest standard.

- The accuracy and precision should be determined by using minimum of five determinations per concentration level excluding blank samples. The average value should be within ±15% of the theoretical value. The LLOQ should be up to ±20%. The coefficient of variance of precision should not exceed 15% and for LLOQ should not exceed 20%. The methods which give the results of accuracy and precision with these above-mentioned values should be acceptable.

- There should be proper demonstration of concentration of analyte in biological matrix with which the accuracy and precision is determined. This can be performed by analyzing replicate sets of QC samples from same biological matrix. This QC sample should be representative of entire concentration range selected for standard curve. From which one concentration within LLOQ, one should be middle one that is middle QC (MQC) and last should be upper limit of standard curve that is High QC (HQC).

- All outliers should be included in reported method validation data and accuracy and precision data. The values of outliers that are determined statistically can also be reported with the calculations of accuracy and precision.

- The storage temperature stability in biological matrix should be determined for analyte. The freeze-thaw stability at minimum of three cycles of two concentrations in triplicates should be studied.

- The ambient temperature stability of analyte should be determined over the time period equal to typical sample preparation, sample handling and analytical run times.

- In case of instrument failure, reinjection reproducibility should be evaluated to determine an analytical run could be reanalyzed.

- For determination of specificity of assay method, a minimum of six concentration of same matrix should be studied. In case of hyphenated techniques like mass-spectrometry based methods, it is not important to study six independent matrices. There should not any compromise to study matrix effect to ensure precision, selectivity and sensitivity in case of Liquid Chromatography-Mass spectrometry (LC-MS) and Liquid Chromatography-Mass spectrometry-Mass Spectrometry (LC-MS-MS) based procedures. The selectivity should be evaluated throughout method development, method validation and it should be continued up to the application of method to actual study samples.

The acceptance/rejection criteria for spiked, matrix-based calibration standards and validation of QC samples should be based on theoretical concentration of analytes. For studying accuracy and precision, the specific criteria should be set in the standard concentration range [13] (**Figure 3**).

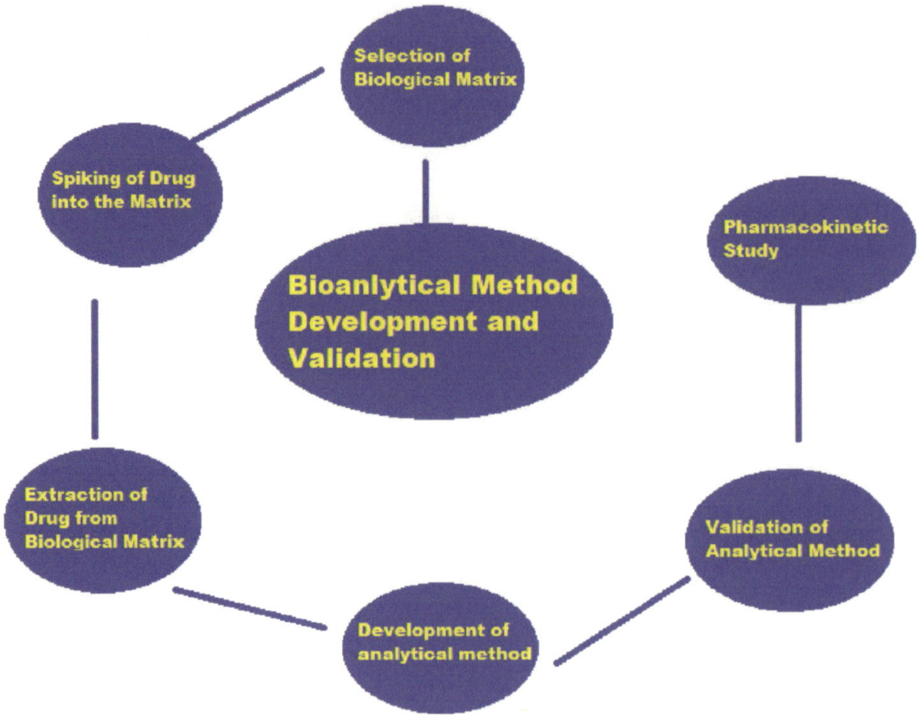

Figure 3.
Sequence showing development and validation of bioanalytical method.

7. Analytical techniques for method development and validation

The various analytical techniques are available for Qualitative and Quantitative analysis, which can be used in above explained types of analytical methods. The chromatographic techniques used as a separation tool and spectroscopic techniques are used for an identification and to obtain structural information. Among all these techniques High performance liquid chromatography, High performance thin layer chromatography, Spectrophotometric techniques and Hyphenated techniques are explained in brief.

7.1 High performance liquid chromatography

When we draw the attention towards the working principle of HPLC, which involves the injection of small sample (Generally in μl) into the stationary phase composed of 3–5-micron tiny particles. The injected components of the sample moved through the abovementioned stationary phase with the mobile phase which is forced through high pressure by the pump.

The HPLC technique is having advantage of High speed, Efficiency, Sensitivity and vastly superior over the simple liquid chromatography. The process of separation of components of sample involves chemical and physical interactions with the stationary phase particles. The separated components are detected at the end of column by the detector in the form of the liquid chromatogram (**Figure 4**).

The HPLC can be applied for separation of non-volatile compounds like Aspirin, Ibuprofen, Acetaminophen and then for separation of salts like Sodium Chloride, Potassium Phosphate. For the separation of proteins like Egg white and Blood

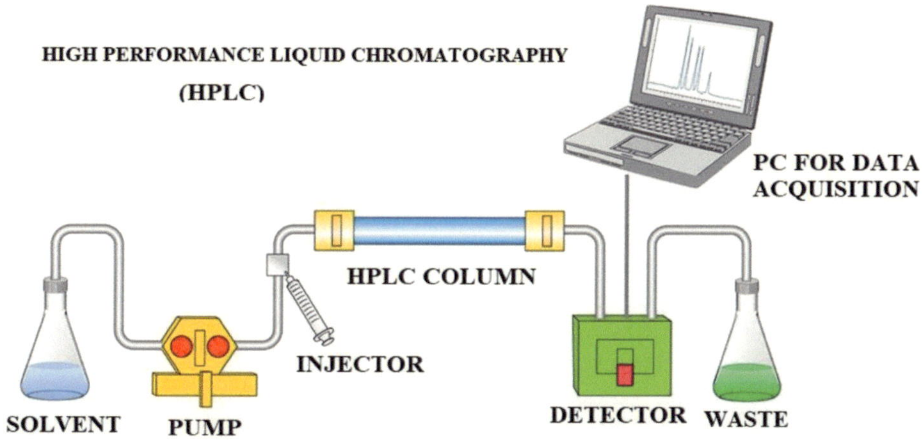

Figure 4.
Flow diagram of working principle of HPLC.

proteins. The HPLC can also be applied for separation of Organic Polymers, Heavy Hydrocarbons, Natural Products, Thermally unstable compounds and Enzymes [14].

For separation of many complex mixtures including biological samples high performance liquid chromatography is the best form of liquid chromatography (HPLC). HPLC is widely used for qualitative and quantitative analysis of different types of pharmaceuticals due to its sensitivity. By using HPLC, one can obtain individual sample with its role in that sample. In 20th century, the HPLC methods were appeared for assay of bulk drugs and later become a principal method of Pharmacopeia. The interaction among the solute molecules and stationary phase, decides the mode of chromatography. HPLC is more versatile technique as various modes are available. By using HPLC, the proper values of precision can obtain with excellent specificity of the methods. Though the specificity, precision and accuracy are obtained with HPLC methods, the system suitability parameters are first analyzed before analyzing these parameters. The more attention should also be providing for high accuracy, precision and specificity. By doing wide literature survey it was observed that HPLC is widely used technique among all chromatographic techniques. One of the reasons for this is detection system of HPLC which can able to detect every component of mixture. The UV detector is most widely used detector for HPLC. The Ultra-violet (UV) detector can analyze various wavelengths simultaneously by giving multiple wavelength programmers on HPLC software. Every component present in mixture which UV can detect that can be obtained by UV detector. A Photodiode array (PDA) detector is one of useful spectroscopic detector. By placing at the image plane of spectrophotometer various wavelength can be scanned simultaneously. For analysis of alcohols, sugars, carbohydrates, fatty acids and polymers, the refractive index detector is used as there is restriction for UV absorption of these compounds. The Refractive Index (RI) detector is one of the lowest sensitivity detectors but it can be applied for trace detection with low noise. For analyzing oxidizable and reducible substances, the electrochemical detector is implemented. In this detector the electrical output obtained by electron flow due to chemical reaction at electrode surface due to presence of above-mentioned compound is used for qualitative and quantitative analysis of these types of samples. Among various detectors available for HPLC, the most sensitive detector is fluorescence detector. The sensitivity of fluorescence detector is 10–1000 times more sensitive as compared with the UV detector. If sample contains any specific fluorescent compound, then it can be easily detected by this detector. For estimation of pharmaceuticals especially fluorescence detector is applied. As most of pharmaceuticals are polar in nature, there analyses are carried out as reverse phase HPLC. In recent years most of the researchers used reverse phase chromatography with UV detection, due to that the results are obtaining with best reliability, analysis, repeatability and sensitivity. Generally, in pharmaceutical industry Octadecyl silyl (ODS) C18 is mostly used stationary phase. Many drugs can be easily obtained in pharmaceutical formulations and biological fluids by using HPLC. Nowadays, HPLC is one of the important tools for solving many problems in pharmaceutical industry. There are certain limitations to HPLC that high price of column, HPLC grade solvents and it is difficult to obtain long term reproducibility due to nature of column packings.

The Liquid Chromatography-Mass spectrometry (LC-MS) is wide choice for quality control and quality assurance in various stages in pharmaceutical industry. The LC-MS can be easily applied for assay of many drugs and pharmaceuticals also applied for analysis of impurities and degradation products. The most hyphenated technique like Liquid Chromatography-Mass spectrometry-Mass spectrometry (LC-MS-MS) is also available for above mentioned work [15].

7.2 High performance thin layer chromatography (HPTLC)

The advancement of Thin layer chromatography (TLC) is the HPTLC that is, High performance thin layer chromatography which an instrumental semi or automatic form of TLC. It is fast working, sensitive and can be able to analyze wide range of samples.

The HPTLC is advantageous to handle a sample with short analysis time for analysis of even complex samples including crude drugs. As automation with the instrument it can be able to analyze entire chromatogram with many parameters without any interruption. The samples can be analyzed simultaneously or independently with standard that shows the reliability of technique. The HPTLC is equipped with high performance adsorbent layers having refined uniform particles, approx. 5 microns in diameter. All processes of experimentation including method development, optimization of various parameters and documentation are performed with standardized methods. The HPTLC can be applied for both qualitative and quantitative analysis of mixtures, as the technique is automated, the quantitative mode is more optimized as compared with the TLC. Also, it can be used for the assay of the compound (**Figure** 5).

The advantages of the HPTLC are as follows:

- Colored samples can be easily separated
- Many samples can be assessed easily on single plate which reduces the cost and time.
- Two-dimensional mode is possible
- Visualizing agent used which are Specific and Sensitive for detection purpose
- Other evaluation technique can be implemented for different samples with different light absorption characteristics.
- Radio labeled compounds can be monitored and microbial activity can be assessed.
- No regeneration and cleanup are required as the technique is disposable.

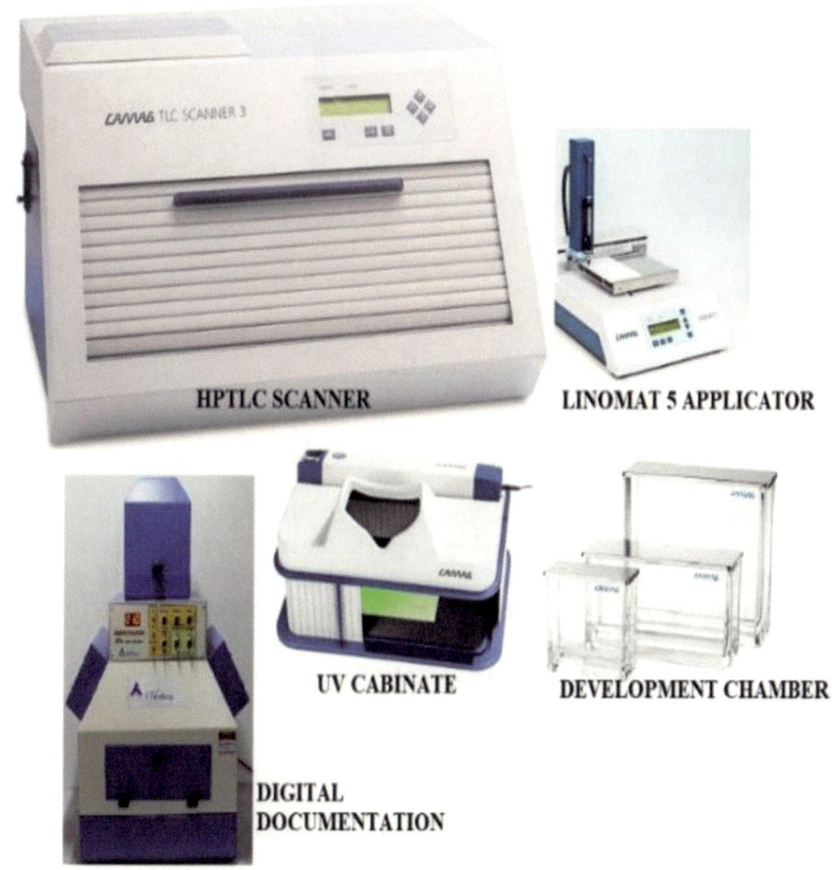

Figure 5.
Diagram showing components of HPTLC.

Development of plate and evaluation of plate are separate processes therefore both can perform as per time available differently [15].

The important advantages of the HPTLC are fast, inexpensive method of analysis. It can prove over high-performance liquid column chromatography when it is performed by skilled person for quantitative analysis. The qualitative and quantitative analysis by the HPTLC with automated sample application and densitometric scanning shows very sensitive and reliable results. The HPTLC has important advantage of providing chromatographic fingerprints which can be stored as an electronic image [16].

7.3 UV-spectrophotometric methods

For the quantification of components present in solution, the UV absorption spectroscopy works on the principle of Beer-Lambert law [17] (**Figure 6**).

According to Beer-Lambert law,

$$A = \log I_0 / I = \varepsilon . c . l$$

where A = Absorbance.

I_0 = intensity of incident light.

I = intensity of emergent light.

ε = molar absorptivity.

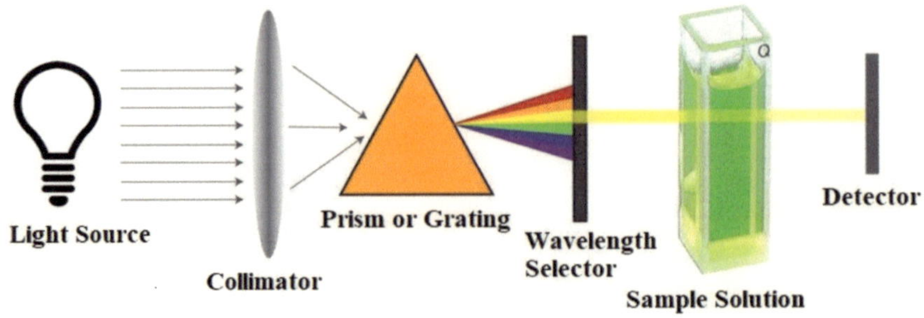

Figure 6.
Flow diagram of working principle of UV spectrophotometer.

c = molar concentration of solute.

I = length of sample cell.

The natural Ultra-Violet (UV) absorption methods and chemical reactions spectrophotometric methods are having importance in pharmacopeia. In these methods quantitative data of reflection or transmission by the analyte as function of wavelength is measured. The method is based on the fact, that functional group of analyte absorbs UV radiation at specific wavelength in a solvent system. The λmax is the term used for maximum absorption of wavelength which is independent of concentration. These methods require less time and less labor consumption. The method also gives best precision. The UV-Visible methods are applied for multicomponent analysis of samples [15].

7.4 Brief introduction of other spectroscopic techniques

The Near Infrared spectroscopy (NIRS) is one of the spectroscopic techniques that can be applied for multicomponent analysis of all types of samples and having advantage of non-destructive technique. For the purpose of raw material testing, quality control of finished product and to monitor the process, the NIR spectroscopy plays important role in recent years. The great advantage of NIR is there no requirement of sample pre-treatment, the use of fiber optic probes and both the chemical and physical parameters can be obtained in single spectrum [15].

The Nuclear Magnetic Resonance (NMR) spectroscopy is one of the advantageous techniques over UV and IR spectroscopy that it can detect the intermediate products like ions, reaction complexes, solvents of chemical reaction. The NMR spectroscopy provides unique information on the structure of intermediate due to which there is no need to restore various hypotheses to explain the mechanism of the process [18]. The mass spectrometry is one of the outstanding techniques in all type's spectroscopies due to its sensitivity, detection limits and its wide range of application. It is widely applied in biochemical problems like proteome, metabolone, drug discovery and metabolism. This technique can also be applied for pollution control, food control, forensic science and natural product or process monitoring. It can also be applied in atomic physics, reaction physics, reaction kinetics, inorganic chemical analysis, ion-molecule reactions and determination of thermodynamic parameters [19].

7.5 Hyphenated techniques

Generally, for the qualitative and quantitative analysis of the samples, the separation technique is combined with the identification technique. In the analysis, chromatographic and electrophoresis are used for separation or isolation of the required components. The quantitative determination or structural information of the sample under study is performed by spectrophotometry. The Hyphenated technique is nothing but the combination of both above said techniques that are separation and spectrophotometric technique. The various hyphenated techniques like Liquid chromatography-Mass spectrometry (LC-MS), Gas chromatography-Mass spectrometry, Liquid Chromatography-Nuclear Magnetic resonance (LC-NMR), Liquid chromatography-Fourier transform infra-red spectroscopy (LC-FTIR), Capillary electrophoresis-Mass spectrometry (CE-MS) are used widely for qualitative and quantitative analysis. There may be a combination of more than one separation or detection technique like Liquid chromatography-Photodiode array-Mass spectrometry (LC-PDA-MS), Liquid chromatography-Mass spectrometry-Mass spectrometry (LC-MS-MS), Liquid chromatography-Nuclear magnetic resonance-mass spectrometry (LC-NMR-MS), Liquid chromatography-photodiode array-nuclear magnetic resonance-Mass spectrometry (LC-PDA-NMR-MS) [20].

Author details

Mahesh Mukund Deshpande
Department of Pharmaceutical Chemistry, Amrutvahini College of Pharmacy, Sangamner, M.S., India

*Address all correspondence to: mahesh_deshpande11@rediffmail.com; maheshdeshpande83@gmail.com

References

[1] Jordan D. An overview of the common technical document (CTD) regulatory dossier. Medical Writing. 2014;**23**(101):101-105. DOI: 10.1179/2047480614Z.000000000207

[2] Analytical procedures and methods Validation for drugs and Biologics, Guidance or Industry, U.S. Department of Health and Human Services, Food and Drug Administration, center for Drug Evaluation and Research (CDER), Centre for Biologics Evaluation and Research (CBER); 2015. pp. 1-15

[3] Robert WL, Laurie G. The Central Role of Analytic Method Development and Validation in Pharmaceutical Development, Life Science Connect 5340 Fryling Road, Suite 300, Erie, PA 165101, 814.897.7700. pp. 1-3. Available from: www.lifescienceconnect.com

[4] ICH Topic Q3A Impurities testing guideline: impurities in new drug substances. The European agency for evaluation of Medical product, Human Medicines Evaluation Unit, CPMP/ICH/142/95. pp. 1-11

[5] ICH topic Q3B (R2) impurities in new drug product, European Medicine Agency CPMP/ICH/2738/99; June 2006. pp. 1-14

[6] Cione AP, Tonhi E, Silva P. Stability Indicating Methods. Brazil: Bioagri Laboratories; 2011. pp. 27-36. DOI: 10.5772/19940

[7] Ngwa G. Forced degradation studies, forced degradation as an integral part of HPLC stability-indicating method development. Drug Development & Delivery. 2010;**10**(5):1-4

[8] Sengupta P, Chatterjee B, Tekade RK. Current regulatory requirements and practical approaches for stability analysis of pharmaceutical products: A comprehensive review. International Journal of Pharmaceutics. 30 May 2018;**543**(1-2):328-344. DOI: 10.1016/j.ijpharm.2018.04.007

[9] Sharma MK, Murugesan M. Forced degradation study: An essential approach to develop stability indicating method. Journal of Chromatography. 2017;**801**:1-3

[10] Deshpande MM, Kasture VS, Mohan M, Chaudhari SR. Practical approach for development and validation of bioanalytical method: A review. Inventi Rapid/Impact: Pharm Analysis & Quality Assurance. 2017;**2017**(1):1-8

[11] Booth B, Stevenson L, Pillutla R, Buonarati M, Beaver C, et al. White paper on recent issues in bionalysis: FDA BMV guidance, ICH M10BMV guidance and regulatory inputs (part 2-recommendations on 2018 FDA BMV guidance, 2019 ICH M10 BMV draft guidance and regulatory agencies input on bioanalysis, biomarkers and immunogenicity). Bioanalysis. 2019;**11**(23):2099-2132

[12] Booth B, Vazvaei F, Fluhler E, Myler H, Woolf E. AAPS, workshop report on ICH M10. The AAPS Journal. 2020;**22**:10. DOI: 10.1208/s12248-019-0398-7

[13] Guidance for Industry Bioanalytical Method Validation, U.S. Department of Health and Human Services Food and Drug Administration Center for Drug Evaluation and Research (CDER) Center for Veterinary Medicine (CVM). BP; May 2001. pp. 8-11

[14] HPLC Basics. Fundamentals of liquid chromatography (HPLC). Courtesy of Agilent Technologies, Inc.; 2007

[15] Siddiqui MR, Alothman ZA, Rahman N. Analytical techniques in

pharmaceutical analysis: A review. Arabian Journal of Chemistry. 2017;**10**:S1409-S1421

[16] Srivastava MM. High-Performance Thin Layer Chromatography (HPTLC). Heidelberg Dordrecht London, New York: Springer. 2011. pp. 9-10. DOI: 10.1007/978-3-642-14025-9

[17] Stauffer E, Newman R. Other Techniques of Analysis and the Future of Fire Debris Analysis. ScienceDirect. Amsterdam: Global Business; 2008

[18] Ionin BI, Erhoy BA. Application of NMR spectroscopy in various fields of organic chemistry. In: NMR Spectroscopy in Organic Chemistry. Physical Methods in Organic Chemistry. Boston, MA: Springer; 1970

[19] Hoffmann ED, Stroobant V. Mass Spectrometry Principles and Applications. 3rd ed. Hoboken, New Jersey: John Wiley and Sons, Ltd. 2007. p. 1

[20] Patel KN, Patel JK, Patel MP, Rajput GC, Patel HA. Introduction to hyphenated techniques and their applications in pharmacy. Pharmaceutical Methods. 2010;**1**(1):2-13

Chapter 9

Cost-Effective Technical Tips for Agarose Gel Electrophoresis of Deoxyribonucleic Acid

Noboru Sasagawa

Abstract

Agarose gel electrophoresis is one of the most fundamental experiment in biochemistry and/or molecular biology, especially in analyzing deoxyribonucleic acid (DNA) or ribonucleic acid (RNA). Many laboratories do agarose gel electrophoresis almost every day. Besides, sometimes we need to prepare tens of agarose gels at a time for training and/or practices of students. In such situations, the more cost-effective way we have, the much more experiments in laboratories/trainings of students we can achieve. Actually, experiments of using agarose can be achieved in a more inexpensive way. In this manuscript, conditions of agarose gel electrophoresis experiment (agarose, buffer, and equipment) are considered, and achievements of such efforts are described.

Keywords: agarose, electrophoresis, buffer, equipment

1. Introduction

In molecular biology and biochemistry, the size of biomolecules (molecular weight of protein, length of nucleic acids, and so on) is an important key information in the experiment. One of the popular methods for size fractionation of such molecules is electrophoresis. Agarose is often used as the gel structure for electrophoresis to fractionate nucleic acids. Although polyacrylamide gel is also usable, agarose gel is the most major compound for electrophoresis of nucleic acids, because of its easy handling [1–3]. A fractionation effect depends on the pore size in the gel. Generally, nucleic acids for research have rather large size (around 20 to several thousand base pairs), and the pore size of agarose gel is enough for such large molecule (nucleic acids) to be fractionated. The pore size of 1% agarose gel is estimated to be around 200 nm [4]. Basically, 0.3–2.0% (weight per volume in buffer) of agarose is used in electrophoresis [5].

In this manuscript, several cost-effective ways of agarose gel electrophoresis of DNA are explained. On the other hand, a modification to develop a quality of agarose gel electrophoresis is reported; adding and mixing graphene oxide powder in agarose gel enhances a separation quality of electrophoresis [6]. Furthermore, several modifications of agarose gel electrophoresis by adding a special reagent in the agarose gel are proposed [7, 8]. The principle of these modified electrophoresis methods is basically the same as the traditional method described here, and the cost-saving method in this manuscript will also be applicable for such modified methods.

Agarose gel electrophoresis is a very popular experiment for training of students in educational institutions [9–11]. Cost-effective methods described here should be good news for such institutions, because running costs cannot be ignored in student training practices.

2. Agarose

Agarose is a kind of carbohydrate macromolecules (polysaccharides), also known as a kind of dietary fibers. Agarose is purified from a certain red seaweed *Rhodophyta*. Polysaccharides from *Rhodophyta* mainly consist of agarose and agaropectin. Agaropectin is not capable of forming gels and has to be removed as impurities from the agarose/agaropectin mixture. The quality of agarose depends on such purification steps, and these steps push up agarose commercially much expensive.

2.1 Quality of agarose

In Asian countries, agar is widely known as an ingredient in foods and/or desserts (e.g. mitsumame, yokan, and so on in Japan). Agar is also a well-known gelling reagent for bacteria medium.

"Agarose" and "agar" are sold as different merchandises, but the origin of them is the same; both are made from the same seaweed. It can be said that agar for bacteria medium is a partially and roughly purified form of sea weeds and is of lower quality than agarose for electrophoresis.

In my experience, agar for bacteria medium is quite suitable for a gel electrophoresis reagent. INA AGAR® BA-30 (Ina Food Industry Co., Ltd. (Nagano, Japan)—Funakoshi Co., Ltd. (Tokyo, Japan)) (**Figure 1**, left) is an agar of which grade is for bacteria, but its quality is very good for electrophoresis. The cost for this reagent goes to about 1/5 of standard agarose for electrophoresis. Moreover, Ina agar S-7 (Ina Food Industry Co., Ltd. (Nagano, Japan)) (**Figure 1**, right) is an agar

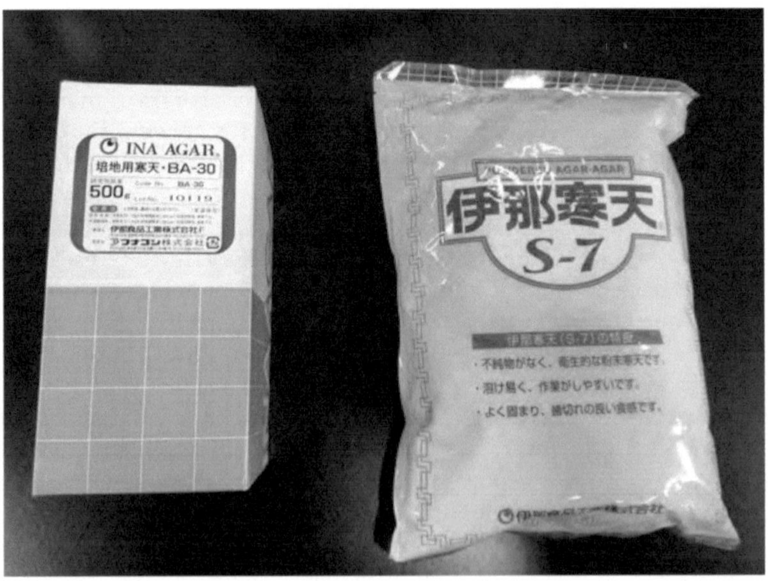

Figure 1.
Agars for not electrophoresis but other use. Left, INA AGAR BA-30 for bacteria medium. Right, INA S-7 agar for cooking.

for cooking, the quality of which is adequate for agarose gel electrophoresis. In this case, the cost is as much as 1/20. Although there is no warranty or trust for results (i.e. a quality test should be done at each package), it is worth doing in each laboratory to test agars for bacteria and/or for cooking. An example of the result by using BA-30 is shown in **Figure 2**, and S7 is shown in **Figure 3**.

Generally, agarose of low purity is more breakable because of its low gel strength. This disadvantage was critical especially when Southern or northern blotting was achieved in the experiment. In recent days, such blotting techniques have given way to the other; for example, polymerase chain reaction (PCR) to see DNA polymorphism and real-time PCR to see gene expressions. The major visualizing way of DNA in agarose gel is to use ethidium bromide (EtBr) or the other DNA intercalators that make fluorescence excited in certain wavelength [12]. Several protocols for staining reagent to intercalate DNA are known; (a) add the reagent in the gel before solidifying, (b) add the reagent in the loading buffer at

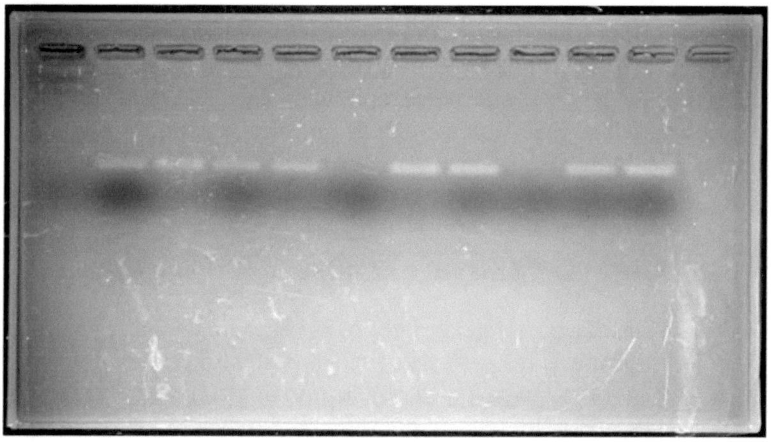

Figure 2.
INA agar BA-30 for agarose gel electrophoresis. 2% weight per volume of agar was applied.

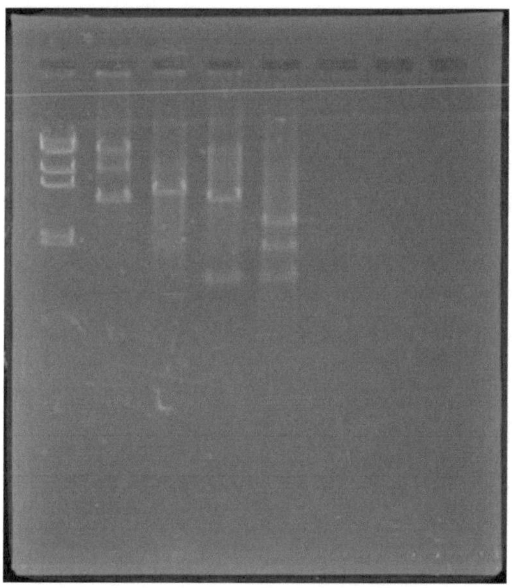

Figure 3.
Result of the electrophoresis by using INA S-7 agar. The gel is 1% weight per volume.

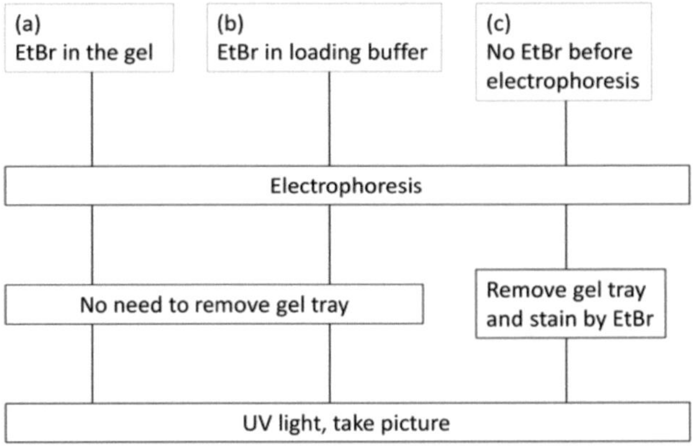

Figure 4.
Scheme for staining and visualizing of DNA in agarose gel electrophoresis.

electrophoresis, and (c) soak the gel in the reagent buffer after electrophoresis (**Figure 4**). In (a) and (b), a photograph of the gel can be taken with a gel tray, when the tray is clear. In such a situation, a low gel strength does not disturb electrophoresis and DNA visualization. Based on these reasons, a low gel strength seems not a too much annoying point when simply doing electrophoresis and taking photographs.

2.2 Recycling and reusing of agarose

It is known that used agarose gel is reusable again and again. Recycling of agarose after electrophoresis is very effective for cost-saving. Several reports are published [13, 14], in which used agarose gels are simply boiled and poured to a gel tray. After cooling to make the recycle gel solid, the recycled gel is enough for applying another electrophoresis. On the other hand, the DNA staining reagent (such as EtBr) still remains in the used agarose gel. EtBr is well known as toxic mutagen [15, 16], so when used and stained gel is boiled, toxic fumes containing

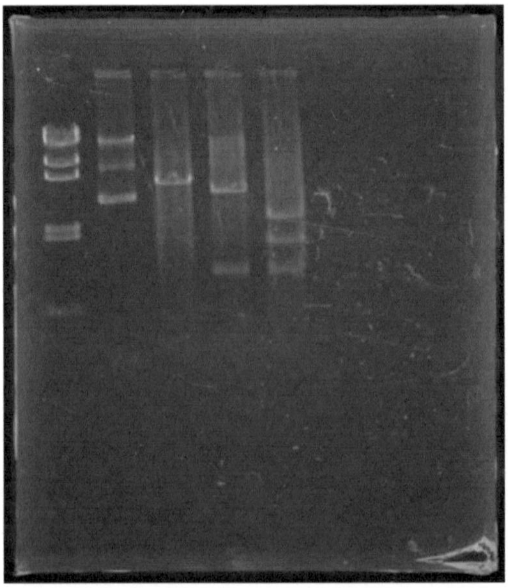

Figure 5.
Recycled gel (1% weight per volume) was applied to the electrophoresis.

EtBr appear. This fume will be hazardous when incorporated through the respiratory system.

To avoid such hazardous fumes, a freeze-and-thaw of used agarose gel is very effective for removing toxic EtBr from the gel [17]. By repeating freeze-and-thaw, the EtBr concentration of used agarose gel dramatically reduces to as much as a negligible level. The result of electrophoresis by recycled agarose is shown in **Figure 5**.

Agarose is hydrolyzed in acidic condition. Therefore, repeating the boiling and melting step in acidic condition might degrade the polymer structure of the agarose. The freeze-and-thaw method mentioned above is free from such a degradation.

3. Electrophoresis buffer

The most standard buffer for agarose gel electrophoresis is TAE buffer (tris, acetic acid, EDTA). TBE (tris, boric acid, EDTA) is the second major buffer. It is said that TBE has an advantage to fractionate small length DNA; in an old sequence analysis, a combination of acrylamide gel and TBE buffer was a standard condition.

When TAE is compared with TBE, the cost of TBE is higher than TAE. This is because of the difference of the price of acetic acid and boric acid.

For RNA electrophoresis, MOPS buffer (MOPS, sodium acetate and EDTA) is another standard, although this buffer is much expensive. Anyhow, daily agarose gel electrophoresis is achieved in a condition of using TAE in standard.

Yet another electrophoresis buffer is SB buffer, which is obtained from sodium borate. The vast majority of SB buffer is the cost, 1/4 of TAE and 1/10 of TBE [18].

In my experience, DNA is well migrated and fractionated in the agarose gel electrophoresis with SB buffer, although small but many air cavities appeared after finishing the electrophoresis. The cavities do not exist when starting the electrophoresis, but they do appear several ten minutes after switching on and/or staining the gel after electrophoresis (**Figure 6**).

3.1 Concentration of buffer

It is a very simple and effective idea of cost-saving that dilution of the buffer is available or not. If 1/2 dilution is available, the cost also will be 1/2. In my experience, 0.5× TAE buffer works fine (**Figure 7**).

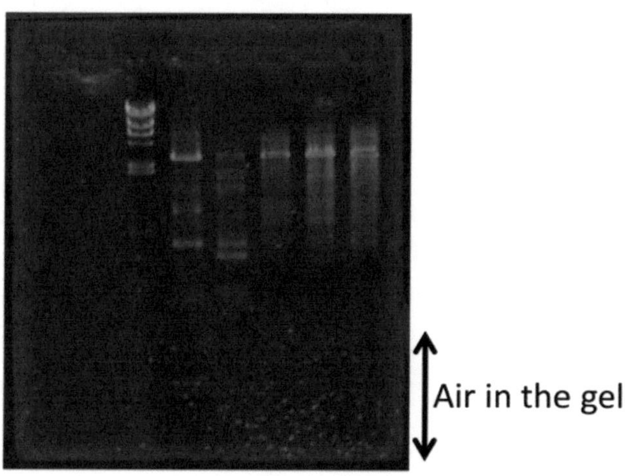

Figure 6.
SB buffer resulted in small but many air cavities in the gel.

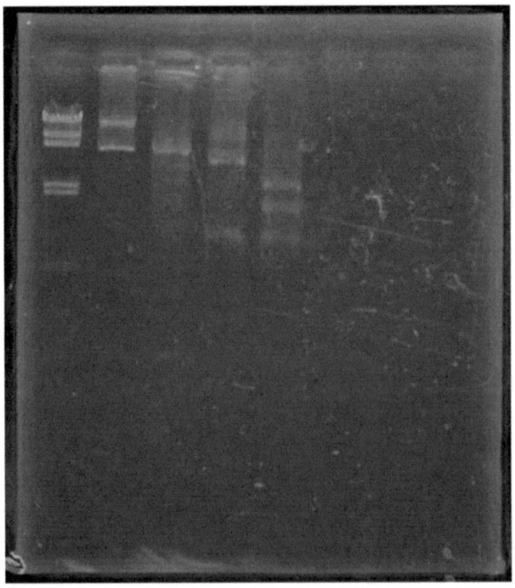

Figure 7.
Agarose gel electrophoresis with 0.5% TAE.

A low concentration of ions in the buffer results in higher resistance in an electric circuit, which leads to heating of the buffer. Therefore, too much dilution of the buffer might result in boiling of buffer and melting of agarose gel.

4. Equipment

There is so much commercial equipment for agarose gel electrophoresis, but unfortunately, they are rather expensive for its purpose; for example, tens of electrophoresis tanks are needed at a time in students' practice, but it is sometimes difficult to buy so many tanks at a time. One of the major reasons of this higher cost is that platinum is used as electrodes in the tank. Platinum is a precious and noble metal, which is very stable and never degraded in electrolysis. The second reason is that the buffer tank of the equipment has a special shape. Generally, the bottom face of the buffer tank has an anti-U-shaped structure (**Figure 8**). The third reason is that the equipment is sold with the special power supply. It seems that no electrophoresis is available without the manufacturer's specified power supply.

4.1 Buffer tank

The anti-U-shaped structure is not always necessary in buffer tank. Basically, a structure of an electrophoresis tank can be much more simple, and we can make it by do-it-yourself (DIY) (**Figure 9**).

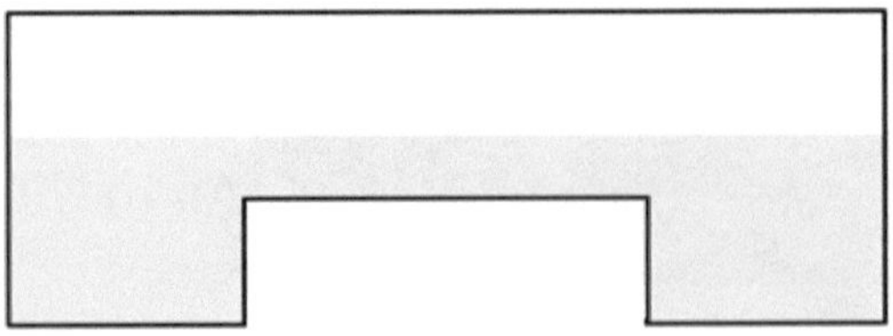

Figure 8.
A horizontal view of typical buffer tank for agarose gel electrophoresis.

Figure 9.
DIY buffer tank for agarose gel electrophoresis. Note that the basket has a flat bottom, unlike a standard commercial buffer tank.

4.1.1 Making DIY buffer tank

A plastic basket in variety store (so-called 100-yen shops, 99 cents store, Dollar store, etc.) is good enough for electrophoresis tank. Plastic tape is put on the basal plane in three- to fourfold repeatedly, which works as a stopper of the gel during electrophoresis (**Figure 9**).

4.1.2 Electrodes

Although carbon stick like a lead of a pencil works as an electrode, stainless steel wire in hardware stores is a good choice of electrodes for agarose gel electrophoresis. No expensive metal is needed; almost the cheapest one will be worth testing. Wireframe of 1-2 mm in diameter leads to a good result. Wires are run at the bottom corner of the tank, simply put by mending tape (**Figure 10**).

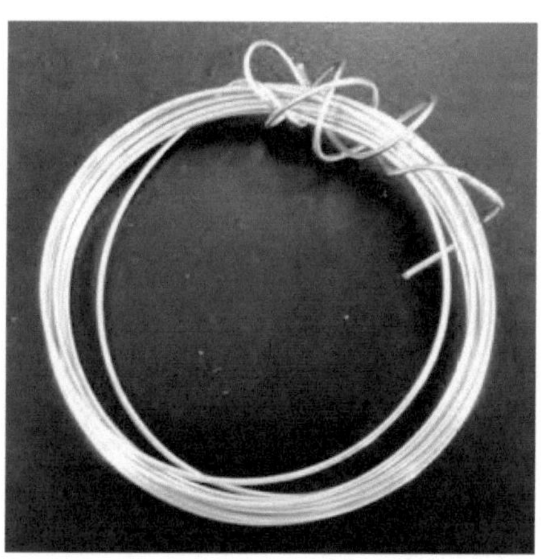

Figure 10.
An inexpensive stainless wire as electrodes of electrophoresis tank. This wire is 1.2 mm in diameter.

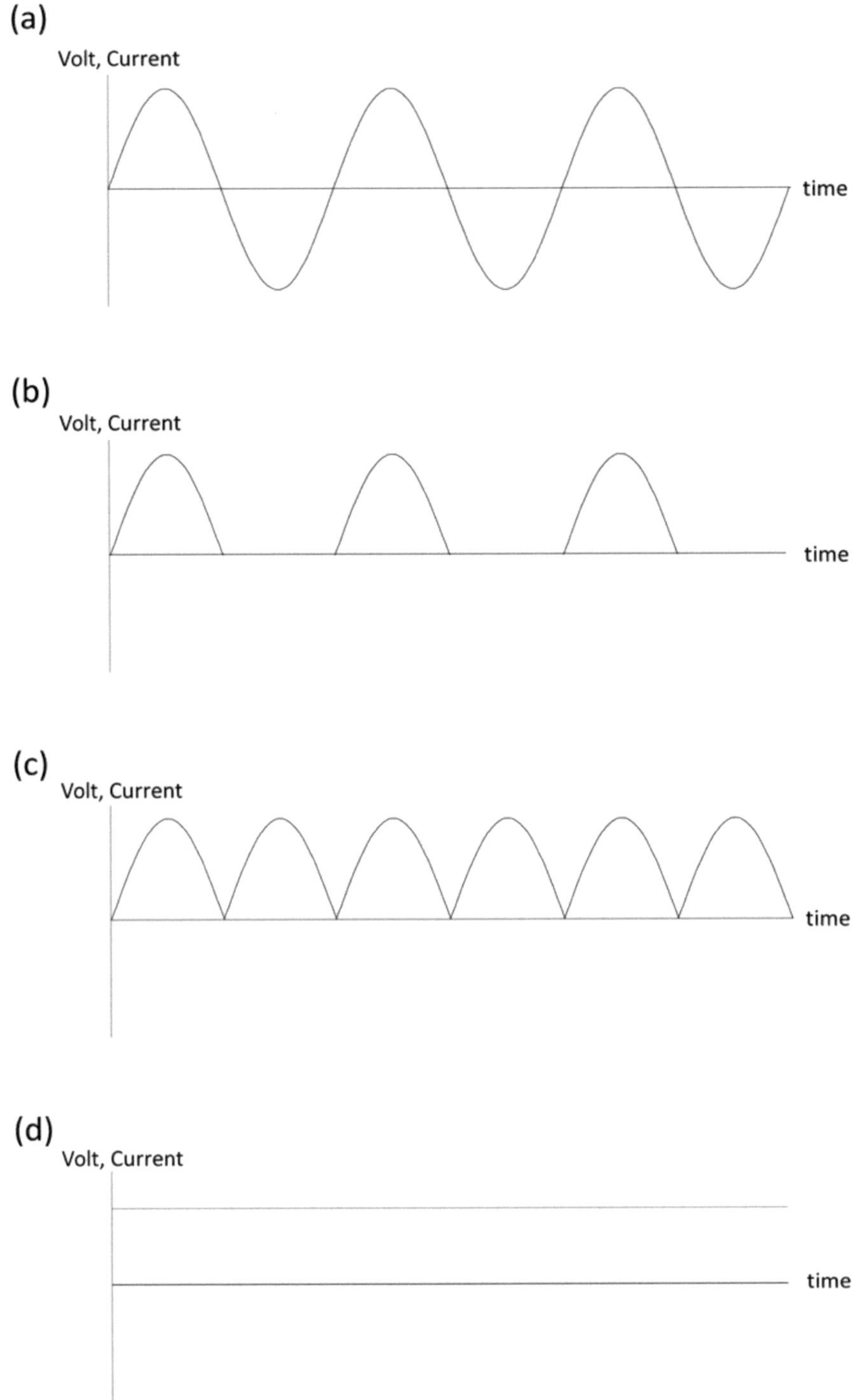

Figure 11.
An alternating current (a), half-wave rectified current (b), full-wave rectified current (c), and a true direct current. (b) and (c) are enough for doing agarose gel electrophoresis, and true direct current like (d) is not needed.

One avoidable notice is that copper is included in the wire as a component of wire. Copper is thought to be toxic, and it is ionized and flow out into the buffer during electrophoresis. Such a wire should be avoided, and it can be easily and clearly determined if copper is ionized because copper ion turns the buffer blue.

4.2 Power supply

Basically, agarose gel electrophoresis is achieved in around 100 volts [19]. In laboratories of molecular biology and biochemistry, power supply for SDS-PAGE is very popular equipment. This supply gives fine and direct current, promising proteins to migrate correctly. Of course, this power supply is also available for use in agarose gel electrophoresis. But actually, such a high-quality direct current is not needed in agarose gel electrophoresis.

Generally, the household electric power is supplied as alternating current (**Figure 11(a)**). This alternating current is not usable as electrophoresis. The current is passed through a diode, and an odd part of the current is picked up (half-wave rectification (**Figure 11(b)**)). A combination of diode enables to make all the alternating current as one direction (full-wave rectification (**Figure 11(c)**)). Half-wave rectified current and full-wave rectified current are a kind of pulsating current, which is not a true direct current (**Figure 11(d)**). It is known that such half-wave or full-wave rectified current is enough for agarose gel electrophoresis [20].

4.2.1 DIY power supply

To make half-wave or full-wave rectified current is not so difficult. **Figure 12** is a diagram showing full-wave rectification from alternating current. Four diodes are needed in this diagram, and they are substituted by one Graetz bridge.

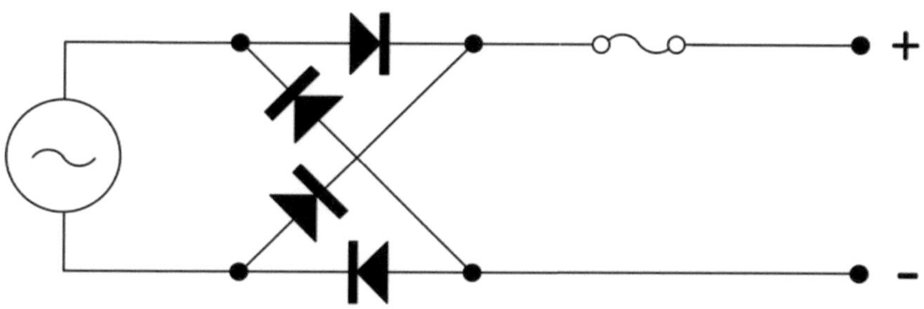

Figure 12.
A simple circuit diagram of full-wave rectification for agarose gel electrophoresis.

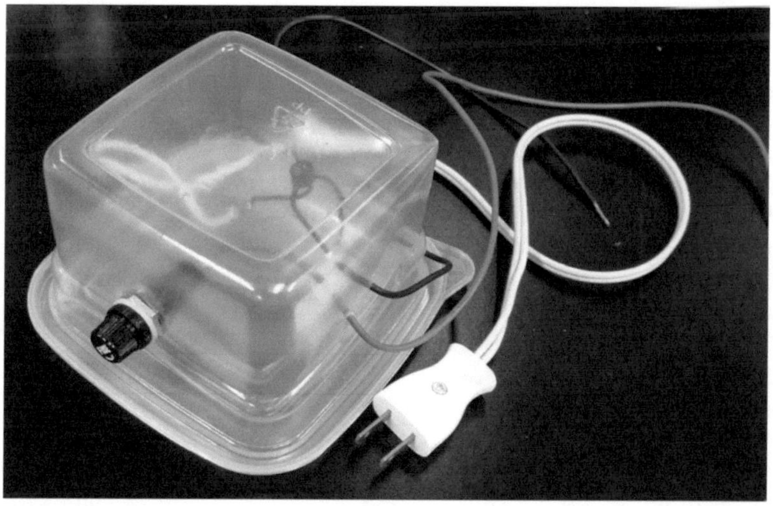

Figure 13.
A DIY power supply based on a diagram of **Figure 12**.

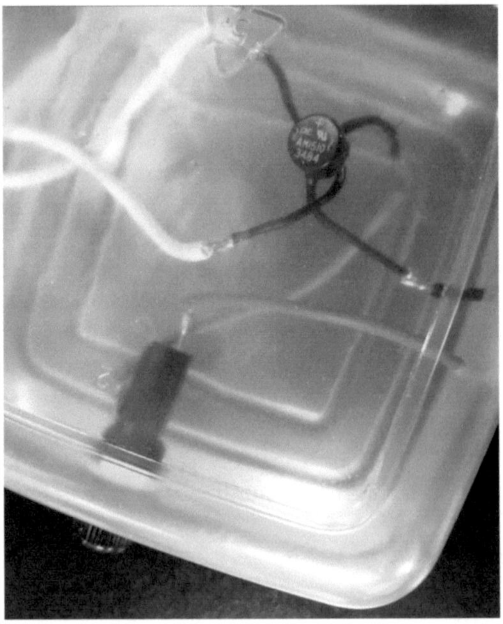

Figure 14.
*A close view of the DIY power supply (**Figure 13**). An inexpensive Graetz bridge (AM1510) is used in the supply.*

Figure 13 is an example of DIY power supply, which is based on the diagram of **Figure 12**. **Figure 14** is a close view of the same DIY power supply, in which only one Graetz bridge is used. A fuse is incorporated in this supply for safety.

5. Concluding remarks

In this manuscript, several technical tips for low-cost agarose gel electrophoresis have been described. The key factor of the tips is agarose (or agar) selection, recycling of agarose, buffer selection, and DIY equipment. Several experiments need a step to recover and isolate fractionated DNA from the agarose gel. In such cases, a high quality of agarose can affect the experiment. Nevertheless, such a high-quality agarose is not always needed for simply checking the band patterns of fractionated DNA. Agarose quality can be changed in its purpose, time, place, and occasion.

Agarose gel electrophoresis is a simple technique. Based on its principle, it can be modified and customized as how much cost you spend to the experiment. Moreover, technical tips described here do not mean downgrading of experiment quality; DNA can migrate and be fractionated as the same way as the standard protocol. The important point is that a calibration test is needed at each reagent and equipment. In my experience, gel strength varies in each product, and concentration of the agar in the gel should be adjusted at each condition.

In this manuscript, the topic has been focused into mainly DNA electrophoresis by agarose gel. RNA is far more sensitive to nuclease (ribonuclease for ribonucleotides) than DNA (deoxyribonuclease for deoxyribonucleotides). This also means that much higher quality of reagents is required for RNA electrophoresis, especially eliminating a contamination of ribonuclease. Moreover, some special technique is required in RNA electrophoresis for denaturation of tertiary structure of single strand RNA. Even though there stand such points to take account of RNA, the buffer tank and power supply in this manuscript will also be able to work in RNA electrophoresis, because of the same principle of the electrophoresis of nucleic acids.

Acknowledgements

I thank Mr. Masayuki Goshima for technical advice and Ms. Haruka Yano and Mr. Yuta Yamada for discussing about experiments.

Conflict of interest

The authors declare no conflict of interest.

Author details

Noboru Sasagawa
Department of Applied Biochemistry, School of Engineering, Tokai University, Kanagawa, Japan

*Address all correspondence to: noboru.sasagawa@tokai-u.jp

References

[1] Sambrook J, Fritsch EF, Maniatis T. Molecular Cloning: A Laboratory Manual. 2nd ed. NY, US: Cold Spring Harbor Laboratory Press; 1989

[2] Barril P, Nates S. Introduction to Agarose and Polyacrylamide Gel Electrophoresis Matrices with Respect to their Detection Sensitivities. Gel Electrophoresis—Principles and Basics. Sameh Magdeldin: IntechOpen; 2012. DOI: 10.5772/38573

[3] Reddy PR, Nomula RN. Gel-Electrophoresis and its Applications. Gel Electrophoresis—Principles and Basics. Sameh Magdeldin: IntechOpen; 2012. DOI: 10.5772/38479

[4] Narayanan J, Xiong JY, Liu XY. Determination of agarose gel pore size: Absorbance measurements vis a vis other techniques. Journal of Physics: Conference Series. 2006;**28**:83-86. DOI: 10.1088/1742-6596/28/1/017

[5] Yılmaz M, Ozic C, Gok I. Principles of Nucleic Acid Separation by Agarose Gel Electrophoresis. Gel Electrophoresis—Principles and Basics. Sameh Magdeldin: IntechOpen; 2012. DOI: 10.5772/38654

[6] Li J, Yang Y, Mao Z, Huang W, Qiu T, Wu Q. Enhanced resolution of DNA separation using agarose gel electrophoresis doped with graphene oxide. Nanoscale Research Letters. 2016; **11**:404. DOI: 10.1186/s11671-016-1609-0

[7] Hegedu E, Kokai E, Kotlyar A, Dombradi V, Szabo G. Separation of 1–23-kb complementary DNA strands by urea–agarose gel electrophoresis. Nucleic Acids Research. 2009;**37**:e112. DOI: 10.1093/nar/gkp539

[8] Harms C, Klarholz I, Hildebrandt A. Two-dimensional agarose gel electrophoresis as a tool to isolate genus- and species-specific repetitive DNA sequences. Analytical Biochemistry. 2000;**284**:6-10. DOI: 10.1006/abio.2000.4693

[9] Lee PY, Costumbrado J, Hsu CY, Kim YH. Agarose gel electrophoresis for the separation of DNA fragments. Journal of Visualized Experiments. 2012;**62**:e3923. DOI: 10.3791/3923

[10] Tan TTM, Tan ZY, Tan WL, PFP L. Gel electrophoresis: DNA science without the DNA! Biochemistry and Molecular Biology Education. 2007;**35**: 342-349

[11] Tweedie JW, Stowell KM. Quantification of DNA by agarose gel electrophoresis and analysis of the Topoisomers of plasmid and M13 DNA following treatment with a restriction endonuclease or DNA topoisomerase I. Biochemistry and Molecular Biology Education. 2005;**33**:28-33

[12] Motohashi K. Development of highly sensitive and low-cost DNA agarose gel electrophoresis detection systems, and evaluation of non-mutagenic and loading dye-type DNA-staining reagents. PLOS One. 2019;**14**: e0222209. DOI: 10.1371/journal.pone.0222209

[13] Palacios G, Giménez C, García ED. Recycling agarose. Plant Molecular Biology Reporter. 2000;**18**:47-49

[14] Seng TY, Singh R, Faridah QZ, Tan SG, Alwee SS. Recycling of superfine resolution agarose gel. Genetics and Molecular Research. 2013; **12**:2360-2367

[15] Lunn G, Sansone EB. Ethidium bromide: Destruction and decontamination of solutions. Analytical Biochemistry. 1987;**162**:453-458

[16] Quillardet P, Hofnung M. Ethidium bromide and safety—Readers suggest

alternative solutions. Trends in Genetics. 1988;**4**:89-90

[17] Sasagawa N. A freeze-and-thaw method to reuse agarose gels for DNA electrophoresis. Bioscience Trends. 2018;**12**:627-629. DOI: 10.5582/bst.2018.01267

[18] Brody JR, Kern SE. Sodium boric acid: A tris-free, cooler conductive medium for DNA electrophoresis. Biotechniques. 2004;**36**:214-216. DOI: 10.2144/04362BM02

[19] Lee SV, Bahaman AR. Discriminatory Power of Agarose Gel Electrophoresis in DNA Fragments Analysis. Gel Electrophoresis—Principles and Basics. Sameh Magdeldin: IntechOpen; 2012. DOI: 10.5772/36891

[20] Kadokami K, Takao K, Saigo KA. A simple, inexpensive "power supply" for multiple electrophoresis. Protein, Nucleic Acid and Enzyme. 1982;**27**: 2108-2111. Japanese

Chapter 10

Application of PCR Technique to Detect Polymorphism of the KRTAP1.1 Gene in Three Sheep Breeds - A Review

Theopoline Omagano Itenge

Abstract

The quality of wool and pelt products depends on the quality of the wool and pelt grown on farm. Genes coding for the proteins involved in the structural components of wool fibre; keratin intermediate filaments (KRTs) and keratin keratin-associated proteins (KAPs) have been extensively researched. The KAPs form a matrix in which the KRTs are embedded. In sheep, KRTAP1.1 (previously B2A) is one of the four genes encoding proteins that make up the KAP1.n family. The ovine KRTAP1.1 gene is clustered with the KRTAP1.3 and KRTAP1.4 genes on chromosome 11. In this chapter, the Polymerase Chain Reaction (PCR) – Applied Fragment Length Polymorphism (AFLP) typing method used to detect polymorphism in the KRTAP1.1 gene is reviewed. Three length variation KRTAP1.1 alleles; named A, B and C, of the lengths 341 base pair, 311 base pair and 281 base pair, respectively have been reported in three sheep breeds; Romney sheep of New Zealand, Merino sheep of New Zealand and Swakara sheep of Namibia. Genetic variation within the KRTs and KAPs can be further exploited to determine as to whether such variation impacts on wool quality. The presence of genetic variation within KRTs and KAPs offers opportunities for the development of gene markers affecting wool and pelt quality traits.

Keywords: genetic variation, KAPs, KRTs, KRTAP1.1, PCR, pelt, sheep, wool

1. Introduction

Domesticated Sheep (*Ovis aries*) are a major source of meat, wool, milk, and pelts/skin or fur, around the s the generic name of hair from animals such as goat, camel, vicuna, alpaca, angora rabbit and yak [1]. Wool grows from follicles on the sheep's skin, similar to the way that hair grows on human skin. Wool fibres are resistant to sunlight, ultraviolet radiation, heat and fire [2]. Because of the unique attributes of wool (excellent insulative properties, breathability and fire resistance [2], the fibre is widely used for clothing, bedding, carpets and other interior textiles.

High quality pelts are produced from Swakara sheep of Namibia, although there are other pelt-producing sheep breeds. Swakara is a fat-tailed sheep that is very hardy and well adapted to arid conditions and mainly kept for the production of pelts [3–6], though it may also be reared for meat and wool. Originating from Uzbekistan in Central Asia, and imported into Namibia in 1907 (as Karakul sheep then) [3], the intensive research and strategic breeding programmes which were subjected to Karakul sheep in Namibia has resulted in the production of a unique breed named Swakara in Namibia [6]. Swakara pelt production is an exclusive industry that produces outstanding quality pelts characterised by short hair, exceptional patterns and better hair texture [7]. The main product produced from Swakara pelt is high quality leather apparel of various colours that is sought after in the fashion industry. Other accessories, such as hand bags, carpets, shoes, car seat covers, belts are processed.

Wool fibre is made up of three main structures: the cortex, cuticle, and in some coarse wools, the medulla [8]. The cortical cells comprise 90% of the wool fibre, and are responsible for the major physical properties of wool fibre [9]. The cortex consists of the microfibrils, made up of keratin intermediate-filament proteins (KRTs) and embedded in a matrix of keratin intermediate-filament-associated proteins (KAPs) [8, 10, 11] through disulphide cross-linkages [12, 13]. The matrix consists of KAPs, and is divided into three groups based on their amino acid compositions; high-sulphur (HS), ultra-high sulphur (UHS) and high-glycine-tyrosine (HGT) KAPs [8, 14].

The HS KAPs occur at relatively high concentrations in the paracortex when compared to the orthocortex [15], and are encoded by five multigenes families referred to as KRTAP1.n, KRTAP2.n, KRTAP3.n, KRTAP11.n and KRTAP13.n [12]. The HS KAPs are highly conserved at both the amino acid and nucleotide sequence levels [16]. However, there is a consecutively repeated decapeptide unit (QTSCCQPTSI), which varies in its frequency between the HS KAPs [17]. This decapeptide occurs between four times in the KRTAP1.1, three times in the KRTAP1.2, twice in the KRTAP1.3 and five times in the KRTAP1.4 protein [18]. The KAP1.n family is composed of four known proteins, referred to as KAP1.1, KAP1.2, KAP1.3 and KAP1.4, previously referred to as B2A, B2B, B2C and B2D, respectively. The genes that code for the proteins making up the KAPl.n family lack introns and they usually occur in gene clusters [19]. The majority of genes coding for the KAPs have a conserved 18-bp sequence that varies slightly, immediately 5′ to the initiation codon [20, 21]. This suggests that KAPs evolved from a common ancestor, and that their expression may have elements in common.

The KRTAP1.1 gene has been mapped to ovine chromosome 11, clustering with other KRTAP genes [22]. KRTAP1.1 has been reported to be polymorphic in Romney sheep breed of New Zealand [18, 23]; Merino sheep breed of New Zealand [24, 25] and Swakara sheep breed of Namibia [26].

Furthermore, the region spanning the KRTAP1.1/KRTAP1.3/KRT33A loci on ovine chromosome 11 has been associated with variation in wool staple strength in Romney sheep [27].

A number of typing methods have been developed to detect polymorphism in genes that code for traits of economic importance in sheep. These methods are used depending on the specific gene being studied, and the resources available to researchers. In this chapter, the PCR – Applied Fragment Length Polymorphism (AFLP) typing method used to detect polymorphism in the KRTAP1.1 gene is reviewed. Other techniques used to identify polymorphism in keratin genes include PCR - Single strand conformational polymorphism (PCR-SSCP) and PCR - Restriction Fragment Length Polymorphism (PCR-RFLP).

Many studies have described genetic variation within genes that code for the KRTs and the KAPs using PCR-SSCP, including those of Gong *et al.*, [28–31], Rogers *et al.*, [18], Itenge-Mweza *et al.*, [24], Chai *et al.,* [32].

Similarly, many studies have described genetic variation within genes that code for the KRTs and the KAPs using PCR-RFLP. The RFLP patterns were obtained by cutting the KAP1.3 locus defining a 598 bp amplimer using *Bsr* I restriction enzyme in sheep breeds by Xu *et al.,* [33], Chen *et al.,* [9], Kumar *et al.*, [34], Mahajan *et al.*, [35] and Meena *et al.,* [36]. Parsons *et al.*, [37] reported a diallelic polymorphism at the KAP6 locus using *Bam*HI PCR-RFLP to give alleles designated A1 (24.5 kb) and A2 (14.1 kb). Rogers *et al.*, [38] reported a di-allelic polymorphism at the KRT33A, formerly known as KRT1.2 defining a 480 bp amplimer in Romney sheep, while Arora *et al.,* [39] found three genotypes (MM, MN, NN) at the KRT33A locus defining 480 bp amplimer in Indian native sheep breeds. Furthermore, Kumar *et al.,* [34] reported three KRT33A genotypes (MM, MN, NN) in Patanwadi and Nali sheep breeds. McLaren *et al.*, [22] reported two alleles at the KRT83, formerly known as KRT2.10 locus using a *Bsr* DI.

2. Polymerase chain reaction (PCR)-agarose gel electrophoresis

2.1 Polymerase chain reaction

Developed by Kary Mullis in 1983, Polymerase Chain Reaction (PCR) is a molecular biology technique that is used to produce relatively large numbers of DNA molecules from very small quantity or poor quality. The reaction involves the amplification of a specific segment of the template DNA. A very important requirement is that the sequence of nucleotides on either side of the sequence of interest must be known, so that primers on either side of the sequence of interest can be designed [40]. Primers are short, single-stranded DNA sequence, typically about 18–30 nucleotides in length, that are used as a starting point in DNA synthesis and define the region of the DNA to be amplified. Primers are also referred to as oligonucleotides. In addition to the template DNA and primers, other ingredients needed for the PCR reactions are the DNA polymerase, all four deoxyribonucleotides (dNTP) and magnesium ion (Mg^{2+}). The DNA polymerase used is usually *Taq* Polymerase, isolated from hot springs bacterium, *Thermus aquaticus*, which can withstand the denaturing temperatures [40]. The cycling can be continued without interruption in PCR machines that are simply programmable water baths that accurately and rapidly can change the water temperature that surrounds the reaction mixture [40]. Up to 96 samples can be processed at a time.

2.2 Agarose gel electrophoresis

Gel electrophoresis is a technique by which charged molecules are separated according to their size, by moving through a gel while an electric current is being applied [41]. Agarose is a macromolecular substance that is derived from the cell walls of a number of genera of red algae, such as *Gelidium* and *Gracilaria* [42]. It can be purified to a whitish granular powder which, when mixed with water and heated, it sets like a jelly. This is called a gel and it is used as a molecular sieve for the DNA molecules that can be characterised by both charge and size [41]. DNA has a negative charge due to the negative charge of its phosphate groups attached to the 5′ carbon of one nucleotide and the 3′ carbon of the next nucleotide. When put in solution and an electric field is applied, DNA fragments move from the negative (black) terminal to the positive (red) terminal because of the net negative charge in solution [41]. The movement of charged molecules is called migration. DNA is loaded into pre-cast wells in the gel and a current applied.

The speed that the DNA travels through a gel is inversely proportional to the size of the DNA. A molecular weight marker (MM) also known as DNA marker or size standard is often included on the gel to give an indication of the fragment size. The fragments in the MM are of a known length, and can therefore be used to approximate the size of the fragments in the samples.

Smaller molecules migrate through the gel more quickly than large DNA molecules, as they are less physically restrained by the gel matrix and therefore travel further than larger fragments that migrate more slowly and will therefore travel a shorter distance. As a result, the molecules are separated by size.

Agarose gel electrophoresis can be affected by the following factors:

- The size of DNA molecule
- The percentage of agarose, which affects the sieving of the DNA molecules.
- The voltage applied during the electrophoresis, which cause the DNA molecules to move.
- The type of agarose
- The electrophoresis buffer.

To make a gel, agarose powder is mixed with a running buffer and heated to a high temperature until all of the agarose powder has melted. The most common gel buffers are TBE buffer (89 mM Tris, 89 mM orthoboric acid, 2 mM Na_2EDTA) and TAE (Tris-acetate-EDTA).The main difference between the two buffers is that TAE buffer has better conductivity than TBE. Therefore, DNA fragments migrate faster in TAE buffer than TBE. Nonetheless, TBE buffer supports better agarose cross-linkage, and better resolution of large DNA fragments is obtained better in TBE buffer and better resolution of smaller DNA fragments in TAE buffer.

The molten gel is left to cool down, and then poured into a gel casting tray before it solidifies. To make wells for the sample, a "comb" is placed at one end of the casting tray. The gel is left to set for about 30 minutes, and the comb is removed. The gel is placed in a gel box, also called electrophoresis tank and can be used immediately, or wrapped in plastic wrap and stored at 4 °C until use [43]. Enough running buffer is added into the tank to cover the surface of the gel. It is important to note that the same running buffer is used as the one used to prepare the gel [43].

The DNA samples are mixed with loading dyes (0.25% bromophenol blue, 0.25% xylene cyanol, 30% glycerol) prior to loading them into the wells of a gel. The loading dyes have three main functions; firstly, they add density to the sample, allowing it to sink into the gel. Secondly, they provide colour and therefore helps to track how far DNA sample has travelled, and finally, they move at standard rates through the gel, allowing for the estimation of the distance that DNA fragments have migrated. Agarose gel electrophoresis is the most effective way of separating DNA fragments of varying sizes ranging from 100 to 25,000 bp [43]. **Table 1** shows the typical agarose gel concentration used to resolve DNA fragments. It is worth noting that base pairs less than 500 are better separated using polyacrylamide gel, with gel percentage between eight to 20%.

After the electrophoresis is complete, the molecules in the gel can be stained with an appropriate dye to make them visible. This may be performed either with Ethidium bromide (EtBr) to a concentration of 0.5–1 μg/ml, silver staining, or coomassie blue dye. Though other methods may also be used to visualise the separation, EtBr is the most common reagent used to stain DNA in agarose gels. It works

DNA fragment (bp)	Agarose gel percentage (%)
1000–30,000	0.5
800–12,000	0.7
500–10,000	1.0
400–7,000	1.2
200–3,000	1.5
100–2,000	2.0

Table 1.
Typical agarose gel concentration for resolving DNA fragments.

by intercalating itself in the DNA molecule in a concentration dependent manner. This allows for an estimation of the amount of DNA in any particular DNA band based on its intensity [43].

To visualise the DNA, the gel is stained with a fluorescent dye that binds to the DNA, and is placed on an ultraviolet transilluminator which will show up the stained DNA as bright bands. EtBr is the most common reagent used to stain DNA in agarose gels. When exposed to ultraviolet light, electrons in the aromatic ring of the EtBr molecule are activated. EtBr works by intercalating itself in the DNA molecule in a concentration dependent manner. This allows for an estimation of the amount of DNA in any particular DNA band based on its intensity.

2.3 Length variation of the KRTAP1.1 gene

2.3.1 KRTAP1.1 primers

KRTAP1.1 primers were designed by Itenge-Mweza *et al.*, [24], designed to amplify a 311 bp fragment of the KRTAP1.1 gene, based on a published gene sequence [10]; GenBank accession number X01610). The primers were: KRTAP1.1up 5'-CAA CCC TCC TCT CAA CCC AAC TCC-3′ and KRTAP1.1dn 5′- CGC TGC TAC CCA CCT GGC CAT A-3′.

2.3.2 Amplification of KRTAP1.1 gene using PCR

PCR amplifications was performed by process discussed by Itenge-Mweza *et al.* [24] using Merino Sheep; Itenge [25], Rogers *et al.*, [18] using Romney sheep and Nyoni *et al.*, [26] using Swakara sheep. Amplification consisted of 1 min denaturation at 95 °C, followed by 30 cycles of denaturation at 95 °C for 1 min, annealing at 65 °C for 1 min and extension at 72 °C for 1 min, with a final extension of 72 °C for 7 min. Amplimers were stored at 4 °C until they were subjected to agarose gel electrophoresis, described in Itenge-Mweza *et al.*, [24] and Rogers *et al.*, [18].

2.3.3 Amplified fragment length polymorphism (AFLP) in KRTAP1.1

KRTAP1.1 amplicons visualised by electrophoresis in 2% SeaKem® LE agarose (FMC Bioproducts) gels, and a photograph was taken for records [24]. A study by Rogers *et al.*, [18] was the first to report length polymorphism in the KRTAP1.1 gene of Romney sheep. Three variants were identified, designated Aα, Aβ and Aγ, (GenBank accession numbers L33885, L33886 and L33887, respectively), and ranging from largest to smallest. The frequency of the Aα, Aβ and Aγ alleles in 19 unrelated Romney sheep was calculated to be 0.13, 0.71 and 0.16, respectively.

A study by Itenge-Mweza *et al,* [24] reported three variants of the KRTAP1.1 gene that varied in length in Merino sheep breed of New Zealand. These were designated as A, B and C (**Figure 1**) and differed in part through variation in the length of a conserved 30 nucleotide repeat sequence; GenBank accession numbers AY835603, AY835603 and AY835605, respectively). The lengths of the coding sequences of alleles A, B and C reported by Itenge-Mweza *et al.*, [24] were 341 bp, 311 bp and 281 bp, respectively, and these compared with the alleles reported by Rogers *et al.,* [18]. Sequence analysis revealed that the length of each of these AFLP bands were 341 bp, 311 bp and 281 bp, respectively (**Figure 2**). Sequence comparison shows that Alleles A and C in Itenge-Mweza *et al.* [24, 25] correspond to those of [17]. On the contrary, Merino allele B (GenBank accession no. AY835604) shows five single nucleotide differences from the sequence by Rogers *et al.*, [18], suggesting a possible fourth KRTAP1.1 allele. Sequencing analysis also confirmed that the length variation detected in Itenge-Mweza *et al.*, [24] was the result of an insertion or deletion of a 30 bp region of the sequence (**Figure 2**).

Three length polymorphism alleles within the KRTAP1.1 gene were identified by Nyoni *et al.*, [26] in Swakara sheep, designated as A, B and C. The band sizes of the alleles (A-C) estimated from the molecular marker included on a 2% agarose gel indicated that they were similar to the sizes reported by Itenge-Mweza *et al.*, [24], of 341 bp, 311 bp and 281 bp, respectively. Sequencing results of the three alleles revealed that the A allele had 100% homology with GenBank accession number AY835603. Allele B had 100% homology with GenBank accession numbers FJ169479 and FJ169479.1, while allele C had 100% homology with GenBank accession number MG641066.

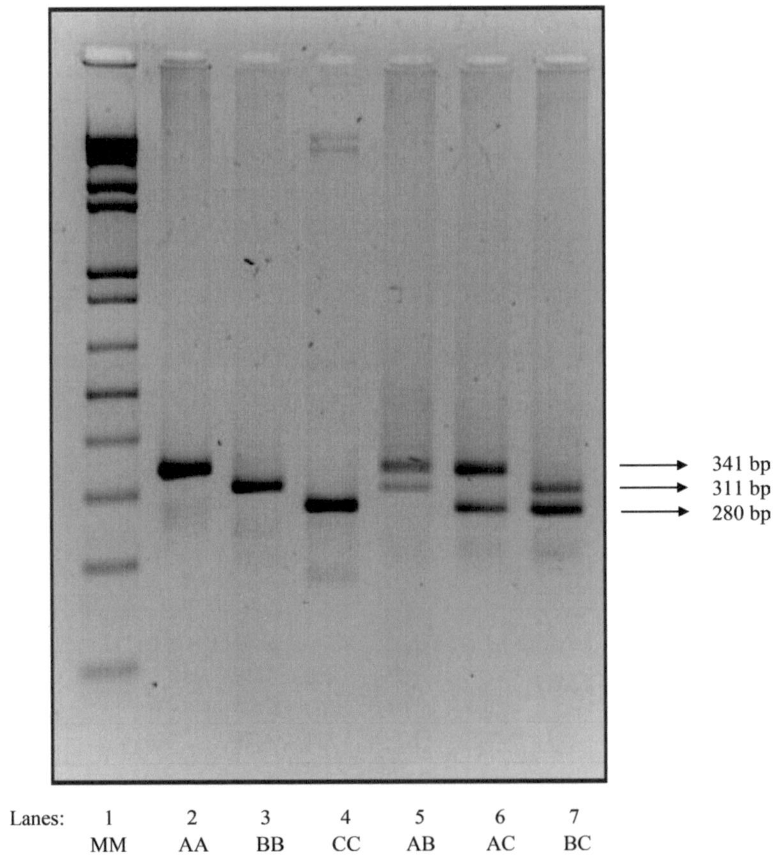

Figure 1.
Length polymorphism within the KRTAP1.1 amplimer, separated in a 2% agarose gel electrophoresis at a constant 10 Vcm^{-1} for 90 min. Lane 1 contains a molecular weight marker (MM) [32].

```
A_(AY835603) CAACCCTCCTCTCAACCCAACTCCTGACACCATGGCCTGCTGTTCCACCA  50
B_(AY835604) --------------------------------------------------  50
C_(AY835605) --------------------------------------------------  50
α_(L33885)   --------------------------------------------------  50
β_(L33886)   --------------------------------------------------  50
γ_(L33887)   --------------------------------------------------  50
X01610       --------------------------------------------------  50

A_(AY835603) GCTTCTGTGGATTTCCCATCTGTTCCACTGGTGGGACCTGTGGCTCCAGT  100
B_(AY835604) --------------------------------------------------  100
C_(AY835605) --------------------------------------------------  100
α_(L33885)   --------------------------------------------------  100
β_(L33886)   -----------------------c--------------------------  100
γ_(L33887)   --------------------------------------------------  100
X01610       --------------------------------------------------  100

A_(AY835603) CCCTGCCAGCAGACCTGCTGCCAGACCAGCTGCTGCCAGCCAACCTCCAT  150
B_(AY835604) ----------c----------------------------------.....  145
C_(AY835605) ----------c----------------------------------.....  145
α_(L33885)   --------------------------------------------------  150
β_(L33886)   ----------c----------------------------------.....  145
γ_(L33887)   ----------c----------------------------------.....  145
X01610       ----------c----------------------------------.....  145

A_(AY835603) CCAGACCAGCTGCTGCCAGCCAACTTCCATCCAGACCAGCTGCTGCCAAC  200
B_(AY835604) .........................-------------------------  170
C_(AY835605) .........................-------------------------  171
α_(L33885)   --------------------------------------------------  200
β_(L33886)   .........................-------------------------  170
γ_(L33887)   .........................-------------------------  171
X01610       .........................-------------------------  170

A_(AY835603) CGATCTCCATCCAGACCAGCTGCTGCCAGCCAACCTCCATCCAGACCAGC  250
B_(AY835604) --------------------------------------------------  220
C_(AY835605) ----..............................----------------  190
α_(L33885)   --------------------------------------------------  250
β_(L33886)   ------t-----------------------------t-------------  220
γ_(L33887)   ----..............................----------------  190
X01610       --------------------------------------------------  220

A_(AY835603) TGCTGCCAGCCAACCTGCCTCCAGACCAGTGGCTGTGAGACGGGCTGTGG  300
B_(AY835604) --------------------------------------------------  270
C_(AY835605) -----------------------------------------c--------  240
α_(L33885)   --------------------------------------------------  300
β_(L33886)   --------------------t-----------------------------  270
γ_(L33887)   -----------------------------------------c--------  240
X01610       --------------------------------------------------  270

A_(AY835603) CATTGGTGGCAGCATTGGCTATGGCCAGGTGGGTAGCAGCG  341
B_(AY835604) -----------------------------------------  311
C_(AY835605) -----------------------------------------  281
α_(L33885)   -----------------------------------------  341
β_(L33886)   ------------------t----------------------  311
γ_(L33887)   -----------------------------------------  281
X01610       -----------------------------------------  311
```

Figure 2.
Comparison of KRTAP1.1 alleles a, B and C (GenBank accession numbers AY835603 - AY835605, respectively) with [11] alleles α, β and γ, (GenBank accession numbers L33885, L33886 and L33887, respectively) and part of the published KRTAP1.1 gene sequence X01610 [32]. Upstream and downstream primers are underlined and the start codon is bolded. Dashes represent similar nucleotide to a allele above and dots represent nucleotide missing in the alleles. Represent chi-like sequences. Represent tandem repeat sequence. It is of note that there is minor sequence variation at the DNA level both within and between the genes.

A major difference between the KRTAP1.1 alleles is the insertion/deletion of a 30 bp nucleotide, which produces a decapeptide (QPTSIQTSCC) in the gene [24]. Interestingly, it was reported by [22, 23, 41] that the inserted/deleted decapeptide is the same as the consecutively repeated decapeptide unit of the other

HS KAPs (KRTAP1.2, KRTAP1.3 and KRTAP1.4. The number of alleles reported in the KRTAP1.2, KRTAP1.3 and KRTAP1.4 genes are two, ten and nine alleles, respectively.

It is not yet clear as to why KRTAP1.1 would have an inserted/deleted decapeptide unit. However, the decapeptide contains two cysteine amino acids. The supply of cysteine is limiting to wool growth [44]. Itenge-Mweza *et al.*, [24] stated that analysis of sheep from diverse breeds may reveal even more KRTAP1.1 alleles, and recommended that a new typing method be developed as the amplified fragment length polymorphism (AFLP) does not differentiate between all of the alleles at the KRTAP1.1 locus in sheep. It was also proposed by Gong *et al.,* [12] that further investigation using other typing methods, rather than AFLP in the KRTAP1.1 is needed, in order that full extent of variation in this gene is revealed, should it exist.

2.4 Length variation of other wool genes

Other than KRTAP1.1, there are other wool genes that have been reported to contain length polymorphism due to a change in the number of repeat sequences. These include trichohyalin (an important wool follicle protein) [45], and the Type II keratins genes [46, 47].

3. Conclusion

Genetic variation within the genes which make up the structural component of wool fibre can be further exploited to determine as to whether such variation impacts on wool quality. The presence of genetic variation within genes that code for the proteins involved in the structural components of wool fibre offers opportunities for the development of gene markers affecting wool quality traits.

Acknowledgements

The author wish to greatly acknowledge the National Commission on Research, Science and Technology (NCRST) for the financial support with the Swakara Project.

Author details

Theopoline Omagano Itenge
University of Namibia, Windhoek, Namibia

*Address all correspondence to: tamushendje@unam.na

References

[1] Höcker, H. (2002) Fibre morphology. In: *Wool: Science and technology*. (Edited by Simpson W. S. and Crawshaw G. H.), pp. 60-79. Woodhead Publishing Limited, Cambridge, England.

[2] Leeder, J.D. (1984). Wool: Nature's Wonder Fibre; Australasian Textiles Publishers: Victoria, Australia.

[3] Bravenboer, B. (2007) *Karakul: Gift from the Arid Land: Namibia 1907-2007*. Karakul Board of Namibia, Windhoek, Namibia.

[4] Campbell, L.J. Evaluation of two indigenous South African sheep breeds as pelt producers. (MSc Thesis). Pretoria, South Africa: University of Pretoria; 2007

[5] Lundie, R. S. (2011) The genetics of colour in fat-tailed sheep: a review. Tropical animal health and production, **43**(7), 1245-1265.

[6] Itenge, T. O. and Shipandeni, M.T.N. (2015) Sale trends of Swakara pelt offered at the Copenhagen fur auction from 1994-2013. Applied Animal Husbandry & Rural Development. **8**:1-5.

[7] Nel, J. A. (1991). Pelt breeds of sheep. In: Karakul. A., Maijala, K. (Eds). Genetic Resources of Pig, Sheep and Goat. World Animal Science. B8. Elsevier Science Publishers B. V., Amsterdam, The Netherlands, pp. 291-299.

[8] Powell, B. C. and Rogers, G. E. (1994) Differentiation in hard keratin tissues: Hair and related structures. In: *Keratinocyte Handbook*. (Edited by Leigh I. M., Lane E. B. and Watt F. M.), pp. 401-436. Cambridge University Press, Cambridge.

[9] Chen, H. Y., Zeng, X. C., Hui, W. Q., Zhao, Z. S., & Jia, B. (2011) Developmental expression patterns and association analysis of sheep KAP8. 1 and KAP1. 3 genes in Chinese Merino sheep. Indian Journal of Animal Sciences, **81**(4):391-396.

[10] Onions, W. J. (1962). *Wool- An introduction to its properties, varieties, uses and production*. Ernest Benn Limited, London, Great Britain.

[11] Powell, B. C. and Rogers, G. E. (1986) Hair keratin: composition, structure and biogenesis, Volume 2. In: *Biology of the Integument.* (Edited by Bereiter-Hahn J., Matoltsy A. G. and Richards K. S.), pp. 695-721. Springer-Verlag, Berlin.

[12] Gong, H., Zhou, H., Forrest, R.H.J., Li, S., Wang, J., Dyer, J. M., Luo, Y. and Hickford, J.G.H. (2016) Wool Keratin-Associated Protein Genes in Sheep—A Review. Gene. **7**(24):1-16; doi:10.3390/genes7060024

[13] Powell, B. C. (1996) The keratin proteins and genes of wool and hair. Wool Technology and Sheep Breeding. **44**:100-118.

[14] Gillespie, J. M. (1990) The proteins of hair and other hard α-keratins. In: *Cellular and Molecular Biology of Intermediate Filament*. (Edited by Goldman R. D. and Steinert P. M.), pp. 95-128, Plenumm Press, New York.

[15] Rogers, G. E. (2006) Biology of the wool follicle: an excursion into a unique tissue interaction system waiting to be re-discovered. Experimental Dermatology. **15**:931-949.

[16] Powell, B. C., Sleigh, M. J., Ward, K. A. and Rogers, G. E. (1983) Mammalian keratin gene families: Organisation of genes coding for the B2 high-sulphur proteins of sheep wool. Nucleic Acids Research. **11**:5327-5346.

[17] Crewther, W. G. (1976) Primary structure and chemical properties of

wool. Proceedings *of the 5th International Wool Textile Research Conference*, Aschen I: 1-101.

[18] Rogers, G. R., Hickford, J.G.H. and Bickerstaffe, R. (1994a) Polymorphism in two genes for B2 high sulphur proteins of wool. Animal Genetics. **25**:407-415

[19] Parsons, Y.M., Piper, L R. and Cooper, D.W. (1994) Linkage relationships between keratin associated protein (KRTAP) genes and growth hormone in sheep. Genomics. **20**:500-502.

[20] Frenkel, M. J., Powell, B. C., Ward, K. A., Sleigh, M. J. and Rogers, G. E. (1989) The Keratin BIIIB gene family: Isolation of cDNA clones and structure of a gene and a related pseudogene. Genomics. **4**:182-191.

[21] Kuczek, E. S. and Rogers, G. E. (1987) Sheep wool (glycine + tyrosine)-rich keratin genes. A family of low sequence homology. *European Journal of Biochemistry*. **166**:79-85.

[22] McLaren, R. J., Rogers, G. R., Davies, K. P., Maddox, J. F. and Montgomery, G. W. (1997) Linkage mapping of wool keratin and keratin-associated protein genes in sheep. Mammalian Genome. **8**:938-940.

[23] Rogers, G. R. Identification of a quantitative trait loci for wool strength in Romney sheep. (PhD Thesis). Christchurch, New Zealand: Lincoln University; 1994

[24] Itenge-Mweza, T.O., Forrest, R.H.J., McKenzie, G.W., Hogan, A., Abbott, J., Amoafo, O. and Hickford, J.G.H. (2007). Polymorphism of the KRTAP1.1, KRTAP1.3 and K33 genes in Merino sheep. Molecular and Cellular Probes. **21**:338-342.

[25] Itenge, T. O. Identification of genetic markers associated with wool quality traits in Merino sheep. (PhD Thesis). Christchurch, New Zealand: Lincoln University; 2007

[26] Nyoni N. F., Itenge T.O., and Shipandeni M.N.T. (2019) Genetic variation in the KRTAP1.1 gene and its potential to supplement Indigenous Knowledge Systems within the Swakara sheep in Namibia. In N.P. Iwuanyanwu and N.O. Vhonani - (Ed.), Proceedings of the 5th Annual Conference of the African Association for the Study of Indigenous Knowledge Systems, (pp. 43-50). Thohoyandou, Limpopo, South Africa: ISBN: 978-0-86808-767-2.

[27] Rogers, G. R., Hickford, J. G. H. and Bickerstaffe, R. (1994b) A potential QTL for wool strength located on ovine chromosome 11. *Proceedings, 5th World Congress on Genetics Applied to Livestock Production, University of Guelph, Guelph, Ontario,* Canada. **21**:291-294.

[28] Gong, H., Zhou, H. and Hickford, J.G.H. (2009). Polymorphism of the ovine keratin-associated protein 1-4 gene (KRTAP1-4). Mol Biol Rep. **37**(7):3377-3380. DOI 10.1007/s11033-009-9925-4

[29] Gong, H., Zhou, H., Yu, Z., Dyer, J., Plowman, J.E., Hickford, J.G.H. (2011) Identification of the ovine keratin-associated protein KAP 1-2 gene (KRTAP1-2). Exp. Dermatol. **20**:815-819.

[30] Gong, H., Zhou, H. and Hickford, J.G.H. (2010) Polymorphism of the ovine keratin-associated protein 1.4 gene (KRTAP1.4). Mol. Biol. Rep. **37**:3377-3380.

[31] Gong, H., Zhou, H., Plowman, J.E., Dyer, J.M. and Hickford, J.G.H. (2012) Search for Variation in the Ovine KAP7-1 and KAP8-1 Genes Using Polymerase Chain Reaction–Single-Stranded Conformational Polymorphism Screening. DNA

Cell Biol. **31**:367-370. https://doi.org/10.1089/dna.2011.1346.

[32] Chai, W., Zhou, H., Forrest, R.H.J., Gong, H., Hodge, S. and Hickford, J.G.H. (2017) Polymorphism of KRT83 and its association with selected wool traits in Merino-cross lambs. Small Ruminant Research. **155**:6-11. http://dx.doi.org/10.1016/j.smallrumres.2017.08.019.

[33] Xu, H.F., Zhao, Z.S., Xue, A.Y., Amina, L.G.L., Ban, Q. and Zhang, J. (2008). Studies on the relation between KAP1.3 gene and 5 breeds of sheep, and with some wool traits. Journal of Shihezi University (Natural Science). **26**(1):60-63.

[34] Kumar, R., Meena, A. S., Kumari, R., Jyotsana, B., Prince, L. L. L., & Kumar, S. (2016) Polymorphism of KRT 1.2 and KAP 1.3 Genes in Indian Sheep Breeds. Indian Journal of Small Ruminants. *22*(1):28-31.

[35] Mahajan, V., Das, A. K., Taggar, R. K., Kumar, D. and Sharma, R. (2017) Association of polymorphic variants of KAP 1.3 gene with wool traits in Rambouillet sheep. Indian Journal of Animal Sciences. **87**(10):1237-1242.

[36] Meena, A. S., Kumar, R., Jyotsana, B., Narula, H. K., & Kumar, S. (2018) Genetic polymorphism of KRT 1.2, KAP 1.3 and THH gene in Magra sheep. Indian Journal of Small Ruminants. **24**(1):27-30.

[37] Parsons, Y. M., Cooper, D. W., Piper, L. R. and Powell, B. C. (1993) A *Bam*HI polymorphism in ovine glycine/tyrosine-rich type II (KRTAP) keratin sequences. Animal Genetics. **24**:218.

[38] Rogers, G. R., Hickford, J. G. H. and Bickerstaffe, R. (1993) *Msp*1 RFLP in the gene for a Type I intermediate filament wool keratin. Animal Genetics. **24**:218.

[39] Arora, R., Bhatia, S. Sehrawat, A., Pandey, A. K., Sharma, R., Mishra, B. P., Jain, A. and Prakash, B. (2008) Genetic Polymorphism of Type 1 Intermediate Filament Wool Keratin Gene in Native Indian Sheep Breeds. Biochem Gene. **46**:549-556. DOI 10.1007/s10528-008-9169-3.

[40] Tamarin, R.H. (1996). Principles of Genetics – Fifth Edition. Wm. C. Brown Publishers, London. pp. 330.

[41] Itenge, T. O. (2012) Identification of polymorphism in the keratin genes (KRTAP3.2, KRTAP6.1, KRTAP7, KRTAP8) and microsatellite BfMS in Merino sheep using polymerase chain reaction-single strand conformational polymorphism (PCR-SSCP) analysis. Chapter publications in: *Electrophoresis*. Edited by Kiumars Ghowsi, pp. 193-220. InTech- ISBN 980-953-307-117-1.

[42] Raven, P.H., Evert, R.F. and Eichhorn, S.E. (1992). Biology of Plants – Fifth Edition. Worth Publishers, Inc. United States of America. pp. 281.

[43] Lee, P.Y., Costumbrado, J., Hsu, C. and Kim, Y.H. (2012). Agarose Gel Electrophoresis for the Separation of DNA Fragments. J Vis Exp. **62**: 3923. doi: 10.3791/3923.

[44] Reis, P. J. (1979) Effects of amino acids on the growth and preperties of wool. In: *Physiological and environmental limitations to wool growth*. (Edited by Black, J. L. and Reis, P. J.), pp. 223-242. University of New England Publications Unit, New South Wales.

[45] Fietz, M. J., McLaughlan, C. J., Campbell, M. T. and Rogers, G. E. (1993) Analysis of the sheep trichohyalin gene: potential structural and calcium-binding roles of trichohyalin in the hair follicle. Journal of Cell Biology. **121**:855-865.

[46] Korge, B. P., Compton, J. G., Steinert, P. M. and Mischke, D. (1992a)

The two size alleles of human keratin 1 are due to a deletion in the glycine-rich carboxyl-terminal V2-subdomain. Journal of Investigative Dermatology. **99**:697-702.

[47] Korge, B. P., Gan, S. Q., McBride, O. W., Mischke, D., and Steinert, P. M. (1992b) Extensive size polymorphism of the human keratin 10 chain resides in the carboxy-terminal V2-subdomain due to variable numbers and sizes of glycine loops. Proceedings of the National Academy of Sciences. **89**:910-914.

Chapter 11

Considerations for Stability of Environmental Samples in Storage for Long-Term Studies

Susan Marie Viet, Maire S.A. Heikkinen and Michael Dellarco

Abstract

It is often advantageous to store collected environmental samples for future retrospective analyses. However, information about sample stability is necessary to determine if there will be analyte loss or gain or degradation under the specified storage conditions and storage period. Failure to evaluate stability could result in inaccurate results and biased exposure assessments. As part of the National Children's Study pilot, we considered which types of environmental samples could be stored for extended periods of time. We conducted an extensive literature review and considered the conduct of long-term stability studies for environmental samples. We present our findings and experience below as guidance for consideration by the environmental research community.

Keywords: Sample stability, Sample storage, Environmental samples, Long-term studies

1. Introduction

For long-term environmental studies, such as for prospective epidemiology studies, it is often advantageous to store collected environmental samples for future retrospective analyses. Traditionally, stored environmental samples have included human tissue and fluids, animal and plant tissues, soils, sediments, and ice cores [1]. Such samples can be used to evaluate results of government policies, health of an animal population, or temporal trends in ecosystems or exposures [2]. In longitudinal studies, this also permits spreading costs of analyses over time – an important consideration as analysis for environmental contaminants can be expensive. Additionally, it provides more flexibility to analyze subsets of samples in nested case–control studies for specific health outcomes or for inclusion of new target analytes or analysis methods. For instance, concern about the presence of pesticides, pharmaceuticals and personal care products (PPCPs) has heightened over the past several years due to the presence of these chemicals in wastewater, groundwater, surface water and drinking water [3–10].

Information about sample stability in long-term studies is critical to determine, if there will be analyte loss or gain or degradation under specified storage conditions and storage period. Failure to evaluate stability could result in inaccurate results and biased exposure assessments due to partial or complete analyte decomposition, chemical transformation, or loss/gain.

The National Children's Study (NCS) was a longitudinal cohort study that aimed to follow 100,000 children from birth until 21 years of age to evaluate the health effects of environmental exposures, including chemical, physical, biological, and psychosocial factors. A variety of different types of samples were collected during the NCS pilot study including environmental samples and biospecimens. The environmental samples that could not be stored because of known instability (e.g., badge samples for air oxidants) were analyzed immediately after collection. Other samples and biospecimens were stored in the NCS repository. The plan was to store the other environmental samples (e.g., water, soil, and dust) and biospecimens (e.g., blood, urine) until children aged, analyzing only samples as requisitioned for children included in subsequent case–control studies.

As part of the NCS pilot, we considered which types of environmental samples could be stored for extended periods of time. Many factors can affect stability of chemical compounds in stored environmental samples over time, such as temperature, humidity, pH, and microbial enzymatic activity of the sample matrix, physicochemical properties of the analytes themselves and their reactions with the matrix, container materials, and other analytes present in the sample [11].

2. Literature survey

The literature was surveyed through 2020 in PubMed, Web of Science and other commercial sources to identify published sample preparation and storage information of contaminants in any environmental matrix. Stability of analytes in analytical standards and in stored biospecimens were not included in this review.

Table 1 summarizes the 63 peer-reviewed articles and 8 reports and book chapters selected after initial screening that discussed stability of anthropogenic chemical compounds in environmental matrices. The most common analytes studied were pesticides and trace elements. Various environmental matrices were considered, including air, water, soil/sediment, dust, and food (plant and animal tissue).

Two major scopes for stability studies were identified: (1) long-term stability in storage (retained samples, environmental specimen banking), and (2) short-term stability of samples in transport conditions. There was no generally accepted procedure for performing a stability study. Stability samples were sometimes prepared and stored all at the same time with some samples removed for analysis at various time periods. Alternatively, samples were prepared and stored at different time points and analyzed all at the same time. The former procedure is subject to day-to-day and longer term variability and changes in analysis procedures; the latter is subject to variability and changes in the preparation and analysis procedures when the time points are far apart.

Another sample stability test procedure, generally used for reference standards and sample transport, is isochronous testing, where samples are prepared at the same time and stored in conditions that offer the best stability, or the least degradation, until they are moved to the storage conditions to be tested for the specified time periods [43]. At the end of the testing period, each sample set is moved back to the original storage until all samples are removed for analysis at the same time. This procedure avoids the challenges of varying sample preparation and analysis conditions.

The determination of stability of an analyte in a sample is measured as the ratio of the concentration measured at time point t, compared to that measured at time point 0. The sample size required to make a decision with specified confidence for each analyte depends on the precision of the analysis method at each concentration level and the amount of change at which samples are considered to have degraded.

Matrix	Analytes	Analysis method(s)	Storage conditions			Results	References
			Temperature	Time	Other		
Air	Endotoxin on PM filters and filter extracts	LAL	4 °C filters, −20°C extracts	3 y		Stored well both in extract and on filter	[12]
	Endotoxin on PM filters and in extracts	LAL	−20°C, 4°C	14 d filters, 24 h extracts	Filter, dessicant	Store both filters and extracts frozen	[13]
	Cat (Fel d 1), dog (Can f 1), mouse (Mus m 1), and mite (DM, Der p 1, TP) allergens	Antigen-specific immunoassay	RT EDC, −20°C and − 80°C extracts	1.5 y	Buffer	EDC 4.8% loss Can f 1 every 30 d, extracts −20°C 1.2% loss DM every 30 d, extract with Tween	[14]
	Ammonia	Indophenol spectrophotometry	RT	0, 1, 2, 4, 6 d	Concentration	Loss 10–17% at 1 d, 40–64% at 6 d, depending on concentration	[15]
	15 volatile organic chemicals (VOC) in SUMMA canister and sorbent tubes	GC-FID, GC-ECD	Not specified	7 d, 30 d		Suitable for compounds tested	[16, 17]
	2,3,3,3-tetrafluoropropene on sorbent	GC-FID	RT, frozen, refrigerated	3, 7, 14, 30 d	sorbent	30 d is acceptable, analysis within 14 d due to migration	[18]
	Formaldehyde, acetaldehyde on sorbent	HPLC	4 °C	4, 12 d		Stable on sorbent for at least 12 d	[19]
	Ortho-phthalaldehyde (OPA) on sorbent	HPLC-MS	Extract refrigerated	5 d		Extract stable for at least 5 d	[20]
	1,2-Dibromo-3-chloropropane (DBCP), 1,3-diClpropene (DCP) on charcoal	GC-ECD	RT, 5.6,-14°C	2, 3, 11 d		Unstable 2 d at RT Stable 1 wk. at −14°C	[21]
	Volatile aliphatic and aromatic solvents on sorbents	GC (detector not specified)	RT, −22°C	7, 14, 28 d	Sorbent	Stable at RT for Chromosorb (except cyclohexanone), better at −22°C for all sorbents	[22]

Matrix	Analytes	Analysis method(s)	Storage conditions			Results	References
			Temperature	Time	Other		
	Ketones on carbon sorbents	GC (detector not specified)	RT, 4°C	1, 7, 30 d	Sorbent, water vapor	Stable 7 days on Anasorb, 30 d on Carboxen 564, except cyclohexanone	[23]
	4-vinyl-1-cyclohexene on charcoal	GC-FID	RT (27–29 °C), 4–6 °C	21 d		Stable 7 days at RT, 21 days refrigerated	[24]
	94 VOC (including terpenes, aromatic, halogenated, and aliphatic compounds) on Tenax GR sorbent	TD/GC–MS-SIM	4°C	1 mo		Stable 1 mo: 87 compounds	[25]
	Hazardous air pollutants (HAPs) in canisters – review of literature	Varied: GC-FID, GC-ECD	RT	Varied: generally 7–14 d, up to 35 d		Stable: 52 HAPs Likely stable: 9 HAPs Likely unstable: 17 HAPs Unknown: 19 HAPs	[26]
(applicable to other matrices)	56 SVOCs (amines, halo ethers, nitrobenzenes, phenols, phthalate esters, polycyclic aromatic hydrocarbons and chlorinated compounds)	GC–MS	Extract purified for analysis: −20°C, 4°C, 22°C	53 days	Vial color (light), fluctuating temperature	SVOCs stable in amber vials, PAHs mostly stable (mon-substituted phenols more affected by temperature), Half-life in clear vials 28–31 d, Constant temperature more important than absolute temperature	[27]
	Polycyclic aromatic hydrocarbons (PAH) on sorbents (XAD2, PYF) and quartz fiber filter	Ref provided	Room temperature, −20°C	10, 20, 30 d	Sorbent, filter	Stable: All at −20°C Stable PUF, RT: PAH Unstable PUF, RT: naphthalene, anthracene, benzo-a-pyrene Stable quartz filter, RT: All but cyclopenta[c,d] pyrene Stable XAD2, RT: All but 2,3-nitrofluoranthrene	[28]

Matrix	Analytes	Analysis method(s)	Storage conditions			Results	References
			Temperature	Time	Other		
	Benzo[a]pyrene on quartz filter	HPLC	−20°C, 20°C	0.5,1,2,4,12 yr	In air, argon	−20°C: Stable for 6 mo, 12% loss at 12 yr. 20°C: Unstable, 50% loss at 12 yr. No effect air vs. argon	[29]
	25 PAHs on PM filters and in extracts	GC–MS-SIM	4°C	7, 30 d filters, 1,5,6 mo extracts		Filters can be stored for 1 mo, extracts for 1 mo, hopanes and steranes for 5 mo	[30]
	Organic solvents on charcoal (ketones, esters, alcohols, aromatic hydrocarbons)	GC-FID/MS	20 °C	1, 2, 4 wk	Carbon sorbent, water, hydroquinone rinse	Unstable: Esters, alcohols, cyclic & aliphatic ketones loss dramatic, more pronounced on one carbon Stable: Toluene, butanol, DMF, styrene, 111-TCA, ethyl acetate Water adsorption affected recovery differently on two carbons Hydroquinone rinse partly effective in improving recovery	[31]
	Cyclic volatile methylsiloxanes (cVMS)	GC–MS	−20°C	14 d	Sorbent type	Stable on ABN Express sorbent, 367–62% loss on ENV+ sorbent	[32]
	30 pharmaceuticals, pesticides on organic-diffusive gradients in thin-films (o-DGT) passive sampler and the polar organic chemical integrative sampler (POCIS)	Not specified	Freezer	o-DGT 18 mo POCIS 6 yr	Sampler type	Average loss 9–14%	[33]

Matrix	Analytes	Analysis method(s)	Storage conditions			Results	References
			Temperature	Time	Other		
	14 Personal care products	GC–MS	4°C	3, 7, 14 d	On sorbent cartridges	Stable 7 d at 4°C	[34]
Water	As(III), As(V)	ICP-MS	6°C	3, 18 mo	pH	Storage in dark, pH < 2, 3 mo	[35]
	As(III), As(V)	ICP-MS	−18°C, 6°C, 20°C	9 d	pH	Storage in dark, 7 d at 6°C	[36]
	As, Se, Sb, Te	HPLC/ICP-MS	−20°C, 3°C, 20°C	30 d	speciation	Stable at 3°C in dark	[37]
	Inorganic Hg, labeled with ^{197}Hg	HPGe	5 °C, 20°C	1, 3, 6, 7, 10 d	Partitioning into particulate, containers	Stable at least for 10 d when acidified and stored in Teflon or glass	[38]
	Perchlorate	IC, LC–MS/MS	4°C, 22°C	638 d		300 d ground water, 90 d surface water	[39]
	44 VOC in wastewater samples	GC–MS	4°C	2–5 wk	Glass vials	Stability varies by VOC: Methylene chloride 2 wks, Chloroform 4 wks.	[40]
	18 bisphenols in wastewater	Not specified	−20°C, 4°C	4 wk		Stable for 4 weeks at either temperature	[41]
	29 PFAS	LC–MS/MS	−20°C, 4°C, 20°C	180 d	Sample type (bottled, surface, effluent)	10 analytes increased or decreased over time, analyte conversion observed in surface and effluent waters	[42]
	Polar pharmaceuticals, pesticides, PFOS, PFOA, caffeine, PBA	LC–MS/MS	4°C, 20°C, 40°C	6 wk		Storage for 6 wk. at 4°C	[43]
	24 pharmaceuticals, pesticides	LC–MS/MS	−20°C	20 mo	In POCIS, in SPE	Small statistically significant losses, add labeled IS	[44]

Matrix	Analytes	Analysis method(s)	Storage conditions			Results	References
			Temperature	Time	Other		
	Pharmaceuticals and pesticides	LC–MS/MS-SRM	4°C	21 d		Steroidal hormones 3 d, others 21 d ok	[45]
	21 organochlorine pesticides on solid-phase extraction (SPE) disks	GC-ECD	−18°C, 4°C	3, 14. 30 d	Sodium azide addition	Stable for at least 30 days at both temperatures, slight improvement for some analytes with sodium azide addition	[46]
	12 pesticides	GC-ECD, HPLC-UV	−20°C, 4°C	3, 30, 90, 180 d	In SPE	Losses in storage for 90 and 180 d storage	[47]
	9 pesticides	GC-NPD	−18°C	119, 319 d	In SPE	119 d recovery 53–155%	[48]
	8 phenylurea pesticides	HPLC	6 °C	21 d	preservatives	21 day storage is ok	[49]
	10 pesticides	GC-MSD	6°C, RT	21 d	In SPE, pH, additives	14 d storage with additives	[50]
	9 organophosphate pesticides	GC-NPD	−20 °C, 4°C, 20°C	60 d	In SPE, drying	60 d at −20°C in dark	[51]
	10 pesticides, 3 atrazine metabolites	LC-APCI-MS, LC-DAD	−20°C, 4°C, RT	1 wk., 3 mo	In SPE	Recovery>90% after storage for 3 mo at −20°C	[52]
	Isoproturon, bentazone, terbuthylazine, alachlor	HPLC-UV	4 °C	14 d, 30 d		Variability due to complexity of matrix	[53]
	OP, OC pesticides, pyrethroids, carbamates	GC-ECD	−17°C, 3°C	2,5,8,14,21,28, 39 d	In SPE, pH	In SPE 6 wk. frozen, otherwise 5 d	[54]
	18 herbicides	LC-APCI-MS	−20°C, 4°C, 20°C	30 d, 60 d	In SPE	Best 60 d at −20°C in dark	[55]
	Phenoxyacid herbicides	LC-DAD-MS	20°C		Water type, bottle type	Half-life river water ~20 d in light, 34–50 d in dark storage; seawater has shorter half-life	[56]

Matrix	Analytes	Analysis method(s)	Storage conditions			Results	References
			Temperature	Time	Other		
	65 Pharmaceuticals and drugs	LC–MS/MS	−20°C	2, 4, 6 wk	In SPE, pH, sample prep	All analytes stored well	[57]
	Pharmaceuticals, PCPs, steroids, hormones	LC/MS/MS	−20°C, 4°C	7, 14, 28 d	Various co-contaminants	Stable 30 d < 6°C, with appropriate preservatives	[7]
	Drugs of abuse	GC–MS/MS	−20°C	1,3,5,7 d, 2,3,12 wk	Preservation, SPE	Stable in SPE at −20°C for at least 3 mo	[58]
	Cocaine	Capillary electrophoresis	8 °C, 25°C, 37°C	1 wk		Store at 8 °C	[59]
	3-mercaptopropionic acid, derivatized	HPLC	4°C	1, 2, 8, 52 wk	buffer	Derivatized samples can be stored at least for 1 y.	[60]
	RDX explosive transformation products	HPLC	4 °C, 23°C, 40°C	45 d	pH, salinity	Stable at 4 °C with 10% sea salts	[61]
Dust/ wipes	Cat (Fel d 1) and mite (Der p 1) allergens	ELISA	−80°C	6, 12, 18, 30 mo		Stable in dust stored in vials for 30 mo	[62]
	Cat (Fel d 1) and mite (Der p 1, Der f 1) allergens, β(1 → 3)-glucan, endotoxin	ELISA, LAL	−20°C	10 mo		Stable in dust stored on vacuum filters stored in plastic bags	[63]
	181 VOCs on activated charcoal cloth (ACC)	GC	−80°C	4 wk	Concentration	Majority stable, exception is n-pentane ~40% loss	[64]
Soil/ sediment	Organophosphate pesticides, organochlorine pesticides, pyrethroids, carbamates	GC-ECD	−17°C, 3°C	2,5,8,14,21,28, 39 d		1 mo frozen	[54]
	4 OP pesticides (diazinon, chlorpyrifos, malathion, fenamiphos) in spiked soil samples	GC	2 °C, Room temp	8 mo		Fenamiphos: Stable Diazinon, CPF, Malathion: Unstable, more so at RT. Store refrigerated or in deep freeze.	[65]

DOI: http://dx.doi.org/10.5772/intechopen.97445

Matrix	Analytes	Analysis method(s)	Storage conditions			Results	References
			Temperature	Time	Other		
	Pesticides	GC–MS	−20°C, 4°C	7, 14, 28, 42, 84, 168 d		Storage for 299 d at −20°C	[66]
	As, Se, Sb, Te	HPLC/ICP-MS	−20°C, 3°C	30 d	speciation	Soil extracts storable 3 d at 3 °C	[37]
	As(III), As(V), monomethylarsonic acid, dimethylarsinic acid in reducing and oxidizing soil	HPLC/ICP-MS of extract	−20°C	1, 2, 3 mo	Extraction solvent	Extracts: MMA, DMA stable, AsIII oxidized to AsV	[67]
	As species in mineral and organic soils	HPLC/ICP-MS	−20°C, 2°C	1 mo	Drying	Storage and drying changes speciation Unstable, freeze-dried samples more stable than wet samples	[68]
	As, Cd, Cr, Cu, Ni, Pb, Zr	FAAS, ZETAAS (Zeeman single-beam AAS), ICP-AES	20 °C, 40°C	24 mo		Storage above 20 °C should be avoided	[69]
	As, Cu, Pb, Zn	ICP-MS	4 °C, RT	90, 183, 284, 392 d	Air drying	Storage at −20°C: 364 709 d	[16]
	Cr(VI)	AAS	−20°C, 4°C	14, 28, 56, 112, 224 d		Storage for 224 d at −20°C	[16]
	Cd, Cr, Cu, Ni, Pb, Zn in irrigation field soil	ICP-AES, ETAAS	−20°C, 4°C, 20°C, 40°C	3, 6, 12, 18 mo	Sterilization	Unstable above 20°C Sterilized samples more stable than non-sterilized	[70]
	C, P, S, Al, Ba, Ca, Mg, Mn, Sr., Zn in forest soil	ICP	−21°C, 3°C, 22°C	1842 d (5 yr)		Increase in all at 22°C due to decreased pH >50 d, store at −20°C or lower	[70]
	Hg	Cold vapor AAS	RT	10, 11, 12, 17 yr	Soil carbon content	Stable	[71]

Matrix	Analytes	Analysis method(s)	Storage conditions			Results	References
			Temperature	Time	Other		
	Methylmercury	Atomic fluorescence detection	30 °C	1 hr., 1 d, 4 d, 7 d	Drying method, soil type	Freeze sample immediately after collection and freeze-dry to prevent bacterial mercury methylation	[72]
	3 PCB, 7 congeners	GC-ECD	−20°C, 4°C	7, 14, 28, 42, 84, 168 d		Storage for 168 d at −20°C	[73]
	17 PAH	GC–MS	−20°C, 4°C	7, 14, 28, 42, 84, 168 d		Storage for 100 d at −20°C	[73]
	3-mercaptopropionic acid, derivatized	HPLC	4°C	1, 2, 8, 52 wk	Buffer	Derivatized samples can be stored at least for 1 y.	[58]
	13 PAH in contaminated soils	HPLC-FID	−20°C, 4°C	6 wk. frozen, 8–10 mo	In dark, additives	Some stable for 8 mo, some only 2 wk. at 4°C (e.g., 3-ring PAH > degradation than 5-ring PAH) Sodium azide-contamination stabilized PAH degradation	[74]
	16 PAH	GC–MS	RT < 30°C	4, 8, 12, 16 mo		Stable in air-dried, ground sediment in shade for 16 mo	[75]
	2 PAH (phenanthrene, pyrene)	GC	Freeze–thaw cycle (−15°C 8 h, 25°C 8 h)	Extraction efficiency at 1, 4, 8, 13, 16, 30 and 120 d	Soil organic content	Competitive effect on extraction efficiency with soil organic content and freeze–thaw cycle	[76]
Plant	As species in needles and mosses	HPLC/ICP-MS	−20°C, 2°C	1 mo	Drying	Storage and drying changes speciation - Unstable, freeze-dried samples more stable than wet samples	[65]

Matrix	Analytes	Analysis method(s)	Storage conditions			Results	References
			Temperature	Time	Other		
	As(III), As(V), monomethylarsonic acid, dimethylarsinic acid in rice	HPLC/ICP-MS	−4°C	1, 2, 3 mo	Extraction solvent	Extracts stable at least for 3 mo.	[67]
	As species in algae	HPLC/ICP-MS	−80°C, −18°C, 4°C, RT	1 yr	Preprocessing	Lyophilization and cryogenic grinding and storage in RT, other temp ok	[77]
	Arsenic species in marine microalgae	LC-ICP-MS	−80°C, −18°C, 4°C, RT	1, 45 d	Processing	Store dried at RT, non-dried at 4°C, freezing results in loss of As	[78]
	19 pesticides in various vegetables, fruits, grains, seeds	HPLC, GC-NPD, GC-ECD	−20°C	1 yr	Hydrolysis, glass bottles	Storage stability referred from hydrolytic behavior: Half-life >10 d at 70°C - stable at least 1 y at −20°C. Half-life <1 d: Unstable	[79]
Animal tissue	As, Se, Sb, Te	HPLC/ICP-MS	−20°C, 3°C	30 d	Speciation	Should be analyzed immediately to avoid transformations	[37]
	Hg-total, Hg-org	CV-AAS	−150°C, −80°C, −30°C, 4°C, 25°C	1 yr		Biobanking: stable after first year	[80]
	As(III), As(V), monomethylarsonic acid, dimethylarsinic acid in fish and chicken	HPLC/ICP-MS	−20°C	3 mo	Speciation	Extracts: Stable for 2 mo, then AsB transformed to DMA	[68]

Matrix	Analytes	Analysis method(s)	Storage conditions			Results	References
			Temperature	Time	Other		
	N-methylcarbamates in beef, duck, chicken liver	HPLC	−4 °C	0.5, 1, 1.5, 2, 3, 4, 5, 6 mo		Store at cryogenic conditions for preparation and storage due to enzymatic activity. Some analytes stable for 6 mo, some not at all	[81]

AAS – Atomic absorption spectrometry, AES – Atomic emission spectrometry, APCI – Atmospheric pressure chemical ionization, DAD – Diode array detector, ECD – Electron capture detection, EDC - Electrostatic dust collector, ELISA - Enzyme-linked immunosorbent assay, FID – Flame ionization detection, GC – Gas chromatography, HPLC – High pressure liquid chromatography, ICP – Inductively coupled plasma, LAL – Limulus amoebocyte lysate, LC – Liquid chromatography, MS – Mass spectrometry, NPD - Nitrogen phosphorous detection, SIM - Selective ion monitoring, SRM – Selected reaction monitorin.

Table 1.
Summary of environmental sample stability studies identified in the literature by matrix and analytes.

The change in concentration over time that indicates significant degradation is often set to 5 or 10 percent, particularly for biological samples. This level is likely too stringent for environmental samples where, for instance, sample processing may itself introduce a reduction of 5 to 10 percent and, therefore, a decision criterion of 20 percent change is most often used.

Despite the widespread practice and numerous and strong benefits of long-term environmental sample storage, we found very little documentation to support preservation of analytes during long-term storage. The findings of the literature review were disappointing in that many questions went mostly unanswered for many sample matrices and analytes of interest, in particular:

- ***What are acceptable long term storage times for various sample types?*** Storage times in the identified studies varied from a few days up to five years. One study extended the storage study to 12 years for air samples [82]. For water, only a few published reports were found on storage stability over 6months, mostly on pesticide stability. These studies do not fully address the compounds or tap water matrix [4, 7, 28, 73, 83–86].

- ***What are acceptable long term storage conditions for various samples types?*** Storage conditions in the identified studies included ambient or room temperature (generally around 20°C), refrigerator (4°C), freezer (−20°C), and/or cryogenic temperatures (−60–80°C). Environmental specimen banks store samples for 50–100 years in cryogenic conditions [79, 80, 87], but do not document the rationale for sample stability under these conditions and time periods. For longer-term storage, some researchers have extracted water contaminants through solid phase extraction (SPE) cartridges and stored frozen for up to a year. Interestingly, colder storage does not always equate to less analyte degradation, as commonly assumed [74].

We also reviewed standard analysis methods for the samples and analytes of interest in the NCS. We found that standard methods are generally focused on regulatory compliance and do not consider long term storage. For example, for ambient vapor and gas sampling, updated storage studies of SUMMA and whole air canisters are needed (US EPA method allows storage of only 30days).

3. NCS stability study experience

The overall objective of the planned NCS storage studies was to determine the stability of target analytes in NCS environmental samples stored in collection containers at specific storage conditions, including reconstitution after thawing and re-freezing of samples, as applicable. For instance, the objective of the tap water study was to evaluate the effect of prolonged storage at −20°C on the stability of the pesticide and pharmaceutical target analytes spiked in analyte-free water. The results of this study were to be used to extrapolate the observed changes in the target analyte concentrations to the stability of those compounds in the NCS study samples.

3.1 Planning

The stability study plans included stability samples similar to environmental field samples, sample processing, analytes or classes of analytes of interest, and analysis methods. The in-depth literature review addressed the prevalence of potential analytes and classes of analytes of interest in an indoor environment, concentrations measured, sampling and analytical methods, reaction or degradation products in these matrices, and storage conditions and stabilities. In addition, experts, agencies and groups with relevant experience, e.g., National Institute of Standards and Technology (NIST), the U.S. Environmental Protection Agency (EPA), the Children's Center Study, National Human Exposure Assessment Survey (NHEXAS), Children's Total Exposure to Persistent Pesticides and Other Persistent Organic Pollutants (CTEPP), and New York State Department of Health/Love Canal project researchers, were contacted for information. This information was used to develop the plans for the preparation and analysis of the stability samples and archiving of both the field and stability samples.

Storage effects due to freeze–thaw cycles, sample pretreatment, and reconstitution were also considered. For example, freeze–thaw cycles can affect the volatile and semi-volatile compound concentrations on a filter or in bulk dust. The pretreatment of samples before storage is performed when metabolism or reaction after collection is known to be of concern. For example, water samples for metals analysis are usually acidified to $pH < 2$ [86], and dust samples are sometimes irradiated to stop microbial growth and metabolism.

3.2 Acquisition of samples

Stability study samples with known concentrations of analytes were planned to be acquired in similar matrices to the NCS samples, i.e., particulate filters, wipes, and vacuum filters. These samples were to include (1) newly-prepared samples that have been spiked with known concentrations of analytes, and (2) pre-tested 'real-world' samples.

Surface dust wipe and air diffusive samples could be made by spiking study sampling media and tap water samples by spiking analyte-free water with known amounts of the target analytes. These kind of samples can be purchased from reference material producers, proficiency testing sample providers, and/or accredited laboratories. To allow for a possible difference in reaction/degradation rate of analytes with their concentration in matrix, each matrix was to be spiked at environmentally relevant concentration levels, based on the literature review, and at concentrations that can be most reliably quantified by the analytical method employed, preferably at concentrations near the midpoint of the instrumental calibration range. Another benefit of this approach was that the concentration of the stability samples could be set to match the concentration used in the laboratory QC samples for increased comparability [7].

Some samples cannot be spiked onto matrices in a manner representative of the corresponding environmental samples, or it would be very difficult to do so. There may be a difference in behavior of analytes in environmental samples and in similar spiked matrices due to pH, microbial enzymatic activity of the sample matrix, physicochemical properties of the analytes themselves and their reactions with the matrix, and other analytes present in the environment. For example, for vacuum filter and deposition plate stability samples, dust from non-participant homes could be obtained, homogenized, and aliquoted before storage.

Each spiked and real-world sample should represent the environmental sample matrix as close as possible to account for inter-analyte and analyte-matrix interactions. If it is not possible to include all analytes of interest in one matrix sample, they can be classified into groups by their chemical characteristics, and include representative analytes of each group deemed most unstable. Because sample homogeneity is very important, for example, surface wipe and dust samples should be aliquoted before spiking and the whole sample used for analysis.

All samples were planned to be collected on exactly the same collection media using the same samplers and stored in the same containers as used in the field study and one manufacturer's lot of each media and containers used, whenever possible. Blank media was to be stored together with the samples for each sample type and analysis time point and unused sample media provided to the analysis laboratories, as required. There should be additional samples stored as back-up for sample loss due to, for instance, filter or container breakage, and for repeat analyses. Additional samples should also be stored to account for the possibility that laboratories or analysis methods may change during the study and simultaneous measurements are necessitated. For each sample type, to the extent possible, samples should be analyzed by one laboratory in order to minimize inter-laboratory bias. For NCS, standard laboratory methods were specified for each analysis, identical to those planned for the field sample analysis.

3.3 Number of samples and analysis time points

Samples are usually analyzed before storage (time point 0) to determine analysis method recovery, concentrations at the beginning of the study (time 0), and to verify the homogeneity of the aliquots. This will take into account any analyte losses during pre-analysis, such as extraction and processing. The frequency of future time points and the number of samples required for statistical significance at each point should be planned carefully before start of the study.

The frequency of time points is chosen to detect the instability of samples and analytes, such as VOC, as early as possible. It is assumed that in most cases, particularly for the organic analytes, the rate of degradation would be highest at the beginning, that is, first order kinetics; this corresponds to a linear change in the log transformed measurement over time [65, 82, 88]. In the NCS stability plans, at each test time t, it is tested statistically with a t-test if a specified percentile of the concentration measurements (the target percentile) will be greater than the specified limit of 80 percent at some future time, assuming stability decreases exponentially (or the log transformed stability decreases linearly) with time. For calculating sample size, it was assumed that the log transformed ratios have a normal distribution and the mean log transformed recovery decreases linearly over time.

As the study proceeds there will be accumulating data from prior test times that can also be used to get a more precise estimate of the slope or fit a nonlinear trend. Calculating samples size when using all prior data is more complicated than calculating sample size for a t-test. The calculations depend on the analysis that might be performed. The number of samples depends on: the alpha and target power for a slope of zero; the standard deviation of the measurement error; the standard deviation of the slope factor; the target percentile (80%); and, the timing of the time points. The test-wise alpha and power for no degradation were set to achieve an overall mean alpha of 5% and an overall power of 95%.

The standard deviation of the measurement error can be estimated from the variation among replicate samples or aliquots, typically measured as a relative standard deviation or, equivalently, a coefficient of variation (CV). The sample sizes were calculated using the error term calculated from the estimated coefficient of variation of the measurement methods. This model was assumed to be adequate for calculating sample sizes and for a variety of sample collection designs, analyses, and distributional or statistical model assumptions. The planned frequency of time points for the NCS was 0, 1, 2, 3, 5, 8, 13, 21, 32, 48, 72, 108, 160, and 240months, altogether 14 time points (or test times). The numbers of planned stability samples varied by sample type and were mostly around 300. The variation was due to the variation of the CV of different analysis methods.

4. Conclusions

Collection and analysis of environmental samples for various analytes in large-scale or longitudinal cohort studies is useful to investigate the contribution of the environment on health outcomes. Storing samples for long periods of time is necessary but expensive in these kinds of studies. Information about preserving sample quality and analyte stability in stored environmental samples is limited. Design and implementation of sample stability studies like those described here are recommended to ensure that samples are stored properly and generate reliable analytical results when required. Publication of environmental sample stability study results will likely provide valuable information to investigators who design and implement large-scale longitudinal environmental health studies.

5. Considerations for the environmental research community

Storage of environmental samples is an important component of large-scale and prospective studies. The environmental research community must address and document the answers to the questions above by conducting and providing data on stability of samples stored over long periods.

- If more stability data are available than presented here, these data should be provided in technical guidelines, study manuals, or published papers.

- Conduct and report on environmental sample storage stability studies, especially for new analytes of interest.

- In the interim, based on our research, a sample stability program should be integrated with sample collection as a part of the quality assurance procedures for the study.

Acknowledgements

This work was conducted under National Institutes of Health, National Institute of Child Health Contract GS-23F-8144H, Order HHSN275201300092U. The authors wish to thank Roslyn Hennessey for initial screening review of articles identified by the literature search, and Barbara Beard for the review of storage stability considerations.

Author details

Susan Marie Viet[1*], Maire S.A. Heikkinen[2] and Michael Dellarco[3]

1 Westat, Rockville, MD, USA

2 Mount Sinai School of Medicine, New York, NY, USA

3 Johns Hopkins Whiting School of Engineering, Baltimore, MD, USA

*Address all correspondence to: susanviet@westat.com

References

[1] Becker, P.R., Gunter, E.W., Schluter, C., Shibata, Y., Wise, S.A. Environmental Specimen Banking. J *Environ Mon* 2006, 8, 776-778.

[2] Rüdel, H., Schröder, W., von der Trenck, K.T., Wiesmüller, G.A. Substance-related Environmental Monitoring: Work Group 'Environmental Monitoring'-Position Paper. *Environ Sci Pollut Res Int* 2009, 16(5), 486-98.

[3] Richardson, S.D., Kimura, S.Y. Emerging environmental contaminants: Challenges facing our next generation and potential engineering solutions. *Environ Technol Innovation* 2017, 8, 40-56.

[4] Reif, A.G., Crawford, J.K., Loper, C.A., Proctor, A., Manning, R., Titler, R. *Occurrence of Pharmaceuticals, Hormones, and Organic Wastewater Compounds in Pennsylvania Waters, 2006-09.* Scientific Investigations Report 2012-5106, U.S. Department of the Interior U.S. Geological Survey with Pennsylvania Department of Environmental Protection, Reston, VA, 2012.

[5] Jelic, A., Gros, M., Ginebreda, A., Cespedes-Sánchez, R., Ventura, F., Petrovic, M., Barcelo, D. Occurrence, Partition and Removal of Pharmaceuticals in Sewage Water and Sludge During Wastewater Treatment. *Water Res* 2011, 45(3), 1165-76.

[6] Phillips, P.J., Smith, S.G., Kolpin, D.W., Zaugg, S.D., Buxton, H.T., Furlong, E.T., Esposito, K., Stinson, B. Pharmaceutical Formulation Facilities as Sources of Opioids and Other Pharmaceuticals to Wastewater Treatment Plant Effluents. *Environ Sci Technol* 2010, 44, 13, 4910-4916.

[7] USEPA. *Stability of Pharmaceuticals, Personal Care Products, Steroids, and Hormones in Aqueous Samples, POTW Effluents, and Biosolids.* EPA-820-R-10-008, Office of Water, U.S. Environmental Protection Agency: Washington, D.C., September, 2010.

[8] Lin, A. and Tsai, Y-T. Occurrence of pharmaceuticals in Taiwan's surface waters: Impact of waste streams from hospitals and pharmaceutical production facilities. *Sci Tot Environ* 2009, 407(12), 3793-3802.

[9] Illinois EPA. *Report on Pharmaceuticals and Personal Care Products in Illinois Drinking Water*. Bureau of Water, Illinois Environmental Protection Agency, June, 2008.

[10] Stackelberg, P.E., Gibs, J., Furlong, E.T., Meyer, M.T., Zaugg, S.D., Lippincott, R.L. Efficiency of conventional drinking-water-treatment processes in removal of pharmaceuticals and other organic compounds. *Sci Tot Environ* 2007, 377(2-3), 255-272.

[11] Dean, J.R., Fitzpatrick, L.J. Pesticides Defined by Matrix. In Handbook of Analytical Separations, Kleibohmer, W., Ed., Elsevier: New York, *Environ Analysis* 2001, 3, 123-173.

[12] Morgenstern, V., Bischof, W., Koch, A., Heinrich, J. Measurements of Endotoxin on Ambient Loaded PM Filters After Long-Term Storage. *Sci Total Environ* 2006, *370*, 574-579.

[13] Spaan, S., Heederik, D.J., Thorne, P.S., Wouters, I.M. Optimization of Airborne Endotoxin Exposure Assessment: Effects of Filter Type, Transport Conditions, Extraction Solutions, and Storage of Samples and Extracts. *Appl Environ Microbiol* 2007, *73*, 6134-6143.

[14] Sander, I., Lotz, A., Zahradnik, E., Raulf, M. Allergen quantification by use of electrostatic dust collectors (EDCs):

Influence of deployment time, extraction buffer, and storage conditions on the results. *Ann Occup Hyg* 2016, *60*, 845-859.

[15] Vikrant, K., Roy, K., Kim, K-H., Bhattacharya, S.S. Insights into the storage stability of ammonia in polyester aluminum bags. *Environ Res* 2019, 177, 108596.

[16] Oliver, K.D., Pleil, J.D., McClenny, W.A. Sample Integrity of Trace Level Volatile Organic Compounds in Ambient Air Stored in SUMMA® Polished Canisters. *Atmospheric Environment* 1986, *20*, 1403-1411.

[17] Oliver, K. D., Cousett, T.A., Whitaker, D.A., Smith, L.A., Mukerjee, S., Stallings, C., Thoma, E.D., Alsoton, L., Colon, M., Wu, T., Henkle, S. Sample Integrity Evaluation and EPA Method 325B Interlaboratory Comparison for Select Volatile Organic Compounds Collected Diffusively on Carbopack X Sorbent Tubes. *Atmospheric Environment* 2017, 163, 99-106.

[18] Mawn, M.P., Kurtz, K., Stahl, D., Chalfant, R.L., Koban, M.E., Dawson, B.J. Analytical method validation for the determination of 2,3,3,3-tetrafluoropropenein air samples using gas chromatography with flame ionization detection. *J Occup Env Hyg* 2013, *10*, 583-589.

[19] Villanueva, F., Colmenar, I., Mabilia, R., Scipioni, C., Cabañas, B. Field evaluation of the Analyst passive sampler for the determination of formaldehyde and acetaldehyde in indorr and outdoor ambient air. *Anal Methods* 2013, *5*, 516-524.

[20] Yamamoto, S., Takeuchi, A., Ishidao, T., Ohkuma, H., Ichiba, M., Hori, H. Development of a measurement method to determine the ceiling exposure concentration of ortho-phthalaldehyde handling workers. *J Occup Health* 2020, 62(1), e12105.

[21] Albrecht, W. N., Hagadone, M. R., Chenchin, K. Charcoal Air Sampling Tube Storage Stability and Desorption Efficiencies of 1,2-Dibromo-3-Chloropropane (DBCP) and 1,3-Dichloropropene (DCP). *Bull Environ Contam Toxicol* 1986, *36*, 629-634.

[22] Gjolstad, M., Bergemalm-Rynell, K., Ljungkvist, G., Thorud, S., Molander, P. Comparison of Sampling Efficiency and Storage Stability on Different Sorbents for Determination of Solvents in Occupational Air. *J Sep Sci* 2004, *27*, 1531-1539.

[23] Prado, C., Alcaraz, M. J., Fuentes, A., Garrido, J., Periago, J. F. Storage Stability of Ketones on Carbon Adsorbents. *J Chromatog.* 2006, *1129*, 82-87.

[24] Kongtip, P., Tangprakorn, B., Yoosook, W., Chantanakul, S. Development of a Sampling and Analysis Method for 4-Vinyl-1-Cyclohexene in Air. *J Occup Health* 2008, 50, 122-129.

[25] Jia, C., Batterman, S., Chernyak, S. Development and Comparison of Methods Using MS Scan and Selective Ion Monitoring Modes for a Wide Range of Airborne VOCs. *J Environ Monit* 2006, *8*, 1029-1042.

[26] Kelly, T.J., Holdren, M.W. Applicability of Canisters for Sample Storage in the Determination of Hazardous Air Pollutants. *Atmospheric Environment* 1995, 29, 2595-2608.

[27] Otim, O., Rocha, J. Disappearance kinetics of 56 toxic semi-volatile organic compounds under common storage conditions. *Environ Sci Process Impacts* 2020, 22(3), 833-847.

[28] Eaton, A., Haghani, A. *Evaluation of Preservation and Holding Time for Pesticides, SOCS, and VOCS.* In Proceedings of the Water Environment

Federation, Proceedings of the Water Environment Federation Technical Exhibition and Conference (WEFTEC), San Diego, California, October 13 - 17, 2007, WEF: Alexandria, VA, 4721-4739.

[29] Chuang, J.C., Wilson, N.K., Lewis, R.G. Methodology of Ambient Monitoring for Polycyclic Aromatic Hydrocarbons. *Fresenius Environ Bull* 1999, *8*, 547-556.

[30] Huang, L., Bohac, S.V., Chernyak, S.M. Composition and integrity of PAHs, nitro-PAHs, hopanes, and steranes in diesel exhaust particulate matter. *Water Air Soil Pollut* 2013, *224*, 1630-1643.

[31] Rudling, J., Bjorkholm, E., Lundmark, B.-O. Storage Stability of Organic Solvents Adsorbed on Activated Carbon. *Annals Occup Hyg* 1986, *30*, 319-327.

[32] Warner, N.A.,Nikiforov, V., Krogseth, I.S., Bjørneby, S.M., Kierkegaard, A., Bohlin-Nizzetto, P. Reducing sampling artifacts in active air sampling methodology for remote monitoring and atmospheric fate assessment of cyclic volatile methylsiloxanes. *Chemosphere* 2020, 255, 126967.

[33] Challis, J.K., Hanson, M.L., Wong, C.S. Pharmaceuticals and pesticides archived on polar passive sampling devices can be stable for up to 6 years. *Environ Toxicol Chem* 2018, 37(3), 762-767.

[34] Ramirez, N., Marce, R. M., Borrull, F. Development of a Thermal Desorption-Gas Chromatography-Mass Spectrometry Method for Determining Personal Care Products in Air. *J Chromatogr* 2010, *1217*, 4430-4438.

[35] Daus, B., Weiss, H., Mattusch, J., Wennrich, R. Preservation of Arsenic Species in Water Samples Using Phosphoric Acid--Limitations and Long-Term Stability. *Talanta* 2006, *69*, 430-434.

[36] Daus, B., Mattusch, J., Wennrich, R., Weiss, H. Investigation on Stability and Preservation of Arsenic Species in Iron Rich Water Samples. *Talanta* 2002, *58*, 57-65.

[37] Lindemann, T., Prange, A., Dannecker, W., Neidhart, B. Stability Studies of Arsenic, Selenium, Antimony and Tellurium Species in Water, Urine, Fish and Soil Extracts Using HPLC/ ICP-MS. *Fresenius. J Anal Che.* 2000, *368*, 214-220.

[38] Ribeiro Guevara, S., Horvat, M. Stability and behaviour of low level spiked inorganic mercury in natural water samples. *Anal Methods* 2013, *5*, 1996-2006.

[39] Stetson, S.J., Wanty, R.B., Helsel, D. R., Kalkhoff, S.J., Macalady, D.L. Stability of Low Levels of Perchlorate in Drinking Water and Natural Water Samples. *Anal Chim Acta* 2006, *567*, 108-113.

[40] Moschandreas, D., Basu, S., Tata, P. Compositing and Storage of Air and Wastewater Samples. *J Environ En.* 1999, *125*, 360-371.

[41] Kovačič, A., Česen, M., Laimou-Geraniou, M., Lambropoulou, D., Kosjek, T., Heath, D., Heath, E. Stability, biological treatment and UV photolysis of 18 bisphenols under laboratory conditions. *Environ Res* 2019, 179(Pt A), 108738.

[42] Woudneh, M.B., Chandramouli, B., Hamilton, C., Grace, R. Effect of Sample Storage on the Quantitative Determination of 29 PFAS: Observation of Analyte Interconversions during Storage. *Environ Sci Technol* 2019, 53(21), 12576-12585.

[43] Gawlik, B.M., Loos, R., Bidoglio, G., Fauler, G., Guo, X., Lankmayr, E.,

Linsinger, T. Testing Sample Stability in Short-Term Isochronous Stability Studies for EU-Wide Monitoring Surveys of Polar Organic Contaminants in Water. *Trends Anal Chem* 2012, 36, 36-46.

[44] Carlson, J.C., Challis, J.K., Hanson, M.L., Wong, C.S. Stability of Pharmaceuticals and Other Polar Organic Compounds Stored on Polar Organic Chemical Integrative Samplers and Solid-Phase Extraction Cartridges. *Environ Toxicol Che.* 2013, *32*, 337-344.

[45] Aboulfadl, K., De, P. C., Prevost, M., Sauve, S. Time-Dependent Integrity During Storage of Natural Surface Water Samples for the Trace Analysis of Pharmaceutical Products, Feminizing Hormones and Pesticides. *Chem Cent J* 2010, *4*, 10.

[46] Barion, A., Balsaa, P., Werres, F., Neuhaus, U., Schmidt, T.C. Stability of organochlorine pesticides during storage in water and loaded SPE disks containing sediment. *Chemosphere* 2018, 210, 57-64.

[47] Senseman, S. A., Lavy, T. L., Mattice, J. D., Myers, B. M., Skulman, B. W. Stability of Various Pesticides on Membranous Solid-Phase Extraction Media. *Environ Sci Technol* 1993, *27*, 516-519.

[48] Anyusheva, M., Lamers, M., Schwadorf, K., Streck, T. Analysis of Pesticides in Surface Water in Remote Areas in Vietnam: Coping With Matrix Effects and Test of Long-Term Storage Stability. *Int J Environ Anal Chem* 2012, *92*, 797-809.

[49] Bassett, M.V., Wendelken, S.C., Dattilio, T.A., Pepich, B.V., Munch, D.J. The Application of Tris Buffer and Copper Sulfate for the Preservation of Phenylurea Pesticides Analyzed Using U.S. EPA Method 532 in the UCMR Survey. *Environ Sci Technol* 2002, *36*, 1809-1814.

[50] Winslow, S.D., Prakash, B., Domino, M.M., Pepich, B.V., Munch, D.J. Considerations Necessary in Gathering Occurrence Data for Selected Unstable Compounds in the USEPA Unregulated Contaminant Candidate List in USEPA Method 526. *Environ Sci Technol* 2001, *35*, 1851-1858.

[51] Sabik, H., Jeannot, R. Stability of Organophosphorus Insecticides on Graphitized Carbon Black Extraction Cartridges Used for Large Volumes of Surface Water. *J Chromatogr* 2000, *879A*, 73-82.

[52] Aguilar, C., Ferrer, I., Borrull, F., Marce, R. M., Barcelo, D. Monitoring of Pesticides in River Water Based on Samples Previously Stored in Polymeric Cartridges Followed by on-Line Solid-Phase Extraction-Liquid Chromatography–Diode Array Detection and Confirmation by Atmospheric Pressure Chemical Ionization Mass Spectrometry. *Analytica Chimica Acta* 1999, *386*, 237-248.

[53] Mouvet, C., Jeannot, R., Riolland, H., Maciag, C. Stability of Isoproturon, Bentazone, Terbuthylazine and Alachlor in Natural Groundwater, Surface Water and Soil Water Samples Stored Under Laboratory Conditions. *Chemosphere* 1997, *35*, 1083-1097.

[54] Lyytikainen, M., Kukkonen, J. V., Lydy, M. J. Analysis of Pesticides in Water and Sediment Under Different Storage Conditions Using Gas Chromatography. *Arch Environ Contam Toxicol* 2003, *44*, 437-444.

[55] Sabik, H., Jeannot, R., Sauvard, E. Stability of Herbicides and Their Degradation Products on Graphitized Carbon Black Extraction Cartridges Used for Large Volumes of Surface Water. *Analusis* 2000, *28*, 835-842.

[56] Dabrowska, D., Kot-Wasik, A., Namiesnik, J. Stability Studies of Selected Phenoxyacid Herbicides in

Water Samples and Determination of Their Transformation Products. *Bull. Environ Contam Toxicol* 2006, *77*, 245-251.

[57] Baker, D.R., Kasprzyk-Hordern, B. Critical evaluation of methodology commonly used in sample collection, storage and preparation for the analysis of pharmaceuticals and illicit drugs in surface water and wastewater by solid phase extraction and liquid chromatography-mass spectrometry. *J Chromatogr* 2011, *1218A*, 8036-8059.

[58] González-Mariño, I., Quintana, J.B., Rodríguez, I., Cela, R. Determination of drugs of abuse in water by solid-phase extraction, derivatization and gas chromatography-ion trap-tandem mass spectrometry. *J. Chromatogr* 2010, *1217A*, 1748-1760.

[59] D'Elia, V., Calcerrada, M., Montalvo, G., Ruiz, C.G. Monitoring of the stability of cocaine and some metabolites in water and oral fluid by a newly developed CE method. *Electrophoresis* 2017, 38(8), 1217-1223.

[60] Salgado, P., Visnevschi-Necrasov, T., Kiene, R.P., Azevedo, I., Rocha, A.C.S., Almeida, C.M.R., Magalhães, C. Determination of 3-mercaptopropionic acid by HPLC: A sensitive method for environmental applications. *J Chromatogr* 2015, 992B, 103-108.

[61] Paquet, L., Monteil-Rivera, F., Hatzinger, P.B., Fuller, M.E., Hawari, J. Analysis of the Key Intermediates of RDX (Hexahydro-1,3,5-Trinitro-1,3,5-Triazine) in Groundwater: Occurrence, Stability and Preservation. *J Environ Monit* 2011, *13*, 2304-2311.

[62] Merritt, A.S., Andersson, N., Almqvist, C. Cat and house dust mite allergen content is stable in frozen dust over time. *Environ Sc. Technol* 2013, *47*, 3796-3799.

[63] Fahlbusch, B., Koch, A., Douwes, J., Bischof, W., Gehring, U., Richter, K., Wichmann, H.-E., Heinrich, J. The effect of storage on allergen and microbial agent levels in frozen house dust. *Allergy* 2003, *58*, 150-153.

[64] Creta, M., Poels, K., Thoelen, L., Vranckx, K., Collaerts, P., Jansen, F., Vangeel, M., Godderis, L., Duca, R-C., Vanoirbeek, J.A. Method to Quantitatively Assess Dermal Exposure to Volatile Organic Compounds. *Ann Work Expo Health* 2017, 61(8), 975-985.

[65] McLay, P. *Organophosphorus Pesticides in Soil: Storage Stability Study.* No. 98-1, Australian Government Analytical Laboratories: Canberra, Australia, April, 1998.

[66] Tulve, N. S., Jones, P. A., Nishioka, M. G., Fortmann, R. C., Croghan, C. W., Zhou, J. Y., Fraser, A., Cavel, C., Friedman, W. Pesticide Measurements From the First National Environmental Health Survey of Child Care Centers Using a Multi-Residue GC/MS Analysis Method. Environ Sci Technol 2006, 40, 6269-6274.

[67] Pizarro, I., Gomez, M., Camara, C., Palacios, M. A. Arsenic Speciation in Environmental and Biological Samples: Extraction and Stability Studies. *Analytica Chimica Acta* 2003, *495*, 85-98.

[68] Huang, J. H., Ilgen, G. Factors Affecting Arsenic Speciation in Environmental Samples: Sample Drying and Storage. *Int J Environ Anal Chem* 2006, *86*, 347-358.

[69] Pueyo, M., Sahuquillo, A., Rigol, A., Lopez-Sanchez, J.F., Rauret, G. A New Quality Control Soil Material for Monitoring Trace Metals in Accidentally Polluted Areas. *Analytica Chimica Acta* 2005, *533*, 41-49.

[70] Ogner, G., Randem, G., Remedios, G., Wickstrom, T. Increase of Soil

Acidity and Concentrations of Extractable Elements by 1 M Ammonium Nitrate After Storage of Dry Soil for Up to 5 Years at 22°C. *Communications in Soil Science and Plant Analysis* 2001, *32*, 675-684.

[71] Navrátil, T., Burns, D.A., Nováková, T., Kaňa, J., Rohovec, J., Roll, M., Ettler, V. Stability of mercury concentration measurements in archived soil and peat samples. *Chemosphere* 2018, 208:707-711.

[72] Kodamatani, H., Maeda, C., Balogh, S.J., Nollet, Y.H., Kanzaki, R., Tomiyasu, T. The influence of sample drying and storage conditions on methylmercury determination in soils and sediments. *Chemosphere* 2017, 173:380-386.

[73] USEPA. *Sample Holding Time Reevaluation*, EPA/600/R-05/124, Office of Research and Development, U.S. Environmental Protection Agency: Washington, DC, 2005.

[74] Rost, H., Loibner, A.P., Hasinger, M., Braun, R., Szolar, O.H. Behavior of PAHs During Cold Storage of Historically Contaminated Soil Samples. *Chemosphere* 2002, 49, 1239-1246.

[75] Fang, L., Wang, W., Liu, H., Qiu, H., Zhao, Y., Lu, B., Wu, Z. Preparation and certification of two freshwater sediments cerified reference materials for polycyclic aromatic hydrocarbons. *Inter J Environ Anal Chem* 2015, *95*, 879-893.

[76] Zhao, Q., Xing, B., Tai, P., Yang, K., Li, H., Zhang, L., Lin, G., Li, P. Effect of Freeze-Thawing Cycles on Aging Behavior of Phenanthrene, Pyrene and Their Mixture in Soil. *Sci Total Environ* 2013, *452-453*, 246-252.

[77] Amaral, C.D.B., Nóbrega, J.A., Nogueira, A.R.A. Investigation of arsenic species stability by HPLC-ICP-MS in plants stored under different conditions for 12 months. *Microchem J* 2014, *117*, 122-126.

[78] Pell, A., Márquez, A., Rubio, R., López-Sánchez, J.F. Effects of sample processing on arsenic speciation in marine microalgae. *Anal Methods* 2013, *5*, 2543-2550.

[79] Pugh, R. S., Becker, P.R., Porter, B.J., Ellisor, M.B., Moors, A.J., Wise, S.A. Design and Applications of the National Institute of Standards and Technology's (NIST's) Environmental Specimen Banking Programs. *Cell Preserv Technol* 2008, 6, 59-72.

[80] Soggia, F., Abelmoschi, M.L., Riva, S.D., Pellegrini, R.D., Frache, R. Antarctic Environmental Specimen Bank-First 5 Years of Experience. *Int J Environ Anal Chem* 2001, 79, 367-378.

[81] Ali, M.S., White, J.D., Bakowski, R.S., Phillippo, E.T., Ellis, R.L. Analyte Stability Study of N-Methylcarbamate Pesticides in Beef and Poultry Liver Tissues by Liquid Chromatography. *J AOAC Int* 1993, *76*, 1309-1316.

[82] Ambe, Y., Mukai, H. Long Term Stability of Benzo[a]Pyrene in Stored Atmospheric Particulate Matter Samples. *Chemosphere* 1997, 34, 2023-2028.

[83] Glassmeyer, S., Shoemaker, J.A. Effects of Chlorination on the Persistence of Pharmaceuticals in the Environment. *Bull Environ Contamin Toxicol* 2005, 74(1), 24-31.

[84] Barber, L.B., Keefe, S.H., Brown, G.K., Taylor, H.E., Antweiler, R.C., Peart, D.B., Plowman, T.I., Roth, D.A., Wass, R.D. *Organic and Trace Element Contaminants in Water, Biota, Sediment, and Semipermeable Membrane Devices at the Tres Rios Treatment Wetlands, Phoenix, Arizona.* Water-Resources Investigations Report 03-4129, U.S. Geological Survey, Denver, CO, 2003.

[85] Shapiro, S.D., Busenberg, E., Plummer, L.N. *The Stability of Chlorofluorocarbons (CFCs) in Ground-Water Samples Archived in Borosilicate Ampoules.* Open-File Report 04-1392. U.S. Department of the Interior, U.S. Geological Survey, Reston, VA, 2005.

[86] McFarland, M., England, S., Hamilton, M.C. *Assessment of the Integrity of Chemicals in Environmental Samples Over an Extended Period of Time.* Environmental Conservation Branch, Environment Canada: North Vancouver, BC, 1995.

[87] Rüdel, H., Weingertner, M. *Guidelines for Sampling and Sample Processing: Storage of Environmental Samples under Cryogenic Conditions*, v 2.0.0. 2008. Federal Environment Bank of the Federal Republic of Germany, Fraunhofer Institute for Molecular Biology and Applied Ecology, Schmallenberg, GE.

[88] Armishaw, P., Millar, R. *Storage Stability of Organochlorine, Organophosphorus and Synthetic Pyrethroid Pesticides in Frozen Homogenised Tomato*. Australian Government Analytical Laboratories: Canberra, Australia, 2000.

Chapter 12

Improving the Technology of Synthesis Absolutized Bioethanol

Sergiy Kurta and Khatsevich Olga

Abstract

The article describes the technology of synthesis of absolute ethyl alcohol, as well as the types, properties and efficiency of water-withdrawing reagents and types, properties and efficiency of reagents (CaO, tetraethoxytitanium), which chemically bind water and dehydrate ethanol. The composition and properties of the absolutized ethanol were studied using the infrared spectroscopy, mass spectroscopy and gas-chromatography. The octane number of gasoline with the addition of absolutized ethanol together with the combustion activator (diethyl ether) was measured as well. It has been shown that the content of even very small amounts (up to 2%) of the combustion activator in absolutized ethanol results in the increase of the octane number and improves running abilities of gasoline.

Keywords: technology, absolutized ethanol, calcium oxide, tetraethoxytitanium, diethyl ether, octane number

1. Introduction

Some methods for obtaining absolute alcohol are well known, namely: (1) by binding water to water by subtracting substances; (2) pressure absolutization on many column rectification (distillation) plants; (3) the so-called azeotropic rectification using methods based on the phenomenon of azeotropism of three-component systems; (4) using methods based on the phenomenon of vapor diffusion through porous partitions or molecular sieves [1, 2]. Using as a desiccant, calcined calcium oxide can achieve an ethanol concentration of 99.8%, using magnesium sawdust—89.95%. With the combined use of sodium alcoholate ethanol can be absolutized to 99.5% [3].

Another way to obtain absolute ethyl alcohol is to separate the mixtures in distillation columns operating at different pressure values. The method is based on the dependence of ethanol content in the mixture on pressure. Typically, the installation consists of two columns. In the first column receive the alcohol strength of 70–95%, which comes in second column under pressure. The pressure in different methods varies from 8 kPa to 7.6 MPa. It is in the second column that absolute alcohol (99.2%) is obtained. The absolute alcohol (99.2%) is obtained in the second column. The disadvantage of these methods is their high energy consumption, because the rectification columns operate under pressure and require of the using of vapor or other high-temperature media to provide the necessary parameters [4–10]. Many industrial distillation plants that produce absolute ethanol are azeotropic using a third component, which forms an azeotrope with the components of the mixture—alcohol and water. Analysis of the literature shows that substances such as benzene, toluene, isooctane, cyclohexane and hexane can form the third component, which is capable of forming azeotropes with ethanol and water. The concentration of alcohol, obtained by this method, is reached to 99.7% [11–21].

Others are the methods built on the application of the phenomenon of adsorption. For this purpose, the alcohol-water mixture should be converted into a vapor phase in which the water molecules are in dissociated form and can be separated by adsorbents that absorb only water molecules from the vapor mixture and can be regenerated for reuse. Such adsorbents include polymer compositions based on dioxols, silica gel, synthetic zeolites and the like. Technologically, they can be made in the form of membranes, which are ceramic tubes, the inner layer of which is covered with a thin layer of zeolite or in the form of molecular sieves. Different composition of synthetic zeolites is known, but aluminum-potassium and aluminum-sodium zeolites have become the most widely used because they have a high adsorption capacity [22–33]. Due to the high cost of synthetic zeolites, ethanol absolutization methods have been developed, using the so-called natural zeolites—clinoptilolite: $(NaK)_4CaAl_6Si_{30}O_{72}\cdot 24H_2O$ and mordenite, which are widespread in the form of ores in certain territories of Ukraine, which is an undeniable advantage of their use [34].

Due to stabilization production of the volume petroleum and decrease in its price and the ban on the use of the tetraethyl lead gasoline in recent years, many countries of the world tend to increase the usage of oxygen-containing compounds as the additions in commercial high-octane gasoline. Methyl (MA), ethyl (EA) and tert-butyl (TBA) alcohols, methyl tert-butyl ether (MTBE), having high octane numbers and low boiling points are quite widely used among the oxygen-containing compounds. The usage of these additions results in an increase of the octane number and the improvement of the oxygen coefficient and the efficiency of the fuel combustion. All types of the fuel with oxygen-generating additives reduce the release of carbon monoxide (CO) and unburnt fuel parts (C) into the atmosphere. Thus, the addition of 10–15% MTBE reduces CO content in exhaust gases by 20%. The use of alcohol-gasoline mixtures as a motor fuel is also promising. The world ethanol production has recently reached 90 million m^3 per year [35]. The main producers are the USA, Brazil and India. It should be noted that more than 80% of produced ethanol is used as a component of motor fuels (fuel ethanol). The rest is used for manufacture of the strong drinks, solvents, the ethyl ethers and esters synthesis and as a raw material for organic synthesis [36].

In many countries of the world, gasoline with 10–15% of various fuel additives is already used. In particular, a mixture of gasoline and ethanol (10–12%) is successfully used in the USA, Canada and Brazil, where its production is based on a national program. In the United States, 80% of produced ethanol is used as a fuel. In France, fuel containing ethanol (5%) is used too. Ukraine consumes around 200 million tons of fuel and energy resources annually and has a deficit of energy resources because the needs for energy consumption are covered only up to 53% by its own resources. The rest is covered by import: 75% of the required volume of natural gas and 85% of crude oil and petroleum products. Such a structure of the energy economy creates Ukraine's dependence on oil and gas exporters and is threatening its energy and national security. Motor transport occupies a leading place in the transportation of goods. The prognosis of the development of Ukraine's motor car park proves the tendency toward a steady increasing of the number of cars and fuel consumption. In 2016, the total motor car park and the park of light commercial vehicles in Ukraine consist of, according to AUTO-Consulting, 9121,000 cars, which consume 32.7 million tons of motor fuel. The alcohol industry of Ukraine fully provides internal needs of alcohol for the manufacture of alcoholic drinks, working only up to 30% of its total capacity [37].

Recently, ethanol production of Ukraine has decreased almost twice from 310 million liters in 1996 to 140 million liters in 2017. Today, 25% of the 80 distilleries with a total capacity of 480 million liters per year operate. To date, in Ukraine there are more than 40 licensed distilleries that can process 900,000 tons of grain per year and obtain 320 million liters of grain ethanol. The annual need of beet molasses is 1.1 million tons, from which you can get other 300 million liters of ethanol. In the structure of realization of rectified ethanol the largest part is occupied by vodka and ethanol from 81 to 94.5% in 2015. In January 2016, the manufacturer "Ukrspirt" produced more than 860,000 l of ethyl alcohol, 85,000 l of strong drinks and 756,000 l of car washers. At the same time, the company work is provided by 16 factories [38].

In recent years, the production of bioethanol has been steadily increasing. From 2000 to 2014 the amount of produced ethanol has increased by five times. The market leaders are: USA (corn), Brazil (sugar cane), Germany and France (sugar beet and grain). Governments of these countries intend to increase the bioethanol production by 70%. If 10% of bioethanol is added to gasoline, emissions of aerosol particles are reduced by 50% and carbon monoxide emissions by 30%. Therefore, there is a Directive 2009/28/EC, which requires up to 10% bioethanol in a motor fuel by 2020. Ukraine signed the Association Agreement with the European Union, so it should undertake the similar obligations. While in the EU this index is gradually approaching 5% in Ukraine, it does not exceed 1% [39].

Macroeconomic analysis has shown that adding 20% of bioethanol to gasoline leads to an annual decrease in gasoline import by $ 400 million, which will result in $ 640 million increase in Ukraine's gross domestic product (GPD). Replacing 20% of heavy distillates with biodiesel will reduce imports by $ 1 billion, which can increase Ukraine's GDP by $ 1.4 billion.

In August 2014, there were 13 enterprises operating in Ukraine that produced alcoholic fuels, 10 of which based on the state alcohol distilleries with a capacity of 23,700 tons/month. However, in August 2014 there was introduced 99 EUR tax on alternative gasoline in favor of importers and network monopolists, resulting in the fact that by the beginning of 2015 only two plants ("Haysinsky" and "Ekoeenergy") with a total capacity of 2700 tons/month worked [39]. The cost of growing rape grain in recent years ranged from 800 to 1200 UAH/t. At the same time, when the seed market price is 3000 UAH/t, the cost of production of biodiesel from the fossil is 11,000–12,000 UAH/t (about 10 UAH/l), which is 2.5 times lower than the existing prices for diesel oil from oil [40]. The dynamics of growth of bioethanol and biodiesel part in the world represented by the author [37] speaks for itself, so the production of bioethanol has exceeded 100 billion liters, and biodiesel amount reaches 30 billion liters.

It is known that for homogenization of gasoline ethanol mixtures the third component is used. This component is aromatic hydrocarbons [41], which are added to ordinary gasoline to increase the octane number. In addition, it is not profitable, because the cost of toluene is by many times higher than the cost of absolute ethanol. As for the preparation of perfect mixtures of bioethanol and biodiesel with gasoline and diesel, only absolute ethanol (99.95%) must be used, therefore the research of new methods and reagents of ethanol absolutization is an actual problem and purpose of our work, which has a practical and economic value.

2. Experimental

The properties of known oxygen-generating additives which activate the motor fuel combustion are presented in **Table 1** [38]. The table shows reagents which do

Indices	MTBE	ETBE	MTAE	DIPE	Methanol	Ethanol	*tert*-Butanol
Research octane number (RON)	112–130	120	105–115	111.6	120	106	108
Motor octane number (MON)	75–115	102	95–105	98.4	90	95	98
Octane number index (RON + MON)/2	110	111	105	105	101	100.5	103
Saturated vapor pressure, MPa (at 293 K)	0.06	0.03	0.02	0.03	0.42	0.13	0.07

Table 1.
Octane number of oxygen generating gasoline additives, which is measured by different methods.

not contain water, including ethyl alcohol and ethers, but in this work it is not described how the ethyl alcohol was absolutized, therefore in different sources it has a different octane number. Consequently, we propose our own method of ethanol absolutization with calcium oxide or tetraethoxytitanium (TET).

Researches of the method of ethanol absolutization by chemically binding water reagents such as calcium oxide and TET are conducted. Alcohols practically do not interact with alkalis, because among the reaction products there is water that decomposes formed alcoholate and a reverse reaction takes place.

$$2CH_3CH_2OH + CaO \rightarrow CH_3CH_2OCaOCH_2CH_3 + 2H_2O\uparrow \rightarrow 2CH_3CH_2OH + Ca(OH)_2 \quad (1)$$

Therefore, fresh prepared (1123 K, 2 h) mineral calcium carbonate, which has the following characteristics, presented in **Table 2**, was used for ethanol absolutization.

After calcination, samples of CaO were cooled in a vacuum desiccator, to reduce air access and to re-carbonize calcium oxide. Only 1 grade with minimum content of impurities was taken for work (**Table 2**).

It was suggested that if you take freshly calcined calcium oxide, then absolutized ethyl alcohol and alcoholates of the corresponding metals can be obtained during an irreversible reaction:

$$CH_3CH_2OH \cdot H_2O + CaO \rightarrow Ca(OH)_2\downarrow + CH_3CH_2OH \quad (2)$$

$$2CH_3CH_2OH + CaCO_3 \rightarrow CH_3CH_2OCaOCH_2CH_3 + 2H_2O + CO_2\uparrow \rightarrow 2CH_3CH_2OH + Ca(OH)_2\downarrow \quad (3)$$

The component content of calcium oxide	Types of CaO obtained from mineral calcium carbonate, $CaCO_3$		
	1	2	3
Active CaO	≤90	≤80	≤70
Active MgO	≥5	≥5	≥5
$CaCO_3$ and $MgCO_3$ carbons	≥3	≥5	≥10
Impurities SiO_2 and Al_2O_3	≥1	≥10	≥15

Table 2.
Properties of calcium oxide.

The insoluble calcium hydroxide in the alcohol was filtered off, and the absolute ethyl alcohol by reactions (4) and (3) was discharged through a direct refrigerator, and the calcium hydroxide remained in the flask in the precipitate form.

Tetraethoxytitanium $(CH_3CH_2O)_4Ti$ as a dehydrator acts in different way.

$$(CH_3CH_2O)_4Ti + 4H_2O \rightarrow 4CH_3CH_2OH + Ti(OH)_4\downarrow \quad (4)$$

During the hydrolysis of TET the titanium tetrahydroxide, which is insoluble in alcohol, precipitates and can be filtered out; the absolutized alcohol is distilled at a special installation under vacuum, or with air dehydrators under normal conditions.

As can be seen from **Table 3** [42], ethanol has the most similar properties with gasoline according to the main indicators (boiling point and octane number). In addition, it is the cheapest and most affordable as a bioethanol—a product of natural raw materials processing, therefore, it was selected for absolutization. The ethanol concentrations were measured by the weight pycnometer method and determined by theoretical data [43]. The volume concentration of alcohol is determined by the areometers (ASP-1, ASP-2) using tabular data. Chromatography of ethanol and gasoline is carried out on by a gas–liquid chromatography, a method for determining oxygen-containing compounds in gasoline (HPCHEM\SEQUENCE \D161124A. S\METHOD\ D4815N.M) by internal standard (ISTD) based on 15 known calibrated substances, including ethanol (**Figures 2–4**). Infrared spectroscopy of ethanol samples is carried out by IR spectroscope (IS 50 FT-IR "NICOLET" Slovensko). Mass spectroscopy of the samples is carried by a mass spectroscope (MX-7304A, AO.SELMI, Sumy, Ukraine, 1 to 210 mass range). Determination of the cetane number of diesel fuel L-0.2-40 and the octane number of petrol A-80, as well as the freezing point of diesel fuel was carried out using the laboratory analyzer "Octane meter" of Shatox sx-100 k type [44]. Also, an assessment of the petrol gasoline by a motor method was carried out at the "OKKO" quality control laboratory in Galich, Ivano-Frankivsk region of Ukraine.

Parameter	Units	Methanol	Ethanol	Butanol	Gasoline
Molecular formula		CH_3OH	C_2H_5OH	C_4H_9OH	C_6H_6
Molecular weight	g/mol	32.04	46.07	74.12	78.11
Melting point	K	175	159	183	278
Boiling point	K	338	351	390	353
Density	g/cm^3	0.791	0.789	0.81	0.879
Saturated vap. pres.	kPa, 298 K	16.2	6.5–7.5	0.8	12.7
Vapor density	g/l, 298 K	1.3	1.9	2.4	3.2
Water solubility	g/100 ml	Unlimited	Unlimited	7.9 g/100 ml	1800 mg/l
Viscosity	s·Pa, 298 K	0.544	1.074	–	0.649
Henry's constant	C_{air}/C_{H_2O}, 298 K	1.087E−4	2.097E−4 to 2.571E−4	–	2.219E−1
Energy density	MJ/l	16	19.6	29.2	–
Octane number RON	–	156	132	104	82.5–98.0
Cetane number MON	–	92	89	78	95

Table 3.
Comparative physico-chemical characteristics of oxygen generating additives—gasoline combustion activators [42].

3. Results and discussion

Table 4 shows the results of the ethanol absolutization using tetraethoxytitanium. As can be seen from the table, absolutized ethanol can be obtained with the ethanol concentration, which increases from 92.5 to 93–96 wt% (or 95.41–99.50 vol%), but does not reach 100 wt%.

Therefore, at the second stage of our research, we used a more reactive dehydrator—calcium oxide which was calcined at 1123 K; its characteristics are presented in **Table 2**. We selected a sample no. 1 with a minimum content of magnesium oxide (MgO—5%), carbonates ($CaCO_3$ + $MgCO_3$–3%) and other impurities (SiO_2 + Al_2O_3–1%). As can be seen from the results presented in **Table 4**, the CaO consumption decreases to 17 wt%, compared with 25 wt% of $(C_2H_5O)_4Ti$, and the concentration of absolutized ethanol increases to 99.5–100.0 wt %, or 99.7–100.0 vol%. In addition, the loss of alcohol decreases. After absolutization with $(C_2H_5O)_4Ti$ these losses reaches 15–25% C_2H_5OH (**Table 3**), while for CaO they decrease to 8–13% relative to the initial C_2H_5OH (**Table 4**). So we can say that calcium oxide as a chemically binding water reagent is much more effective dehydrator. As can be seen from **Figure 1**, the ethanol concentration after treatment with

Initial ethanol amount, SSU 4221:200 95 vol%, 92.5 wt %		The amount of ethanol obtained after absolutization		Ethanol yield after absolutization	The amount of Ti $(OC_2H_5)_4$		Ethanol density after absolutization	Ethanol concentration after absolutization	
V, ml	*m*, g	*V*, ml	*m*, g	wt%	*m*, g	%	g/cm^3	vol%	wt%
100	82.15	85	69.07	84.08	26.65	25	0.8126	95.41	93.0
100	82.16	85.1	70.72	86.08	27.55	25.11	0.8065	95.85	93.5
100	82.10	77.59	62.1	75.14	29.15	26.20	0.8014	97.49	96.0
100	82.10	77.59	62.1	75.14	29.15	26.20	0.8014	99.50	96.0

Table 4.
Properties of ethanol after absolutization with tetraethoxytitanium.

The initial ethanol amount, 95 vol%, 92.5% mas.		The amount of ethanol obtained after absolutization		Ethanol yield after absolutization	The amount of CaO		Ethanol density after absolutization	Ethanol concentration after absolutization	
V, ml	*m*, g	*V*, ml	*m*, g	wt%	*m*, g	%	g/cm^3	vol%	wt%
100	82.15	96	75.77	92.23	18	17.9	0.7893	100.0	100.0
100	82.16	93	73.42	89.36	17.8	17.8	0.7894	99.97	99.95
100	82.10	93	73.42	89.43	17.5	17.5	0.7895	99.93	99.89
100	82.10	90	71.69	87.32	17	17.1	0.7899	99.72	99.51

Table 5.
Properties of ethanol after absolutization by CaO (90%).

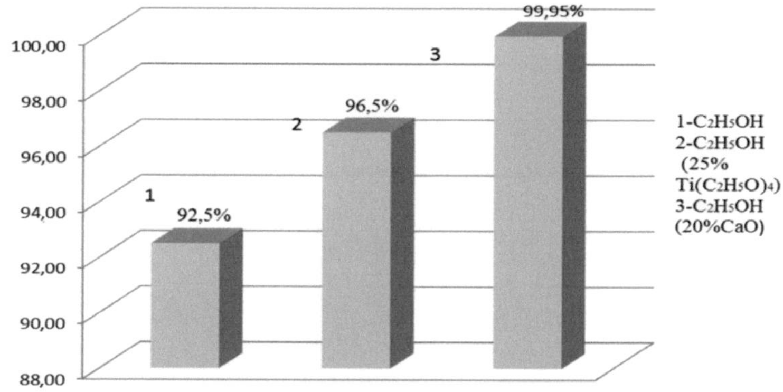

Figure 1.
Dependence of the ethyl alcohol concentration after absolutization with tetraethoxytitanium (25% $(C_2H_5O)_4Ti$) and calcium oxide (20% CaO).

calcined calcium oxide (17.5% CaO) reaches 99.95 wt%, while the absolutization with tetraethoxytitanium increases the concentration of ethanol only to 96.5 wt%.

At the same time, the non-absolutized 95% ethanol contains an admixture of MTBE (0.09531 wt %). In this case, the low-boiling impurity in the non-absolute ethanol is absent (**Figure 2**).

According to the chromatogram (**Figure 2**), the concentration of non-absolutized ethanol reaches 97.8499 vol % against 95.0%, as it was measured by the weight pycnometer method (**Tables 4** and **5**). At the same time, the concentration of ethanol absolutized with calcium oxide (**Figure 3**) is 97.5724 vol %. Taking into account 2.3432% of unidentified low-boiling impurity (T_b = 308–310 K) and 0.03433% of MTBE (according to the chromatogram in **Figure 3**), the total concentration of absolutized ethanol is 99.95 vol %, as it was determined by the pycnometric method (**Tables 4** and **5**).

Taking into account the chromatogram data in **Figure 3** and its interpretation, we determine the qualitative composition of one of the unknown components (X = 2.3432%), which appears in the chromatogram of ethyl alcohol (99.95%), absolutized with CaO for 2.8 min. At the same time, the chromatographic analysis of 99.99% ethyl alcohol absolutized by industrial method [43] (**Figure 4**), shows 100.03 vol % of ethanol. The difference of 0.03% might be a mistake of the device or an alcohol supplier analysis method.

Unfortunately, this impurity is not identified by graded chromatograms of alcohols and ethers, therefore, an IR spectral analysis of this alcohol is performed, which

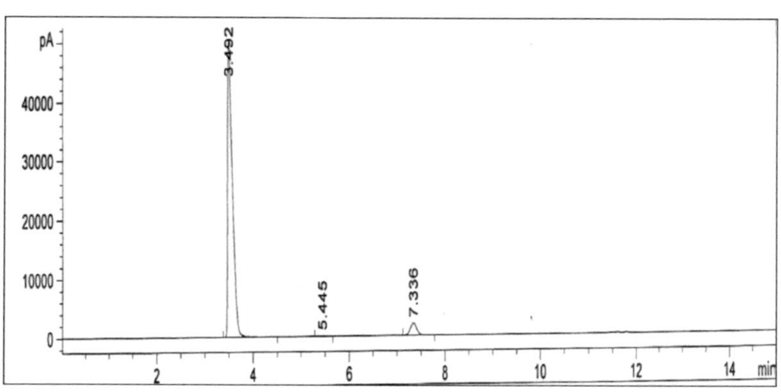

Figure 2.
Chromatogram of the initial, non-absolutized ethanol (95.0 vol%) with corresponding interpretation.

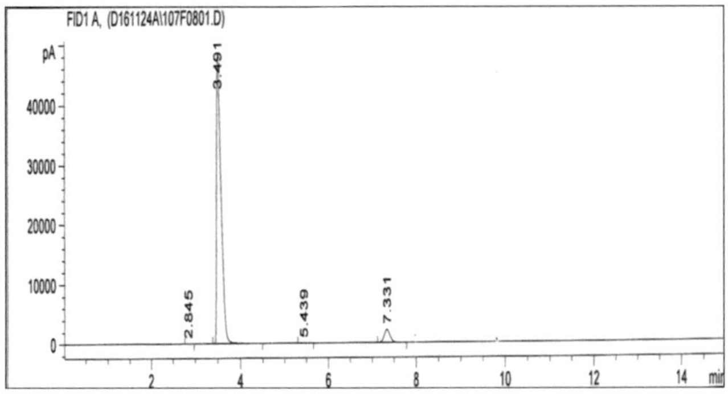

Figure 3.
Chromatogram of the ethanol (99.95 vol%), absolutized with calcium oxide with corresponding interpretation.

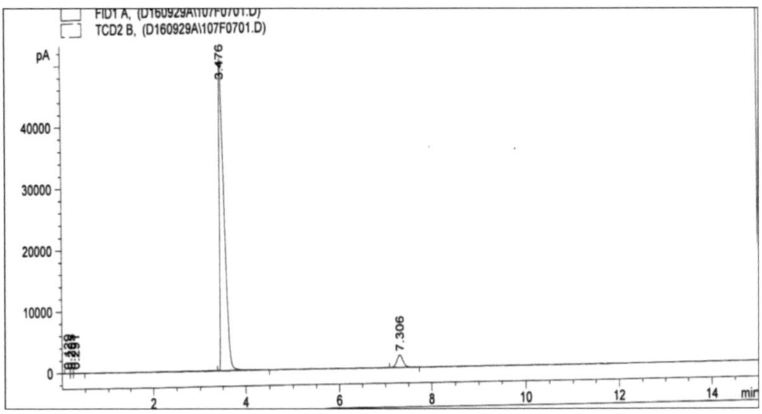

Figure 4.
Chromatogram of the ethanol (99.99 vol%), absolutized by industrial method [43] with corresponding interpretation.

is compared with the IR spectra of the original (95%) and pure absolute (100%) ethyl alcohol (**Figure 5**). As can be seen from **Figure 5**, the IR spectra of all three samples do not differ significantly from each other. Only in two absorption regions 3320–3330 and 1630–1640 cm^{-1}, there is a noticeable difference in the absorption intensity of the fluctuations of the corresponding groups, which significantly decrease from the maximum in the original non-absolutized 95% C_2H_5OH (curve 1, **Figure 5**) to a minimum in absolute 100% C_2H_5OH (curve 3, **Figure 5**).

First of all, the amount of water in samples of ethanol decreases from 4–5% in the initial non-absolutized alcohol (*curve* 1, **Figure 5**) to 0.1–0.05% in the absolutized with CaO ethanol (*curve* 2, **Figure 5**) and up to 0% in the absolutized 100% C_2H_5OH (*curve* 3, **Figure 5**). Therefore, the absorption intensity in the region of 3320–3330 cm^{-1}, which corresponds to the valence fluctuations of the hydroxyl groups of the H_2O molecules, decreases by 20% (according to the spectrograms in **Figure 5**). This is noticed in the transition from the non-absolutized 95% C_2H_5OH (*curve* 1, **Figure 5**) to the absolutized with CaO ethanol (*curve* 2, **Figure 5**) and to the absolute 100% ethanol (*curve* 3, **Figure 5**). In addition, in the spectrograms of the absolutized with CaO 99.95% C_2H_5OH (*curve* 1, **Figure 5**) and absolute 100% ethanol (*curve* 3, **Figure 5**) absorption disappears in the region of deformation oscillations 1630–1640 cm^{-1} of double bonds C=C or C=O [10] compared with non-absolutized 95% alcohol (*curve* 1, **Figure 5**), which can be identified as the presence of ketones, aldehydes, complex esters and ethers, and corresponds to the

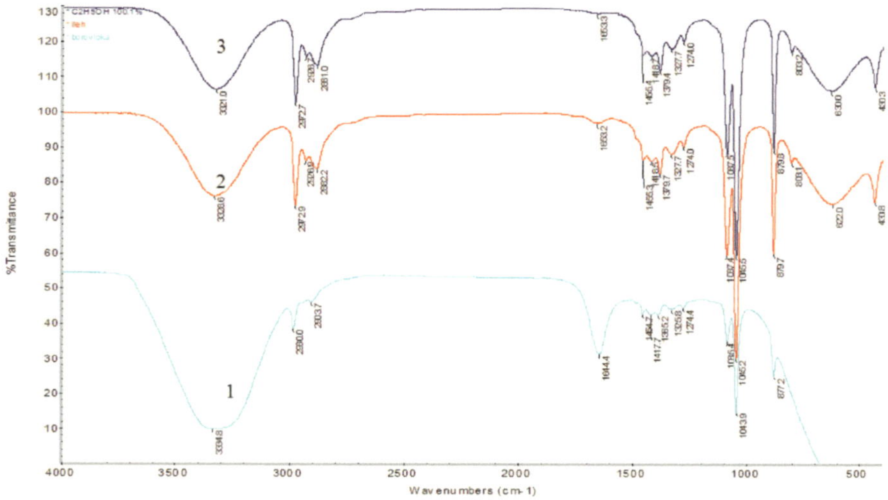

Figure 5.
Infrared spectra of different ethyl alcohols: 95.0 vol% non-absolutized C_2H_5OH (1); 99.95 vol% C_2H_5OH absolutized with CaO (2) and 100% C_2H_5OH of industrial product (3).

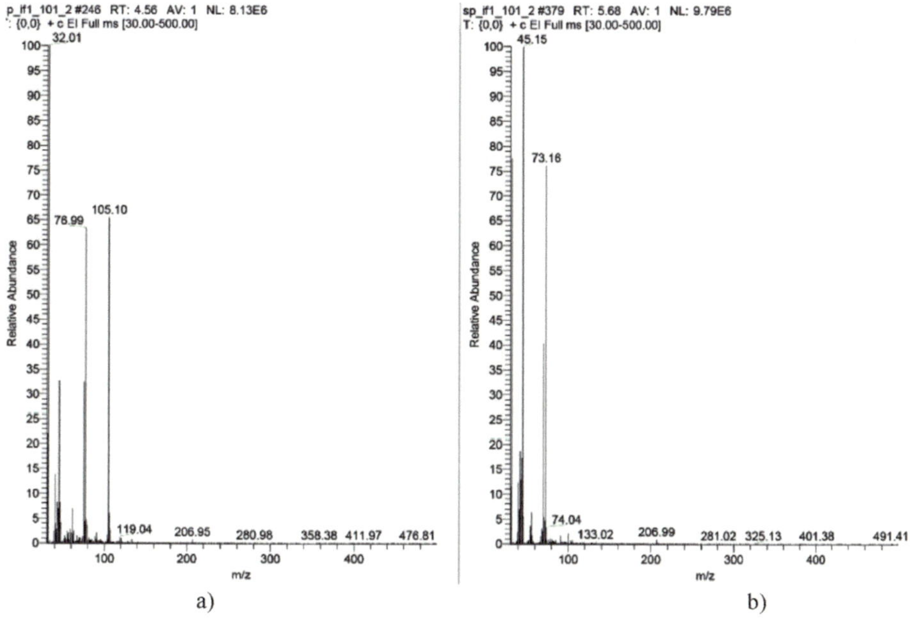

Figure 6.
Mass spectra of initial 95% ethanol (a) and absolutized with CaO 99.95% ethanol (b).

presence of MTBE in ethyl alcohol, which is identified in chromatograms of non-absolutized 95% and absolutized with CaO 99.95% ethyl alcohol (**Figures 2** and **3**) in the range of 0.034–0.095%.

For identification and more proper determination of the composition and structure of the impurity (X = 2.3432%) in ethanol, which is identified in the chromatogram of the absolutized with CaO ethanol at 2.845 min (**Figure 5**), we carried out a mass spectrometric analysis of our samples of non-absolutized 95% and absolutized with CaO 99.95% ethanol by the mass spectroscope MX-7304A, AO.SELMI shown in **Figure 6**.

As can be seen in the mass spectrogram of non-absolutized 95% ethyl alcohol (**Figure 6a**), radical composition of the impurity is formed by the electron action of the mass spectrometer, there can be three types of free radicals with molecular weights such as 32.01 m.u.—C_2H_5• = 29 m.u.; 76.99 m.u.—$(CH_3)_3O$• = 73 m.u.; 105.1 m.u.—$C(CH_3)_3OOCH_2$• = 103 m.u. They may be formed by the action of electrons of the mass spectrometer on ethyl alcohol C_2H_5OH and methyl *tret*-butyl ether $C(CH_3)_3OOCH_2$, which are identified in the chromatograms of the initial non-absolutized 95% alcohol (**Figure 2**). At the same time, for absolutized with calcium oxide 99.95% ethanol (**Figure 6b**) in the composition of the impurity radicals formed by the action of electrons of the mass spectrometer, there can be only two types of free radicals with molecular weights such as 45.15 m.u.—C_2H_5O• = 44 m.u. and 73.16 m.u.—$(C_2H_5)_2O$• = 74 m.u. They can be formed by the electron influence of the mass spectrometer on ethyl alcohol C_2H_5OH and diethyl ether $(C_2H_5)_2O$, which are identified in a chromatogram of the 99.95% ethanol absolutized with CaO (**Figure 3**) [44, 45].

Taking into account that we know the retention time of the impurity (X = 2.3432% at 2.845 min), which is not identified in the chromatogram of 99.95% ethanol, absolutized with CaO (**Figure 3**), therefore, we have constructed dependence the retention time of three kinds of alcohol: *i*-propanol, ethanol, methanol and diethyl ether (DEE) depending on the boiling points of this substances in the chromatogram, shown in **Figure 7**.

The analysis of the curves shows that the diethyl ether retention time at the boiling point of 307.6 K (34.6°C) is 2.85 min, which coincides with the chromatogram data (2.845 min). In this way, we confirm the assumption that during the absolutization of 95% ethanol with calcium oxide, ethyl alcohol containing 2.3432% of diethyl ether and 97.5724% C_2H_5OH can be obtained.

We checked how 2.3432% of diethyl ether influences the ethanol octane number and the gasoline octane number. Previously, the physico-chemical characteristics of some oxygen-generating additives for gasoline, shown in **Figure 8**, were analyzed. As can be seen from **Figure 8**, the density and molecular mass of diethyl ether are close to those of gasoline A-95. The boiling point of the diethyl ether 307.6 K is lower than that of gasoline. It is approaching the boiling point of the first fraction of gasoline, which is 307.6 K. These properties are especially useful for accelerated engines of gasoline cars in winter, when ignition is hampered by a lowered ambient

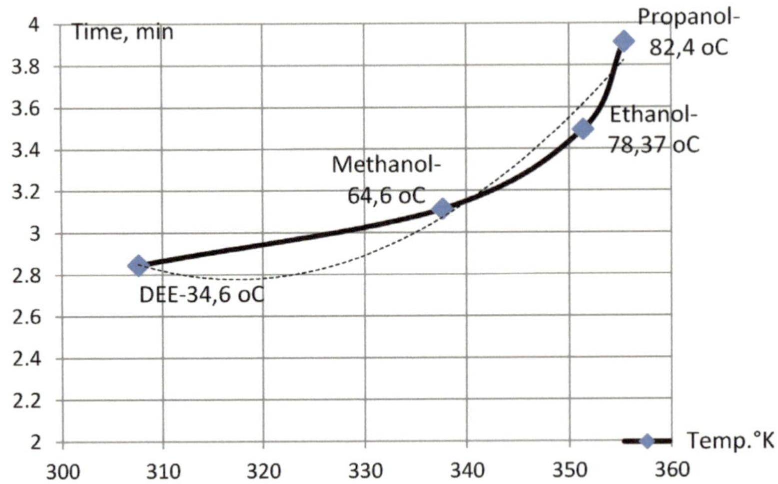

Figure 7.
*Dependence of retention time (min) of i-propanol (355.4°K), ethanol (351.37°K), methanol (337.6°K), and diethyl ether (DEE) (307.6°K), and the boiling points of these substances from the chromatogram (**Figure 3**).*

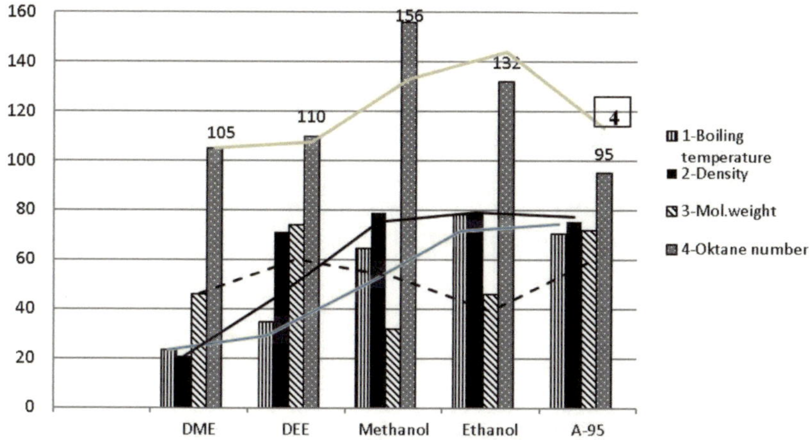

Figure 8.
Physico-chemical characteristics (1, boiling temperature °C; 2, density g/sm³ × 10; 3, molecular weight; 4, octane number) of oxygen-generating additives (DME—dimethyl ether, DEE—diethyl ether, methanol, ethanol) in comparison with the characteristics of gasoline A-95.

temperature (below 263 K). Thus, the diethyl ether presence improves the gasoline combustion efficiency, which we have checked by adding 10–95% absolutized (97.5%) alcohol containing 2.34% of diethyl ether, which is presented in **Figure 9** [46].

Octane numbers of A-80 gasoline with absolute ethanol additives (gasoline-ethanol mixture) are counted according to the authorship formula which is derived from experimental data [47]. These octane numbers are compared with the octane numbers of gasoline, in which ethanol with diethyl ether is added.

$$\text{ONGE} = [26.44 - 0.29(\text{ON0})]\ln \text{Ce} + [1.32(\text{ON0}) - 29.49] \quad (5)$$

where ONGE—octane number of gasoline-ethanol mixture; ON0—octane number of initial gasoline; Ce—ethanol content [48].

Obtained data of the octane number dependence is performed in **Figure 9** (*curve* 1) and compared with the octane number of gasoline-ethanol mixtures to which

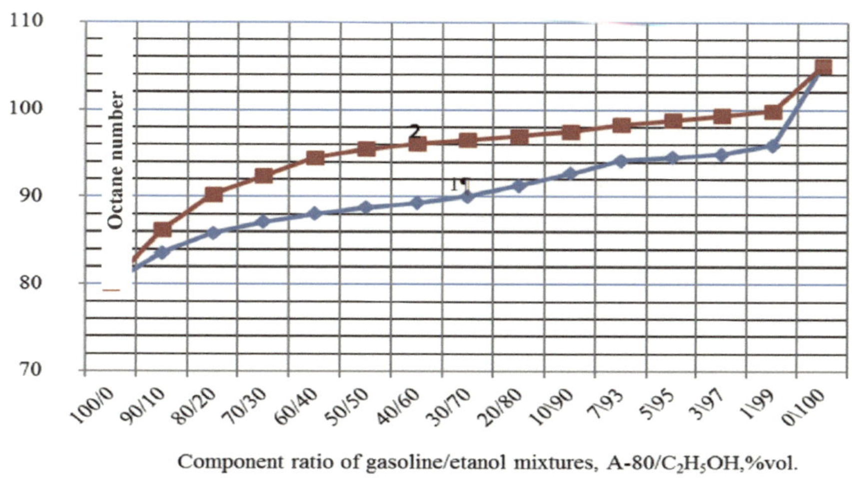

Figure 9.
Dependence of the octane number of gasoline A-80 on the amount and kind of added ethanol: theoretically calculated ON (1) of gasoline-ethanol mixtures from the formula [46] and experimentally obtained ON (2) of gasoline-ethanol mixtures with the addition of absolutized ethyl alcohol containing 2.34% of diethyl ether.

ethanol containing 2.34% of diethyl ether was added (*curve* 2). Thus, it was shown that ethyl alcohol containing diethyl ether can more effectively increase the octane number of gasoline A-80 [49], The content of 20–40% of such ethanol in gasoline A-80 increases its octane number to 91–95 U. While conventional 99.95% ethanol increases the octane number of gasoline A-80 only to 85–88 U. In the case of the maximum possible absolutized ethanol content up to 80–90% in gasoline A-80, the octane number reaches 91–93 U, and the introduction of the same amount of 99.95% ethanol containing 2.34% of diethyl ether in gasoline A-80, the octane number reaches 97–97.5 U [50].

4. Conclusions

The properties and high ability of calcium oxide and tetraethoxytitanium as dehydrators for chemical binding of water and absolutization of 96% ethyl alcohol to 99.95% have been investigated. The dependence of the properties and composition of the absolute alcohol on the dehydrator nature was shown. Infrared spectroscopy, mass spectroscopy, and gas-liquid chromatography confirmed the presence of 2–2.34% of diethyl ether in ethanol, which is absolutized by freshly prepared calcium oxide. It was determined that the octane number of gasoline A-80 with the addition of 20–90% of the absolutized ethanol containing 2.34% of the combustion activator—diethyl ether—increases gasoline octane number more significantly than with the addition of the same amount of an ordinary absolute 99.95% alcohol without diethyl ether and reaches 95–97 U, whereas the octane number of gasoline with ordinary absolute ethanol reaches only 88–95 U. We confirmed literary data that the content of even very small amounts of ethers, as combustion activators, combined with the absolute ethyl alcohol leads to an improvement in the gasoline running abilities [45].

Author details

Sergiy Kurta* and Khatsevich Olga
Department Chemistry, Precarpathian National Vasyl Stefanyk University, Ivano-Frankivsk, Ukraine

*Address all correspondence to: kca2014@ukr.net

References

[1] Стебников ВН. Перегонка и ректификация спирта. Изд. 2-е, 1969. 450с

[2] Frolkova AK, Raeva VM. Bioethanol dehydration. State of the art. Theoretical Foundations of Chemical Engineering. 2010;**44**(4):545-556

[3] Гитис СС, Глаз АИ. Практикум по органической химии. Москва: Высшая школа. 1991. 153с

[4] Львов СВ. Некоторые вопросы бинарных и многокомпонентных смесей. Москва: Изд. Академии наук СССР. 1960. С. 13

[5] Arifeen N et al. Process design and optimization of novel wheat-based continuous bioethanol production system. Biotechnology Progress. 2007; **23**:1394-1403

[6] Dehydrating alcohol and the like: pat. 1676700 USA: Classifications B 01D 3/003; publ. 10.07.1928

[7] Способ получения спирта этилового абсолютированого: пат. 2449979 РФ: МПК С 07С 31/08, С 07 С 29/74; заявл. 06.04.2010; опубл. 10.05.2012, Бюл. №15

[8] Торшин АВ. Повышение качества этилового спирта путем совершенствования технологии эвапюрации: дис.канд.техн.наук: 05.08.01/ Воронежский государственный университет инженерных технологий. Воронеж, 2016. 182 с

[9] Короткова ТГ. Обоснование и разработка инновационных технологий пищевого спирта,абсолютированного этанола и биоэтанол: дис.д-ра. техн. наук: 05.18.01/Кубанский государственный технолог. Универ.. Краснодар, 2013. 576 с

[10] Windadgo S, Seider WD. Azeotropic distillation. AlChE Journal. 1996;**41**(1):96-130

[11] Starula S, Oprea F, Mihaescu D. Separation and purification of anhydrous ethanol by azeotropic distillation using an entertainer. Revista de Chemie. 2005;**56**(5):544-548

[12] Font A, Asensi JC, Ruiz F, Gomis V. Application of isooctane to the dehydration of ethanol. Design of the column sequence too obtain absolute ethanol by heterogeneous azeotropic distillation. Industrial and Engineering Chemistry Research. 2003;**42**:140-144

[13] Gomiz V, Pedrasa R, Frances O, Asensi JC. Dehydration of ethanol using azeotropic distillation. Industrial and Engineering Chemistry Research. 2007; **46**:4572-4576

[14] Gomiz V, Font A, Pedrasa R, Saquete MP. Isobaric vapor–liquid and vapor-liquid-liquid equilibrium data for the system water-ethanol-cyclohexane. Fluid Phase Equilibrium. 2005;**235**:7-10

[15] Gomiz V, Font A, Pedrasa R, Saquete MP. Homogeneity of the water +ethanol +toluene azeotrope at 101.3 kPa. Fluid Phase Equilibrium. 2008;**266**:8-13

[16] Gomiz V, Font A, Pedrasa R, Saquete MP. Isobaric vapor-liquid equilibrium data for water ethanol hexane system. Fluid Phase Equilibrium. 2007;**259**:66-70

[17] Черепов ЕВ. Технология обработки современного роизводства абсолютированого спирта и биоэтанола: дис.канд. техн. наук: 05.18.01, 05.18.12/ Майкопский государ. Технолог. Унив-т, Краснодар, 2011. 167 с

[18] Черепов ЕВ, Лобода АВ, Короткова ТГ. Технология производства биоэтанола и абсолютизированного спирта для пищевой и медиц. промышл. *Известия вузов. Пищевая технология*. 2010. Т5–6. С. 47–50

[19] Технологія спирту. Маринченко В. О. та ін..; за ред. В. О. Маринченка. Вінниця, 2003. 364 с

[20] Короткова ТГ. Основы межфазного равновесия моделирования разделения спиртово-углеводородных смесей с двойными и тройными азеотропами. *Извесття вузов. Пищевая технология*. 2010. №4. С. 77–81

[21] Киреев ВА. Курс физической химии. Москва: Изд. Химия, 1975. 776 с

[22] Глинка НЛ. Общая химия: научн. пособие. Изд. 30-е, испр. Москва, 2003.728 с

[23] Спосіб отримання спирту етилового зневодненого: пат. 105707 Україна: МПК F26D 3/ 347, B01D 61 / 36, C07C 31/ 08. № а 201300030; заявл. 02.01.2013; опубл. 10.06.2014, Бюл. №11

[24] Спосіб виробництва спирту абсолютованого або паливного етанолу: пат.50390 Україна: МПК 6 C12F 3/ 00,C12P 7 / 00. № а 200129262; заявл. 29.12.2001; опубл.15.10.2002, Бюл. №10

[25] Спосіб зневоднення водно-спиртового розчину: пат.64538 Україна: МПК C 10L 1 / 18, C10L 1 / 182, C10L10 / 10, C10L 1 / 30, B01D 15 / 00. № а 2003065808; заявл. 24.06.2003; опубл.15.15.2006, Бюл. №15

[26] Новый способ обезвоживания спирта и устройство для его осуществления: пат. 2400282 РФ: МПК B01D 15 / 00, B01D 53 / 00; заявл. 16.10.2007; опубл.27.09.2010, Бюл. №10

[27] Спосіб получения спирта абсолютированого пат. 2265473РФ: B01D 3/14, C07C 31/08; заявл. 09.07.2004;опубл.10.12.2005, Бюл. №12

[28] Спосіб зневоднення етанолу: пат.64538 Україна: МПК C07C 31/08. № а 201207593; заявл. 20.06.2012; опубл.25.12.2012, Бюл. №24

[29] Regenerative molecular siebe adsorbents used for alcohol dehydration: pat. 847680B2 USA: Classifications B 0120 34 / 08. № a 20100240524A1; app. 23.09.2010; publ. 02.07.2013

[30] Dehydration process using membranes with hydrophobic coating: pat. 847680B2 USA: Classifications B01D 53 / 00, B01D 53 / 362, B01D 53 / 628. № a 20090057224A1; app. 05.03.2009; publ. 30.07.2013

[31] Спосіб зневоднення концентрованих водно – спиртових розчинів: пат.12591Україна: МПК C07C 29 / 00. № u 200508015; заявл. 12.08.2005; опубл. 15.02.2006, Бюл. №2

[32] Спосіб дегідратації етилового спирту: пат.12591Україна: МПК C07C 07/13. № u 20042978; заявл. 12.04.2002; опубл. 15.11.2002, Бюл. №4

[33] Спосіб зневоднеення спирту: пат.7622Україна: МПК C07C 31 / 08. № u 201207593; заявл. 20.06.2012; опубл.25.12.12, Бюл. №24

[34] Спосіб зневоднеення етанолу: пат.106112 Україна: МПК C07C 7/13. № а 201209885; заявл. 15.08.2012; опубл. 25.07.15, Бюл. №14

[35] Bray V, Shutsky I. Dopovidi Nats. Vol. 6. Kiev, Ukraine: Acad. Nauk Ukrainy; 2016. p. 71

[36] Available from: http://www.eniscuola.net/en/mediateca/etanol-and-biodiesel-global-production/

[37] Shchokin A, Kolesnik Y, Novak A. Tsigank: Electron. Zh. "ESCO". 2003;**5**:21

[38] Бойченко СВ, Бойченко МС, Личманенко ОГ, Кабан СМ. Вплив добавок аліфатичних спиртів на властивості бензинів. Наукоємні технології. 2015;**25**(1):86-92

[39] Available from: http://agravery.com/uk/posts/show/deregulacia-apk-

dopomoze-ukraini-zrobiti-proriv-u-virobnictvi-bioetanolu-pavlenko

[40] Boichenko S. Innovative chemmotological thought as an integrated system of knowledge. Chemistry & Chemical Technology. 2014;**8**:349

[41] Dutchak V. PhD thesis. Lviv: Lviv Polytech. Nats. Univ.; 2008

[42] Krahl J, Knothe G, Munack A, Ruschel Y, Schröder O, Hallier E, et al. Comparison of exhaust emissions and their mutagenicity from the combustion of biodiesel, vegetable oil, gas-to-liquid and petrodiesel fuels Fuel 886. 2009. pp. 1064-1069. Available from: www.elsevier.com/locate/fuel

[43] Available from: http://www.itrcweb.org

[44] CAS No.: 64-17-5. Available from: https://russian.alibaba.com/product-detail/ethyl-alcohol-99-99

[45] Starchevsky V, Ribun V, Kurta S, Khatsevich O. Properties and composition of absolutized by chemically ethanol and their effect on the gasoline octane number. Chemistry & Chemical Technology. 2018;**8**(3): 346-354

[46] Bratychak M. Osnovy Promyslovoi Naftokhimii. Lviv: Lviv Polytech. Nats. Univ.; 2008

[47] Kousoulidou M, Dimaratos A, Karvountzis-Kontakiotis A, Samaras Z. Combustion and emissions of a common-rail diesel engine fueled with HWCO. Journal Energy Engineering, Special Issue on Innovative Technologies on Combustion of Biofuels in Engines: Issues and Challenges. 2014; **140**:A4013001

[48] Patrylak L, Patrylak K, Okhrimenko M, et al. Comparison of power-ecological characteristics of diesel engine work on mixed diesel fuels on the basis of ethyl esters of rapeseed and sunflower oils. Chemistry & Chemical Technology. 2015;**9**:383

[49] Boichenko S, Iakovlieva A, Vovk O, et al. Traditional and alternative jet fuels: Problems of quality standardization. Journal of Petroleum & Environmental Biotechnology. 2013;**4** (3). DOI: 10.4172/2157-7463.1000146

[50] Kurta S, Ribun V, Fedorchenko S. Dewaxing of motor fuels is the complex method of increasing the octane and cetane numbers of gasoline and diesel. Deutscher Wissenschaftsherold. German Science Herald; 2017;(3):81-92. DOI: 10.19221/2017321